Directed Model Checks for Regression Models from Survival Analysis

Dissertation
zur Erlangung des Doktorgrades Dr. rer. nat.
der Fakultät für Mathematik und Wirtschaftswissenschaften
der Universität Ulm

vorgelegt von

Axel Gandy
aus
Passau

2005

Bibliografische Information Der Deutschen Bibliothek

Die Deutsche Bibliothek verzeichnet diese Publikation in der Deutschen Nationalbibliografie; detaillierte bibliografische Daten sind im Internet über http://dnb.ddb.de abrufbar.

ISBN 3-8325-1144-X

Logos Verlag Berlin
Comeniushof, Gubener Str. 47,
10243 Berlin
Tel.: +49 030 42 85 10 90
Fax: +49 030 42 85 10 92
INTERNET: http://www.logos-verlag.de

Amtierender Dekan:	Prof. Dr. Ulrich Stadtmüller
1. Gutachter:	Prof. Dr. Uwe Jensen
2. Gutachter:	Prof. Dr. Volker Schmidt
3. Gutachter:	Prof. Odd O. Aalen, PhD
Tag der Promotion:	23. Dezember 2005

To my parents

Contents

Chapter 1

Introduction

The proportional hazards model is the most frequently used regression model from survival analysis. However, since its introduction by Cox (1972), many other regression models have been proposed. How can one check whether a given regression model is suitable for a particular application? Several model checks have been suggested to answer this question. However, they are mostly ad hoc suggestions that are constructed for a certain model and it is usually not clear against which alternatives they are particularly powerful.

The aim of this thesis is to develop model checks for a wide class of regression models from survival analysis. These model checks can be adjusted to increase the power against certain alternatives.

In survival analysis, a group of individuals experiencing events over time is observed. The aim of regression models is to relate these events to certain covariates. A classical application is concerned with patients that undergo a certain type of surgery. Hereby, an event is the death of a patient. One is interested in knowing how covariates like the age of a patient or a certain medication influence the length of survival. However, it is often the case that some patients are still alive when the study ends and a statistical analysis is made. For these patients, it is only known that they survived up to a certain time. This circumstance, called censoring, is typical for survival analysis.

Survival analysis has its origins in biostatistics, but models from survival analysis have also been successfully used in different areas, e.g. reliability theory and financial mathematics. In reliability theory, the individuals may be machines or motors and the events may be failures or the individuals may be pieces of software and the events the submissions of bug reports. In financial mathematics, companies may be observed and the events are bankruptcies or insurance holders may be observed and the events are cancellations of contracts. In the above medical example, each individual experiences at most one event. The modern formulation of survival analysis allows for more than one event per individual, which is the reason that a modern term for survival analysis is event history analysis. For the ith individual, a stochastic process $N_i(t)$ counts the number of events up to time t. Regression models are designed to estimate if and how certain covariates affect the occurrence of events described by the counting process N_i. Models are usually defined by the so-called intensity which we introduce in the following. The Doob-Meyer decomposition guarantees that under integrability conditions, there are predictable, increasing processes Λ_i such that

$$N_i(t) - \Lambda_i(t)$$

is a local martingale. If the paths of the compensator Λ_i are absolutely continuous with respect to Lebesgue measure then a predictable process λ_i such that

$$\Lambda_i(t) = \int_0^t \lambda_i(s)\,\mathrm{d}s$$

is called the intensity of N_i.

1.1 Some Models from Survival Analysis

Several suggestions have been made how to model the intensity λ_i for the ith individual. The most prominent model is the proportional hazards model introduced by Cox (1972), which specifies that the intensity has the following form:

$$\lambda_i(t) = \lambda_0(t)\exp(\boldsymbol{Z}_i(t)\boldsymbol{\beta})R_i(t), \tag{1.1}$$

where the observable covariates $\boldsymbol{Z}_i$ are $k_{\boldsymbol{\beta}}$-dimensional row vectors of predictable stochastic processes, the deterministic function λ_0 and the vector $\boldsymbol{\beta} \in \mathbb{R}^{k_{\boldsymbol{\beta}}}$ are unknown, and the so-called at-risk indicators R_i are observable stochastic processes which indicate whether an individual is at risk or not by taking the values 0 or 1. Vectors are column vectors unless indicated otherwise. In the classical setup with at most one event per individual, $R_i(t)$ is 0 iff the event or the censoring has occurred prior to t. The Cox model is a semiparametric model, i.e. it contains an unknown finite-dimensional parameter and an unknown function.

Aalen (1980) introduced the following additive model:

$$\lambda_i(t) = \boldsymbol{Y}_i(t)\boldsymbol{\alpha}(t), \tag{1.2}$$

where the observable covariates $\boldsymbol{Y}_i$ are row vectors of predictable stochastic processes and $\boldsymbol{\alpha}$ is an unknown deterministic vector-valued function. Note that the at-risk indicator is included in $\boldsymbol{Y}_i$. The Aalen model is a nonparametric model, i.e. it only contains unknown functions.

Parametric models have been put forward as well. For example, in Andersen et al. (1993, Example VII.6.1), the following multiplicative model is considered:

$$\lambda_i(t) = a(t, \boldsymbol{\gamma}) r(\boldsymbol{Z}_i(t)\boldsymbol{\beta}) R_i(t),$$

where the observable covariates $\boldsymbol{Z}_i$ are row vectors of predictable stochastic processes, r and a are some known functions, $\boldsymbol{\gamma}$ and $\boldsymbol{\beta}$ are finite-dimensional parameters, and R_i are the at-risk indicators. Basically, this model is a parametric version of a generalized Cox model.

Various generalizations of these models have been proposed, e.g. a semiparametric restriction of the Aalen model (McKeague and Sasieni, 1994):

$$\lambda_i(t) = \boldsymbol{Y}_i^c(t)\boldsymbol{\alpha}^c + \boldsymbol{Y}_i^v(t)\boldsymbol{\alpha}^v(t), \tag{1.3}$$

a general additive-multiplicative hazard model (Lin and Ying, 1995):

$$\lambda_i(t) = \left[g(\boldsymbol{Y}_i^v(t)\boldsymbol{\beta}) + \lambda_0(t) h(\boldsymbol{Y}_i^c(t)\boldsymbol{\alpha}^c)\right] R_i(t),$$

a sum of a Cox and an Aalen model (Martinussen and Scheike, 2002):

$$\lambda_i(t) = \left[\boldsymbol{Y}_i^v(t)\boldsymbol{\alpha}^v(t) + \lambda_0(t)\exp(\boldsymbol{Y}_i^c(t)\boldsymbol{\alpha}^c)\right] R_i(t),$$

and a product of a Cox and an Aalen model (Scheike and Zhang, 2002):

$$\lambda_i(t) = \boldsymbol{Y}_i^v(t)\boldsymbol{\alpha}^v(t)\exp(\boldsymbol{Y}_i^c(t)\boldsymbol{\alpha}^c),$$

where the observable covariates $\boldsymbol{Y}_i^c$ and $\boldsymbol{Y}_i^v$ are row vectors of predictable stochastic processes, g and h are known functions, $\boldsymbol{\alpha}^c$ and $\boldsymbol{\beta}$ are unknown vectors, λ_0 is an unknown function, $\boldsymbol{\alpha}^v$ is an unknown vector-valued function, and R_i is the at-risk indicator.

The Cox model has been extended to time-dependent coefficients as well (Murphy and Sen, 1991; Martinussen et al., 2002):

$$\lambda_i(t) = \lambda_0(t)\exp(\boldsymbol{Z}_i(t)\boldsymbol{\beta}(t)) R_i(t),$$

where λ_0, $\boldsymbol{Z}_i$, and R_i are as in the classical Cox model (1.1) and the unknown deterministic function $\boldsymbol{\beta}$ is allowed to depend on time.

Different techniques are used to estimate the parameters of the above models. Generally, finite-dimensional parameters can be estimated consistently at a rate of $\sqrt{n}$, where n is the number of individuals. Rather than estimating the unknown functions in the models directly, integrals over the functions are estimated. For example in the Cox model, $\int_0^t \lambda_0(s)\,\mathrm{d}s$ is estimated. Usually, these estimators are $\sqrt{n}$-consistent as well. Estimators for the functions themselves are derived by smoothing the estimators of the integrated functions. The smoothed estimators usually converge at a rate of less than $\sqrt{n}$.

Details about these models and about some of the various other models that have been suggested, can be found for example in Andersen et al. (1993), Hougaard (2000), and Therneau and Grambsch (2000).

1.2 The Test Statistic

In applications, one faces the question if a particular model has a good fit or if maybe another model should be used. It is natural to employ goodness-of-fit tests for this purpose. However, for some of the above models no goodness-of-fit tests have been introduced so far. The tests that have been introduced are specific tests for one model. This is mainly due to the fact that the asymptotic distribution of the test statistic depends on which particular estimator is used for parameters of the null hypothesis.

We present goodness-of-fit tests that can be used for any of the above models. The main idea is to ensure that the asymptotic distribution of our test statistic does not depend on which estimators are used for parameters from the null hypothesis. For our test to work, we only need to know that the estimators converge at certain rates.

All models specified in Section 1.1 can be written in the following form:

$$\lambda_i(t) = f(\boldsymbol{X}_i(t), \boldsymbol{\alpha}^v(t), \boldsymbol{\alpha}^c), \tag{1.4}$$

where f is a known function, $\boldsymbol{X}_i$ are observable stochastic processes, $\boldsymbol{\alpha}^v$ is an unknown vector-valued function and $\boldsymbol{\alpha}^c$ is a finite-dimensional parameter. This is the class of models we shall be dealing with. It contains the most important models from survival analysis. However, it does not cover all models from survival analysis; e.g. frailty models, which include unobserved random effects, are not covered. The problem with frailty models is that with respect to the filtration generated by the observable processes, the interaction between covariates and parameters in the intensity may be more complicated than (1.4).

The basic idea for our test is as follows. By definition,

$$M_i(t) = N_i(t) - \int_0^t \lambda_i(s)\,\mathrm{d}s$$

are mean zero local martingales, meaning heuristically that they should 'fluctuate' around 0. Replacing λ_i by the estimated intensity $\hat{\lambda}_i(\cdot) = f(\boldsymbol{X}_i(\cdot), \widehat{\boldsymbol{\alpha}^v}(\cdot), \widehat{\boldsymbol{\alpha}^c})$, where $\widehat{\boldsymbol{\alpha}^v}$ and $\widehat{\boldsymbol{\alpha}^c}$ are some estimators of $\boldsymbol{\alpha}^v$ and $\boldsymbol{\alpha}^c$, we get the so-called martingale residuals

$$\widehat{M}_i(t) = N_i(t) - \int_0^t f(\boldsymbol{X}_i(s), \widehat{\boldsymbol{\alpha}^v}(s), \widehat{\boldsymbol{\alpha}^c})\,\mathrm{d}s.$$

If (1.4) holds true, f is continuous, and $\widehat{\boldsymbol{\alpha}^v}$ and $\widehat{\boldsymbol{\alpha}^c}$ are close to $\boldsymbol{\alpha}^v$ and $\boldsymbol{\alpha}^c$, then $\widehat{M}_i$ should still 'fluctuate' around 0, but $\widehat{M}_i$ need not be mean zero local martingales. The test statistic we use is basically a weighted average of the martingale residuals $\widehat{M}_i$. More precisely, we use the stochastic process $n^{-1/2}\sum_{i=1}^n \int_0^t c_i(s)\,\mathrm{d}\widehat{M}_i(s)$, where the weights c_i are some predictable stochastic processes. In other words, we use the test statistic $T(\widehat{\boldsymbol{\alpha}^v}, \widehat{\boldsymbol{\alpha}^c}, t)$, where

$$T(\boldsymbol{\alpha}^v, \boldsymbol{\alpha}^c, t) = n^{-\frac{1}{2}} \sum_{i=1}^{n} \int_0^t c_i(s) \left[\mathrm{d}N_i(s) - f(\boldsymbol{X}_i(s), \boldsymbol{\alpha}^v(s), \boldsymbol{\alpha}^c) \, \mathrm{d}s \right].$$

If the null hypothesis holds true, i.e. we have $\lambda_i(\cdot) = f(\boldsymbol{X}_i(\cdot), \boldsymbol{\alpha}_0^v(\cdot), \boldsymbol{\alpha}_0^c)$ for some $\boldsymbol{\alpha}_0^v$, $\boldsymbol{\alpha}_0^c$, then

$$T(\boldsymbol{\alpha}_0^v, \boldsymbol{\alpha}_0^c, t) = n^{-\frac{1}{2}} \sum_{i=1}^{n} \int_0^t c_i(s) \, \mathrm{d}M_i(s)$$

is a mean zero local martingale and, under regularity conditions, it converges weakly to a mean zero Gaussian process with independent increments as the number of individuals n tends to infinity. This can be shown by a central limit theorem for martingales.

The difference $T(\widehat{\boldsymbol{\alpha}^v}, \widehat{\boldsymbol{\alpha}^c}, t) - T(\boldsymbol{\alpha}_0^v, \boldsymbol{\alpha}_0^c, t)$ usually does not vanish as $n \to \infty$. Indeed, by a Taylor expansion

$$\begin{aligned} &T(\widehat{\boldsymbol{\alpha}^v}, \widehat{\boldsymbol{\alpha}^c}, t) - T(\boldsymbol{\alpha}_0^v, \boldsymbol{\alpha}_0^c, t) = \\ &= \int_0^t \left(\frac{1}{n} \sum_{i=1}^{n} c_i(s) \frac{\partial}{\partial \boldsymbol{\alpha}^v} f(\boldsymbol{X}_i(s), \boldsymbol{\alpha}^v, \boldsymbol{\alpha}_0^c) \Big|_{\boldsymbol{\alpha}^v = \boldsymbol{\alpha}_0^v(s)} \right) n^{\frac{1}{2}} (\widehat{\boldsymbol{\alpha}^v}(s) - \boldsymbol{\alpha}_0^v(s)) \, \mathrm{d}s \\ &\quad + \left(\int_0^t \frac{1}{n} \sum_{i=1}^{n} c_i(s) \frac{\partial}{\partial \boldsymbol{\alpha}^c} f(\boldsymbol{X}_i(s), \boldsymbol{\alpha}_0^v(s), \boldsymbol{\alpha}^c) \Big|_{\boldsymbol{\alpha}^c = \boldsymbol{\alpha}_0^c} \mathrm{d}s \right) n^{\frac{1}{2}} (\widehat{\boldsymbol{\alpha}^c} - \boldsymbol{\alpha}_0^c) \\ &\quad + R, \end{aligned} \tag{1.5}$$

where R is the remainder term which will vanish asymptotically under some regularity conditions. The other terms on the right hand side typically do not vanish. To derive the asymptotic distribution of $T(\widehat{\boldsymbol{\alpha}^v}, \widehat{\boldsymbol{\alpha}^c}, t)$, one would have to use Slutsky-type arguments and consider the joint asymptotic distribution of $T(\boldsymbol{\alpha}_0^v, \boldsymbol{\alpha}_0^c, t)$, $n^{1/2}(\widehat{\boldsymbol{\alpha}^v}(\cdot) - \boldsymbol{\alpha}_0^v(\cdot))$, and $n^{1/2}(\widehat{\boldsymbol{\alpha}^c} - \boldsymbol{\alpha}_0^c)$, which depends on the particular estimators used and may not be known.

The key idea of our approach is to simplify the above by imposing orthogonality conditions on the weights $\boldsymbol{c}$ to guarantee that $T(\widehat{\boldsymbol{\alpha}^v}, \widehat{\boldsymbol{\alpha}^c}, t)$ and $T(\boldsymbol{\alpha}_0^v, \boldsymbol{\alpha}_0^c, t)$ are asymptotically equivalent. This ensures that $T(\widehat{\boldsymbol{\alpha}^v}, \widehat{\boldsymbol{\alpha}^c}, t)$ converges to a mean zero Gaussian process with independent increments.

To get the first term in (1.5) to vanish, we require for all s, $\boldsymbol{\alpha}^v(s)$, and $\boldsymbol{\alpha}^c$ that

$$\sum_{i=1}^{n} c_i(s) \frac{\partial}{\partial \boldsymbol{\alpha}^v} f(\boldsymbol{X}_i(s), \boldsymbol{\alpha}^v, \boldsymbol{\alpha}^c) \Big|_{\boldsymbol{\alpha}^v = \boldsymbol{\alpha}^v(s)} = \mathbf{0}. \tag{1.6}$$

To get the second term in (1.5) to vanish for each t, we require for all s, $\boldsymbol{\alpha}^v(s)$, and $\boldsymbol{\alpha}^c$ that

$$\sum_{i=1}^{n} c_i(s) \frac{\partial}{\partial \boldsymbol{\alpha}^c} f(\boldsymbol{X}_i(s), \boldsymbol{\alpha}^v(s), \boldsymbol{\alpha}^c) = \mathbf{0}. \tag{1.7}$$

If our test does not use the whole path of the stochastic process $T(\widehat{\boldsymbol{\alpha}^v}, \widehat{\boldsymbol{\alpha}^c}, \cdot)$ we can relax (1.7). For example, if our test is only based on $T(\widehat{\boldsymbol{\alpha}^v}, \widehat{\boldsymbol{\alpha}^c}, \tau)$, for some $\tau > 0$, then condition (1.7) can be relaxed and replaced by

$$\int_0^\tau \frac{1}{n} \sum_{i=1}^{n} c_i(s) \frac{\partial}{\partial \boldsymbol{\alpha}^c} f(\boldsymbol{X}_i(s), \boldsymbol{\alpha}^v(s), \boldsymbol{\alpha}^c) \, \mathrm{d}s = \mathbf{0}, \tag{1.8}$$

for all $\boldsymbol{\alpha}^v$ and $\boldsymbol{\alpha}^c$.

Of course, the requirements on the weights complicate things a bit, since weights c_i that satisfy these conditions may have to depend on $\boldsymbol{\alpha}^v$ and $\boldsymbol{\alpha}^c$. Plugging in estimators destroys the predictability of c_i, but fortunately the asymptotic results do not change.

Many other goodness-of-fit tests have used weighted sums of the difference between the counting process N_i and its estimated compensator $\hat{\Lambda}_i$ as basis. Besides working with a more general model, what sets our approach apart from classical approaches is the following: Most goodness-of-fit tests implicitly restrict or transform the weights to get rid only of some of the terms on the right hand side of (1.5). For example in tests for the Cox model (1.1), usually the weights are transformed to get rid of the first term in (1.5) but not the second term. As a consequence, the asymptotic distributions of these tests are more complicated and the tests are only applicable for one particular model when used with specific estimators. In contrast, our test can be used with any kind of estimator that is consistent at a certain rate.

If f is affine-linear in some of the parameters, then, under some further conditions, (1.6)-(1.8) do not depend on the particular value of these parameters. Hence, we may choose weights that do not depend on them. In this case, we will see that we do not need to estimate them at all, since they are not required to compute the test statistic or the estimator of its variance.

1.3 Directing the Test

A problem with many goodness-of-fit tests for nonparametric or semiparametric models is that they may exhibit a low power against many alternatives and often it is not clear against which alternatives the tests are particularly sensitive. We aim to improve this in the following way:

Suppose we want to check a certain model. We use this model as null hypothesis of our test. The alternative hypothesis is given by all intensities not contained in the null hypothesis. We assume that we have a second model, called 'competing model', which is given by another set of intensities. For example, the competing model can be one of the models mentioned in Section 1.1. Often, we use a Cox model (1.1) as competing model. The competing model and the null hypothesis need not be separated - they may share certain intensities. In the terms of Vuong (1989), this means that they may be overlapping or nested. For example, the null hypothesis and the competing model can be of the same type, e.g. both can be Cox models, but with different covariates.

We choose the weights c_i to make the test powerful if the competing model holds true and the null hypothesis fails. Since the competing model is not the alternative hypothesis, a rejection of the null hypothesis is no conclusive evidence in favor of the competing model.

In order to direct the test, we need to use parameters of the competing model and the weight process $\boldsymbol{c}$ will depend on those parameters. All parameters besides $\boldsymbol{\alpha} = (\boldsymbol{\alpha}^v, \boldsymbol{\alpha}^c)$ on which $\boldsymbol{c}$ depends will be denoted by $\boldsymbol{\beta}$. In addition to containing parameters from competing models, some components of $\boldsymbol{\beta}$ will also be used to satisfy the orthogonality condition (1.8). The combined parameter $(\boldsymbol{\alpha}, \boldsymbol{\beta})$ will be denoted by $\boldsymbol{\theta}$. Therefore, we often write $T(\widehat{\boldsymbol{\theta}}, t)$ for the test statistic, where $\widehat{\boldsymbol{\theta}}$ is an estimator for $\boldsymbol{\theta}$.

A simple approach to satisfy (1.6) and one of (1.7) or (1.8) is to define $\boldsymbol{c}$ by an (unweighted) orthogonal projection of an arbitrary process $\boldsymbol{d}$. Setting $\boldsymbol{d}$ to an estimate of the difference between the intensity under the competing model and the intensity under the null hypothesis already leads to a good power.

Since the asymptotic distribution is relatively simple, we can improve the power. In fact, we will show how to choose optimal weights against fixed alternatives with respect to the approximate Bahadur efficiency and against local alternatives with respect to the Pitman efficiency. This leads to weighted orthogonal projections.

Under certain conditions, the variance of our test statistic $T(\widehat{\boldsymbol{\theta}}, t)$ may converge to 0. In that case, the usual standardized versions of $T(\widehat{\boldsymbol{\theta}}, t)$ may not converge in distribution and if they converge, it is not clear what the limiting distribution is. In particular, the variance may converge to 0 if the null hypothesis and the competing model are overlapping. We propose several approaches how to deal with this problem. Among them is a test for an equivalent condition for the convergence of the variance to 0. We will only discuss this test for the special case of the Cox model (1.1) as null hypothesis.

1.4 Outline

The thesis is structured as follows.

In Chapter 2, we review some models from survival analysis and discuss properties of their estimators. After that, we present some previous approaches to test goodness-of-fit. Most of these tests use the Cox model as null hypothesis.

In Chapter 3, we consider a relatively simple model in order to demonstrate some important ideas without being distracted by too many technical details. Indeed, we present tests for Aalen's additive risk model (1.2). We will show that in this case our approach leads to consistent tests.

Chapter 4 contains some preparations for the general results. In Section 4.1, we discuss convergence rates of estimators if the model for which they are designed may not be true. In Section 4.2, we present a smoothing method which we need as a preparation for checking more general models in the following chapters.

In Chapter 5, we consider the case of the time-dependent orthogonalization of the weights c_i by conditions (1.6) and (1.7). In this case, our test does not use the information that $\boldsymbol{\alpha}^c$ does not depend on time. Therefore we work with the simplified model:

$$\lambda_i(t) = f(\boldsymbol{X}_i(t), \boldsymbol{\alpha}^v(t)),$$

for which, of course, only condition (1.6) is needed.

Chapter 6 is devoted to the study of the test statistic $T(\widehat{\boldsymbol{\alpha}^v}, \widehat{\boldsymbol{\alpha}^c}, \tau)$ under the conditions (1.6) and (1.8). Now it is important to distinguish between $\boldsymbol{\alpha}^v$ and $\boldsymbol{\alpha}^c$ and hence, the full model (1.4) is used. The proofs of this chapter are considerably simplified by using the results of Chapter 5.

As already mentioned, the variance of the test statistic may converge to 0 as $n \to \infty$. In Chapter 7, we suggest several approaches how to deal with this problem. Among them is a suggestion for a sequential test for which we show the asymptotics.

In Chapter 8, several special cases for our general model are considered and results of simulation studies for most of these special cases are presented. The special cases considered are the Aalen model (1.2), the semiparametric restriction of the Aalen model (1.3), a generalized version of the Cox model (1.1) and parametric models.

In Chapter 9, we apply our tests to three datasets. Two datasets are from biostatistics: the PBC dataset (Fleming and Harrington, 1991), and the Stanford heart transplant data (Miller and Halpern, 1982). Another dataset is from software reliability (Gandy and Jensen, 2004).

Chapter 10 contains several remarks about aspects of our test: tests against several competing models, possible extensions to other models, relation to estimation in semiparametric models, and a case in which the power of our test is the same as if the null hypothesis was a simple hypothesis. Chapter 11 contains some concluding remarks.

In Appendix A, some technical results from the theory of stochastic processes are collected. Some facts about orthogonal projections can be found in Appendix B. After that there is a list of symbols and abbreviations, a list of conditions, the bibliography and a summary in German.

The basic ideas of the least squares projection used in Chapter 3 have already been published in Gandy and Jensen (2005c). The ideas of Chapter 5 for the special case of Cox models as well as the sequential procedure described in Sections 7.1-7.4 are submitted for publication (Gandy and Jensen, 2005b). The least squares projections used in Chapter 6 in the special case of the semiparametric restriction of the Aalen model (1.3) which are explicitly given in Subsection 8.2.1 are accepted for publication (Gandy and Jensen, 2005a). The present thesis generalizes and unifies these papers. The example of the software reliability data used in Section 9.1 has been published in Gandy and Jensen (2004).

Chapter 2

A Review of Models and Checks from Survival Analysis

In the last decades, the field of survival analysis has become very broad. Despite the large number of publications in this subject, the key reference still is the book by Andersen et al. (1993). Other books specialize on a certain topic, e.g. Fleming and Harrington (1991) on the Cox model, Therneau and Grambsch (2000) on extensions of the Cox model, Bagdonavicius and Nikulin (1998) on accelerated testing, and Hougaard (2000) on multivariate survival analysis. By now, many applied books on survival analysis are available, e.g. Cox and Oakes (1984), Parmar and Machin (1995), Hosmer and Lemeshow (1999), Kleinbaum (1996), Miller (1998), Elandt-Johnson and Johnson (1999), Smith (2002), Lee and Wang (2003), Klein and Moeschberger (2003) and Tableman and Kim (2004), some of which are new editions or reprints of older books.

In this chapter, we introduce the counting process approach to survival analysis which has its origins in the work by Odd Aalen, see e.g. Aalen (1977, 1978, 1980). Using the counting process setup, we introduce some regression models. Furthermore, we review some model checks from the literature.

In Section 2.1, we consider the classical example in survival analysis, a clinical trial. We show how the counting process comes into play. In Section 2.2, the basic counting process setup is given, which will also be used in later chapters. Section 2.3 introduces some further notation, which will be used throughout the thesis. Section 2.4 contains some models from survival analysis and some standard estimators in these models. We provide considerably more detail than in Section 1.1. In Section 2.5, we review model checks in survival analysis. The models for which the most checks have been proposed is the Cox model.

2.1 A Classical Example: Survival Analysis in a Clinical Trial

Consider the following example: In a clinical trial, patients are observed and one is interested in the length of survival after a certain surgery. Patients enter the study when the surgery is performed. In Figure 2.1, a timeline for a typical clinical study can be seen. Individuals enter the study over time. Some of them experience the event one is interested in. This can be relapse or it can be death. Some individuals do not experience the event: This can be because the study ends or because they moved and were lost to follow up. However, it may also be that they died of other causes. For those individuals it is only known that they were observed for some time without experiencing the event. The phenomenon that some individuals are no longer observed after a certain point in time is called right censoring.

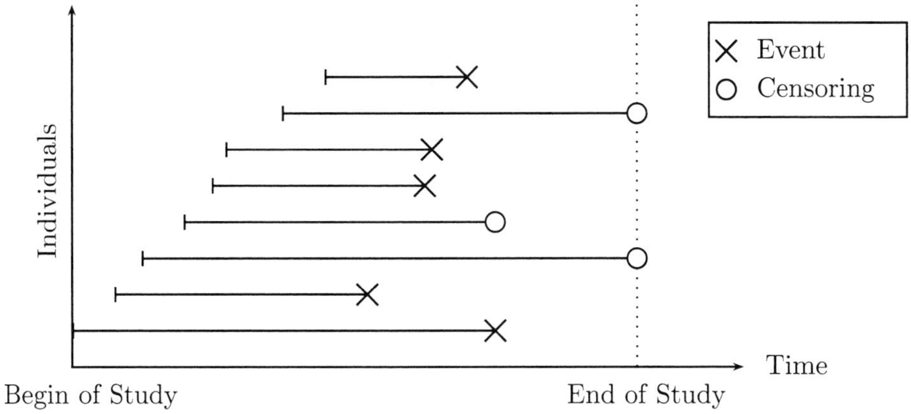

Figure 2.1: Timeline of a clinical trial.

For analyzing this type of data, usually a different time-scale is chosen. One considers the time since individuals entered the study. This is illustrated in Figure 2.2.

So, usually one assumes that for each individual i, there are two random times: T_i, the time of the event and C_i, the time of censoring, i.e. the time after which no more events are observed. One observes only the minimum X_i of C_i and T_i, i.e. $X_i = C_i \wedge T_i$, and whether an event occurs via the indicator variable $\delta_i = I\{T_i = X_i\}$. The connection to counting processes is as follows: Let $N_i(t) = \delta_i I\{X_i \leq t\}$. If the ith individual does not experience the event then

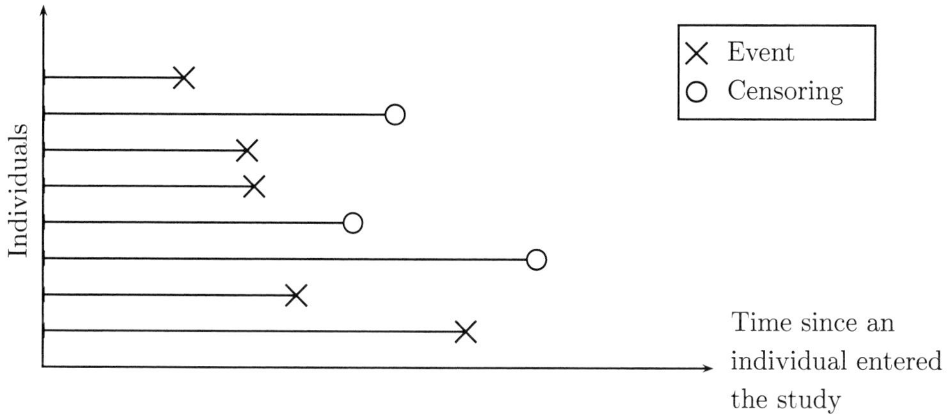

Figure 2.2: Usual timeline for models in survival analysis.

$N_i(t) = 0$ for all t. Otherwise, N_i has precisely one jump from 0 to 1 at time T_i, see Figure 2.3. Hence, N_i is a counting process, albeit a very simple one. The advantage of considering counting processes is that one can define models by the intensity λ_i of the counting process N_i and then use martingale theory for the difference between the counting process and the cumulated intensity $\int_0^t \lambda_i(s)\,\mathrm{d}s$.

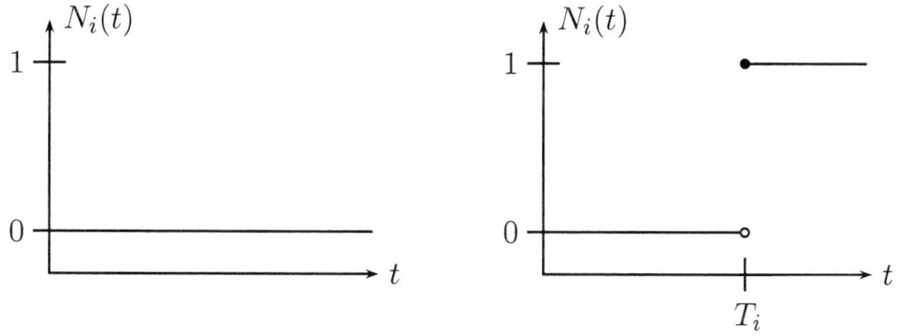

Figure 2.3: Counting processes for classical survival analysis with at most one event per individual. Left: no event, Right: event at time T_i.

2.2 General Setup

It is not necessary to restrict oneself to counting processes having at most one jump. We consider the following setup:

Suppose that we observe events during the finite time interval $[0, \tau]$, where $0 < \tau < \infty$. We assume that we have an underlying probability space $(\Omega, \mathcal{F}, \mathrm{P})$

on which all stochastic processes are defined. Furthermore, let $(\mathcal{F}_t), t \in [0, \tau]$, be a filtration to which properties such as adapted or predictable pertain. $\mathrm{E}[\cdot]$ denotes expectation with respect to P.

We observe an n-variate counting process $\boldsymbol{N} = (N_1, \dots, N_n)^\top$, whose elements have no common jumps, meaning that $\boldsymbol{N}(0) = \boldsymbol{0}$, the paths of the N_i are piecewise constant with jumps that are upward of size one and $\Delta N_i(t) \Delta N_j(t) = 0$ for all $i \neq j$ and all $t \in [0, \tau]$. We assume that $\boldsymbol{N}$ is adapted and admits an intensity $\boldsymbol{\lambda} = (\lambda_1, \dots, \lambda_n)^\top$, i.e. $\boldsymbol{\lambda}$ is a locally bounded predictable process and $\boldsymbol{M}(t) = (M_1(t), \dots, M_n(t))^\top := \boldsymbol{N}(t) - \int_0^t \boldsymbol{\lambda}(s)\, \mathrm{d}s$ is a vector of local martingales. The integrated intensity is denoted by $\boldsymbol{\Lambda}(t) = (\Lambda_1(t), \dots, \Lambda_n(t))^\top := \int_0^t \boldsymbol{\lambda}(s)\, \mathrm{d}s$. An example for a sample path of N_i, Λ_i, and M_i can be seen in Figure 2.4.

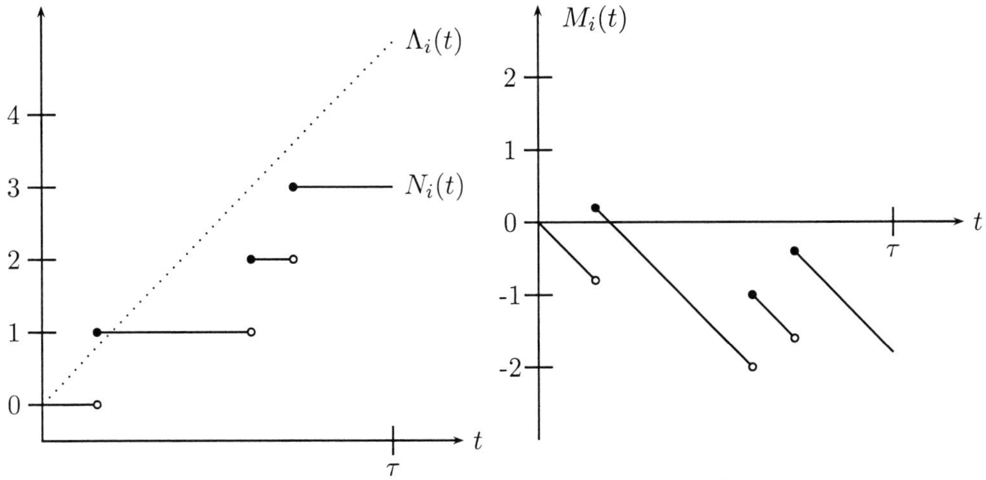

Figure 2.4: Example of sample paths for N_i, Λ_i, and M_i (assuming constant intensity $\lambda_i(t)$).

2.3 Notation

We will mostly be concerned with the limiting behavior of the test statistic T mentioned in the introduction, as the dimension of the counting process becomes large, i.e. n tends to infinity. Since this is basically the only limiting behavior we consider, we suppress indicating this when we mention limits, i.e. limits in this thesis will always be as $n \to \infty$. Stochastic convergence is denoted by $\xrightarrow{\mathrm{P}}$ and convergence in distribution is denoted by $\xrightarrow{\mathrm{d}}$. We use the convention $0/0 = 0$. The minimum of a and b is denoted by $a \wedge b$.

We always write matrices and vectors in bold face $(\boldsymbol{A}, \boldsymbol{x})$ and if we refer to elements of a matrix or a vector we denote the elements by A_{ij} or x_i. Furthermore, $\boldsymbol{A}_i$ denotes the i-th row of $\boldsymbol{A}$, i.e. if $\boldsymbol{A}$ has k columns, $\boldsymbol{A}_i = (A_{i1}, \dots, A_{ik})$.

The transpose of $\boldsymbol{A}$ is denoted by $\boldsymbol{A}^\top$. Usually, matrices are denoted by upper case letters and vectors by lower case letters, but this is no general rule. Vectors are column vectors unless indicated otherwise. $\boldsymbol{0}$ (resp. $\boldsymbol{1}$) denotes the vector, row vector, or matrix whose components are all 0 (resp. 1). The identity matrix is denoted by $\boldsymbol{I}$. The dimensions of $\boldsymbol{0}$, $\boldsymbol{1}$, and $\boldsymbol{I}$ will be clear from the context. As a shorthand, for vectors $\boldsymbol{x}$, $\boldsymbol{y} \in \mathbb{R}^n$, we use the elementwise product $\boldsymbol{xy} := (x_i y_i)_{i=1,\dots,n}$, the elementwise division $\boldsymbol{x}/\boldsymbol{y} := (x_i/y_i)_{i=1,\dots,n}$, and for a function $f : \mathbb{R} \to \mathbb{R}$, we let $f(\boldsymbol{x}) := (f(x_i))_{i=1,\dots,n}$. In particular, $\boldsymbol{x}^k := (x_i^k)_{i=1,\dots,n}$, for integers k. If a matrix $\boldsymbol{A} \in \mathbb{R}^{k\times k}$ is not invertible then $\boldsymbol{A}^{-1}$ is defined to be the $k \times k$ matrix with all elements equal to 0.

For vectors, $\|\cdot\|$ denotes the Euclidean norm. For matrices $\boldsymbol{A}$, the matrix norm induced by the Euclidean norm is denoted by $\|\boldsymbol{A}\|$, i.e. $\|\boldsymbol{A}\| = \sup \|\boldsymbol{Ax}\|$, where the sup is over all vectors $\boldsymbol{x}$ such that $\|\boldsymbol{x}\| = 1$. For sequences (X_n), $n \in \mathbb{N}$, of random variables we say $X_n = O_{\mathrm{P}}(1)$ if for each $\epsilon > 0$ there exists $K > 0$ such that $\sup_{n\in\mathbb{N}} \mathrm{P}(|X_n| > K) < \epsilon$. If (a_n), $n \in \mathbb{N}$, is another sequence of real valued random variables, we write $X_n = O_{\mathrm{P}}(a_n)$ if $X_n/a_n = O_{\mathrm{P}}(1)$. If $(\boldsymbol{X}_n)$, $n \in \mathbb{N}$, is a sequence of random vectors or random matrices, $O_{\mathrm{P}}(\cdot)$ is defined in the same way with $|\cdot|$ replaced by $\|\cdot\|$. For sequences (X_n), $n \in \mathbb{N}$, and $(a_n), n \in \mathbb{N}$, of random variables, we say $X_n = o_{\mathrm{P}}(a_n)$ if $X_n/a_n \xrightarrow{\mathrm{P}} 0$.

For a Borel-measurable set $A \subset \mathbb{R}^p$, the set of all bounded, measurable mappings from $[0,\tau]$ into A is denoted by $\mathrm{bm}(A)$.

We also introduce some non-standard notation which will ease the presentation considerably. For matrices $\boldsymbol{A} \in \mathbb{R}^{n\times a}$, $\boldsymbol{B} \in \mathbb{R}^{n\times b}$ and vectors $\boldsymbol{x} \in \mathbb{R}^n$ with $a, b, n \in \mathbb{N}$ we define

$$\overline{\boldsymbol{A}} = \frac{1}{n}\boldsymbol{A}^\top \boldsymbol{1}, \quad \overline{\boldsymbol{AB}} = \frac{1}{n}\boldsymbol{A}^\top \boldsymbol{B}, \quad \overline{\boldsymbol{AxB}} = \frac{1}{n}\boldsymbol{A}^\top \mathrm{diag}(\boldsymbol{x})\boldsymbol{B},$$

where $\boldsymbol{1} = (1,\dots,1)^\top \in \mathbb{R}^n$ and $\mathrm{diag}(\boldsymbol{x})$ denotes the diagonal matrix with the elements of $\boldsymbol{x}$ on its diagonal. More generally, suppose we have $k \in \mathbb{N}$ matrices $\boldsymbol{A}^{(1)} \in \mathbb{R}^{n\times a_1}, \dots, \boldsymbol{A}^{(k)} \in \mathbb{R}^{n\times a_k}$. Then we define for all $b_j \in \{1,\dots,a_j\}, j = 1,\dots,k$,

$$\overline{\boldsymbol{A}^{(1)}\cdots\boldsymbol{A}^{(k)}}_{b_1\dots b_k} := \frac{1}{n}\sum_{i=1}^{n}\prod_{j=1}^{k} A^{(j)}_{ib_j} = \frac{1}{n}\sum_{i=1}^{n} A^{(1)}_{ib_1}\cdots A^{(k)}_{ib_k}.$$

We call $\overline{\boldsymbol{A}^{(1)}\cdots\boldsymbol{A}^{(k)}}$ *product mean*. Of course, we interpret n-variate vectors as $n \times 1$-dimensional matrices. The matrices and vectors used in the product mean are allowed to be random and may also depend on further parameters which will be indicated in parentheses, e.g. if $\boldsymbol{A}$ and $\boldsymbol{B}$ depend on $t \in [0,\tau]$ then $\overline{\boldsymbol{AB}}(t) = \frac{1}{n}\boldsymbol{A}(t)^\top \boldsymbol{B}(t)$.

If $\boldsymbol{A}^{(1)}, \ldots, \boldsymbol{A}^{(k)}$ depend on $\theta \in \Theta$ where Θ is some measurable space we say that

$$\overline{\boldsymbol{A}^{(1)} \cdots \boldsymbol{A}^{(k)}} \text{ converges } \textit{uip} \text{ (uniformly in probability) on } \Theta$$

if there exists a deterministic, measurable, bounded function $\overrightarrow{\boldsymbol{A}^{(1)} \cdots \boldsymbol{A}^{(k)}} : \Theta \to \mathbb{R}^{a_1 \times \cdots \times a_k}$ such that for all $b_j \in \{1, \ldots, a_j\}$, $j = 1, \ldots, k$,

$$\sup_{\theta \in \Theta} \left| \overline{\boldsymbol{A}^{(1)} \cdots \boldsymbol{A}^{(k)}}_{b_1 \ldots b_k}(\theta) - \overrightarrow{\boldsymbol{A}^{(1)} \cdots \boldsymbol{A}^{(k)}}_{b_1 \ldots b_k}(\theta) \right| \xrightarrow{\mathrm{P}} 0.$$

This includes the assumption that the left hand side is a real valued random variable. The limit of $\overline{\boldsymbol{A}^{(1)} \cdots \boldsymbol{A}^{(k)}}$ will always be denoted by $\overrightarrow{\boldsymbol{A}^{(1)} \cdots \boldsymbol{A}^{(k)}}$.

2.4 Models and their Properties

The purpose of this section is to introduce some models from survival analysis and some commonly used estimators. For our purposes, it is of interest that these estimators are usually $\sqrt{n}$-consistent.

2.4.1 Cox-Type Models

Cox's proportional hazards model (Cox, 1972; Andersen and Gill, 1982) is the most widely used model in survival analysis and event history analysis. It specifies that the intensity $\boldsymbol{\lambda} = (\lambda_1, \ldots, \lambda_n)^\top$ of a counting process $\boldsymbol{N} = (N_1, \ldots, N_n)^\top$ can be written as

$$\lambda_i(t) = \lambda_0(t) \exp(\boldsymbol{Z}_i(t)\boldsymbol{\beta}) R_i(t), \tag{2.1}$$

where the observable row vector of covariates $\boldsymbol{Z}_i$ is a $k_{\boldsymbol{\beta}}$-dimensional predictable stochastic process, $R_i \in \{0, 1\}$ is the at-risk indicator, $\boldsymbol{\beta} \in \mathbb{R}^{k_{\boldsymbol{\beta}}}$ is an unknown regression parameter and the deterministic, nonnegative baseline λ_0 is unspecified. An extension to (2.1) is discussed by Prentice and Self (1983), where

$$\lambda_i(t) = \lambda_0(t) r(\boldsymbol{Z}_i(t)\boldsymbol{\beta}) R_i(t) \tag{2.2}$$

for a known function $r : \mathbb{R} \to [0, \infty)$.

The parameter $\boldsymbol{\beta}$ is usually estimated by a maximum partial likelihood approach (Cox, 1972; Andersen and Gill, 1982), i.e. $\hat{\boldsymbol{\beta}}$ is the maximizer of $C(\boldsymbol{\beta})$ where

$$C(\boldsymbol{\beta}) := \sum_{i=1}^{n} \int_0^{\tau} \log(\rho_i(\boldsymbol{\beta}, s)) \, \mathrm{d}N_i(s) - \int_0^{\tau} \log(n\overline{\boldsymbol{\rho}}(\boldsymbol{\beta}, s)) n \, \mathrm{d}\overline{\boldsymbol{N}}(s), \tag{2.3}$$

and $\rho_i(\boldsymbol{\beta}, t) = R_i(t) r(\boldsymbol{Z}_i(t)\boldsymbol{\beta})$. The standard Breslow estimator $\hat{\Lambda}_0(t)$ of $\Lambda_0(t) = \int_0^t \lambda_0(s) \, \mathrm{d}s$ is given by

$$\hat{\Lambda}_0(t) = \int_0^t \frac{\mathrm{d}\overline{\boldsymbol{N}}(s)}{\overline{\boldsymbol{\rho}}(\widehat{\boldsymbol{\beta}}, s))}. \tag{2.4}$$

Under regularity conditions, it is known that $\sqrt{n}(\widehat{\boldsymbol{\beta}} - \boldsymbol{\beta}_0)$ converges to a normal distribution, and that the Breslow estimator $\hat{\Lambda}_0(t)$ is $\sqrt{n}$-consistent, see e.g. Andersen and Gill (1982); Prentice and Self (1983). Alternative approaches to estimation of parameters in the Cox model can be found in Sasieni (1993).

The approach used to derive the asymptotics for the Cox model does not depend much on the particular form of ρ_i. If ρ_i is twice differentiable with respect to $\boldsymbol{\beta}$ then the proofs of Andersen and Gill (1982) or Prentice and Self (1983) can be modified to show convergence of $\sqrt{n}(\widehat{\boldsymbol{\beta}} - \boldsymbol{\beta})$ and of $\sqrt{n}(\widehat{\Lambda}_0(t) - \Lambda_0(t))$. This basically holds true whenever we speak of Cox models. Therefore, we usually speak of Cox-type models and mean models with intensity of type

$$\lambda_i(t) = \lambda_0(t)\rho_i(\boldsymbol{\beta}, t), \tag{2.5}$$

where ρ_i is an observable nonnegative stochastic process indexed by $\boldsymbol{\beta} \in \boldsymbol{\mathcal{X}}_{\boldsymbol{\beta}} \subset \mathbb{R}^{k_{\boldsymbol{\beta}}}$ and $t \in [0, \tau]$, which is twice differentiable with respect to $\boldsymbol{\beta}$.

2.4.2 Aalen's Additive Model

Aalen (1980) introduced a nonparametric model, in which the covariates and the counting process are linked together by the assumption that the intensity can be written as

$$\lambda_i(t) = \boldsymbol{Y}_i(t)\boldsymbol{\alpha}(t), \tag{2.6}$$

where $\boldsymbol{Y}_i$ are row vectors of predictable stochastic processes containing the observable covariates and $\boldsymbol{\alpha}$ is an unknown deterministic vector-valued function. In matrix-notation, we can rewrite (2.6) as follows:

$$\boldsymbol{\lambda}(t) = \boldsymbol{Y}(t)\boldsymbol{\alpha}(t)$$

where $\boldsymbol{Y}(t) = (\boldsymbol{Y}_1(t)^\top, \dots, \boldsymbol{Y}_n(t)^\top)^\top$.

The standard estimator for $\boldsymbol{A}(t) = \int_0^t \boldsymbol{\alpha}(s)\,\mathrm{d}s$ is the least squares estimator $\widehat{\boldsymbol{A}}(t)$ defined by

$$\widehat{\boldsymbol{A}}(t) = \int_0^t \boldsymbol{Y}^-(s)\,\mathrm{d}\boldsymbol{N}(s),$$

where $\boldsymbol{Y}^-(s)$ is a generalized inverse of $\boldsymbol{Y}(s)$. If $\boldsymbol{Y}(s)$ has full column rank then $\boldsymbol{Y}^-(s) = (\boldsymbol{Y}(s)^\top\boldsymbol{Y}(s))^{-1}\boldsymbol{Y}(s)^\top$. In Aalen (1980), it is shown that $\sqrt{n}(\widehat{\boldsymbol{A}}(t) - \boldsymbol{A}(t))$ converges to a mean zero Gaussian martingale.

To obtain an estimator $\widehat{\boldsymbol{\alpha}}$ of $\boldsymbol{\alpha}$ itself, one can use kernel smoothers as follows:

$$\widehat{\boldsymbol{\alpha}}(t) = \frac{1}{b}\int_0^\tau K\left(\frac{t-s}{b}\right)\,\mathrm{d}\widehat{\boldsymbol{A}}(s),$$

where $b > 0$ is the so-called bandwidth and the kernel $K : \mathbb{R} \to \mathbb{R}$ satisfies $\int_{\mathbb{R}} K(s)\,\mathrm{d}s = 1$. An example for K is the Epanechnikov kernel $K(t) = \frac{3}{4}(1 - t^2)$

for $|t| \leq 1$ and $K(t) = 0$ for $|t| > 1$. In Section 4.2, we modify this kernel smoothing approach slightly to get rates of convergence of the derivative of $\widehat{\boldsymbol{\alpha}}$ and to take care of boundary effects.

In Huffer and McKeague (1991) and in McKeague (1988), weighted least squares estimators for $\boldsymbol{A}$ are discussed. These weighted estimators can lead to a greater efficiency of the estimator. However, the estimation of the weights make the resulting estimator more complicated. The weighted estimator is $\sqrt{n}$-consistent as well. Further discussion of Aalen's model can be found in Aalen (1989, 1993) and in Andersen et al. (1993).

2.4.3 A Semiparametric Additive Risk Model

The following restriction of the Aalen model has been suggested by McKeague and Sasieni (1994):

$$\lambda_i(t) = \boldsymbol{Y}_i^c(t)\boldsymbol{\alpha}^c + \boldsymbol{Y}_i^v(t)\boldsymbol{\alpha}^v(t), \tag{2.7}$$

where the observable covariates $\boldsymbol{Y}_i^c$ and $\boldsymbol{Y}_i^v$ are row vectors of predictable stochastic processes, $\boldsymbol{\alpha}^c$ is an unknown vector, and $\boldsymbol{\alpha}^v$ is an unknown vector-valued function.

To motivate the model, McKeague and Sasieni (1994) point out that in the Aalen model only a limited number of covariates can be handled with small or medium size samples. To get around this problem one can use (2.7) and assume that some covariates have time-independent influence.

Simple estimators of the parameters $\boldsymbol{\alpha}^c$ and $\boldsymbol{A}^v(t) = \int_0^t \boldsymbol{\alpha}^v(s)\,\mathrm{d}s$ are the least squares estimators $\widehat{\boldsymbol{\alpha}^c}$ and $\widehat{\boldsymbol{A}^v}(t)$ suggested by McKeague and Sasieni (1994), which are defined as follows:

$$\widehat{\boldsymbol{\alpha}^c} = \left(\int_0^{\mathcal{T}} \overline{\boldsymbol{Y}^c(\boldsymbol{Q}^{\boldsymbol{Y}^v}\boldsymbol{Y}^c)}(s)\,\mathrm{d}s\right)^{-1} \frac{1}{n}\int_0^{\mathcal{T}} \boldsymbol{Y}^c(s)^\top \boldsymbol{Q}^{\boldsymbol{Y}^v}(s)\,\mathrm{d}\boldsymbol{N}(s),$$

$$\widehat{\boldsymbol{A}^v}(t) = \int_0^t \boldsymbol{Y}^{v-}(s)\left(\mathrm{d}\boldsymbol{N}(s) - \boldsymbol{Y}^c(s)\widehat{\boldsymbol{\alpha}^c}\,\mathrm{d}s\right),$$

where $\boldsymbol{Y}^c(t) = (\boldsymbol{Y}_1^c(t)^\top, \ldots, \boldsymbol{Y}_n^c(t)^\top)^\top$, $\boldsymbol{Y}^v(t) = (\boldsymbol{Y}_1^v(t)^\top, \ldots, \boldsymbol{Y}_n^v(t)^\top)^\top$, $\boldsymbol{Y}^{v-}(t)$ is a generalized inverse of $\boldsymbol{Y}^v(t)$ and $\boldsymbol{Q}^{\boldsymbol{Y}^v}(t)$ is the orthogonal projection matrix onto the space orthogonal to the columns of $\boldsymbol{Y}^v(t)$. McKeague and Sasieni (1994) also discuss weighted versions of these estimators. For weighted and unweighted estimators, McKeague and Sasieni (1994) show that $\sqrt{n}(\widehat{\boldsymbol{\alpha}^c} - \boldsymbol{\alpha}^c)$ converges to a mean zero normal vector and $\sqrt{n}(\widehat{\boldsymbol{A}^v}(t) - \boldsymbol{A}^v(t))$ converges to a mean zero Gaussian process, where $\boldsymbol{\alpha}^c$ and $\boldsymbol{A}^v(t)$ are the 'true' values from the model.

A simplified version of (2.7) with $\boldsymbol{Y}_i^v(t) = R_i(t)$, where $R_i(t)$ is the at-risk indicator, has been considered by Lin and Ying (1994).

2.4.4 Parametric Models

There exists a huge amount of literature on parametric models for survival analysis, see e.g. the bibliographic remarks in Andersen et al. (1993, Chapter VI.4). In Andersen et al. (1993, Chapter VI), the following general model is considered:

$$\lambda_i(t) = \eta_i(t, \boldsymbol{\theta}), \tag{2.8}$$

where the observable parameter-dependent stochastic process η_i is predictable and $\boldsymbol{\theta}$ is an unknown, finite-dimensional vector.

Usually, the parameter $\boldsymbol{\theta}$ is estimated by a maximum partial likelihood approach. The log partial likelihood used for this is given e.g. in Andersen et al. (1993, p. 402) as

$$C(\boldsymbol{\theta}) = \int_0^\tau \sum_{i=1}^n \log(\eta_i(t, \boldsymbol{\theta}))\, \mathrm{d}N_i(t) - \int_0^\tau \sum_{i=1}^n \eta_i(t, \boldsymbol{\theta})\, \mathrm{d}t. \tag{2.9}$$

Under regularity conditions, the maximizer $\widehat{\boldsymbol{\theta}}$ of C has the usual properties of maximum likelihood estimators. In Andersen et al. (1993, Chapter VI.1.2), conditions for $\sqrt{n}(\widehat{\boldsymbol{\theta}} - \boldsymbol{\theta}_0)$ to be asymptotically normal are given, where $\boldsymbol{\theta}_0$ is the 'true' parameter. For our purposes, it is mainly interesting to note that $\widehat{\boldsymbol{\theta}}$ is $\sqrt{n}$-consistent. An extension of the maximum partial likelihood approach is to consider M-estimators, see Andersen et al. (1993, Chapter VI.2).

An example for (2.8) is the parametric Cox model:

$$\lambda_i(t) = a(t, \gamma) \exp(\boldsymbol{Z}_i \boldsymbol{\beta}) R_i(t), \tag{2.10}$$

where a is some known function, $\boldsymbol{Z}_i$ are row vectors of covariates, R_i is the at-risk indicator and γ, $\boldsymbol{\beta}$ are unknown finite-dimensional parameters. The parametric Cox model has been treated in Hjort (1992).

2.4.5 Further Models

Recall the following models already mentioned in the introduction: the general additive-multiplicative hazard model (Lin and Ying, 1995):

$$\lambda_i(t) = \left(g(\mathbf{Y}_i^v(t)\boldsymbol{\beta}) + \lambda_0(t) h(\mathbf{Y}_i^c(t)\boldsymbol{\alpha}^c)\right) R_i(t),$$

a sum of a Cox and an Aalen model (Martinussen and Scheike, 2002):

$$\lambda_i(t) = \left[\mathbf{Y}_i^v(t)\boldsymbol{\alpha}^v(t) + \lambda_0(t) \exp(\mathbf{Y}_i^c(t)\boldsymbol{\alpha}^c)\right] R_i(t), \tag{2.11}$$

and a product of a Cox and an Aalen model (Scheike and Zhang, 2002):

$$\lambda_i(t) = \mathbf{Y}_i^v(t)\boldsymbol{\alpha}^v(t) \exp(\mathbf{Y}_i^c(t)\boldsymbol{\alpha}^c), \tag{2.12}$$

where the observable covariates $\boldsymbol{Y}_i^c$ and $\boldsymbol{Y}_i^v$ are row vectors of predictable stochastic processes, g and h are known functions, $\boldsymbol{\alpha}^c$ and $\boldsymbol{\beta}$ are unknown vectors, λ_0 is an unknown function, $\boldsymbol{\alpha}^v$ is an unknown vector-valued function, and R_i is the at-risk indicator. In the respective papers that introduce these models, $\sqrt{n}$-consistent estimators of the parameters are given. Martinussen and Scheike (2002) also discuss an extended version of (2.11), in which some of the components of $\boldsymbol{\alpha}^c$ are allowed to depend on time. For further information on (2.12), see Scheike and Zhang (2003).

As mentioned in the introduction, the Cox model has also been generalized to include time-dependent regression parameters $\boldsymbol{\beta}(t)$:

$$\lambda_i(t) = \lambda_0(t)\exp(\boldsymbol{Z}_i(t)\boldsymbol{\beta}(t))R_i(t), \tag{2.13}$$

where λ_0, $\boldsymbol{Z}_i$, and R_i are as in the classical Cox model (1.1) and $\boldsymbol{\beta}$ is allowed to depend on time. The model has been considered by Zucker and Karr (1990), Murphy and Sen (1991), Grambsch and Therneau (1994), Pons (2000), Martinussen et al. (2002), Cai and Sun (2003), Winnett and Sasieni (2003). In Scheike and Martinussen (2004), the model (2.13) is restricted by requiring that some components of $\boldsymbol{\beta}(t)$ are constant over time. In the aforementioned papers, several $\sqrt{n}$-consistent estimators of the parameters are given. In Martinussen et al. (2002), the following extension of the Cox model is discussed:

$$\lambda_i(t) = \exp\left(\boldsymbol{Y}_i^v(t)\boldsymbol{\alpha}^v(t) + \boldsymbol{Y}_i^c(t)\boldsymbol{\alpha}^c\right)R_i(t), \tag{2.14}$$

where $\boldsymbol{Y}_i^v$, $\boldsymbol{Y}_i^c$, $\boldsymbol{\alpha}^v$, $\boldsymbol{\alpha}^c$, and R_i are as before.

All models introduced so far fit our framework, i.e. they can be written as

$$\lambda_i(t) = f(\boldsymbol{X}_i(t), \boldsymbol{\alpha}^v(t), \boldsymbol{\alpha}^c),$$

where f is a known function, $\boldsymbol{X}_i$ are observable stochastic processes, $\boldsymbol{\alpha}^v$ is an unknown vector-valued function, and $\boldsymbol{\alpha}^c$ is a finite-dimensional parameter. However, there are several other classes of models, which do not fit our framework (1.4). The following models are some examples:

Frailty models have been introduced since the beginning of the 90s. In these models, some covariates are not observable. Instead, a fixed parametric class of distributions is assumed for them. Some frailty models can be found in Nielsen et al. (1992), Murphy (1995), and Kosorok et al. (2004). As mentioned in the introduction, in frailty models, the interaction between covariates and parameters in the intensity with respect to the filtration generated by the observable processes, may be more complicated.

In Dabrowska (1997), the following model is considered:

$$\lambda_i(t) = \alpha(t, X_i(t))\exp(\boldsymbol{Z}_i(t)\boldsymbol{\beta})R_i(t),$$

where $\alpha(\cdot,\cdot)$ is an unknown deterministic function and $\boldsymbol{\beta}$ is an unknown vector of regression coefficients. The covariates are the one-dimensional stochastic process X_i and the row vector of stochastic processes $\boldsymbol{Z}_i$. As always, R_i is the at-risk indicator.

2.5 Various Approaches to Goodness-of-Fit: A Review

We survey several approaches for model checks, mainly considering those that lead to formal tests. However, since the number of model checks in survival analysis is very large, a complete list is beyond the scope of this thesis.

2.5.1 Checks for the Cox Model

The Cox model is by far the most studied model in survival analysis and thus it is not surprising that most suggestions for model checks in survival analysis are checks for the Cox model. Already in the monograph by Andersen et al. (1993), 23 pages are devoted to checks of the Cox model and considerably more checks have been proposed since. We only sketch some important approaches. For further approaches and worked examples we refer to Fleming and Harrington (1991) and Andersen et al. (1993). Several graphical model checks can be found in Therneau and Grambsch (2000).

First, we describe several approaches that use the so-called martingale residuals as a starting point. By definition, if the Cox model (2.1) holds true then

$$M_i(t) = N_i(t) - \int_0^t \lambda_0(s)\exp(\boldsymbol{Z}_i(s)\boldsymbol{\beta})R_i(s)\,\mathrm{d}s$$

are mean zero local martingales and thus should 'fluctuate' around 0. These martingales can be estimated by the martingale residuals

$$\widehat{M}_i(t) = N_i(t) - \int_0^t \rho_i(\widehat{\boldsymbol{\beta}}, s)\,\mathrm{d}\hat{\Lambda}_0(s) = N_i(t) - \int_0^t \frac{\rho_i(\widehat{\boldsymbol{\beta}}, s)}{\overline{\rho}(\widehat{\boldsymbol{\beta}}, s)}\,\mathrm{d}\overline{\boldsymbol{N}}(s),$$

where $\rho_i(\boldsymbol{\beta}, s) = R_i(s)\exp(\boldsymbol{Z}_i(s)\boldsymbol{\beta})$, $\widehat{\boldsymbol{\beta}}$ is the maximum partial likelihood estimator, and $\hat{\Lambda}_0$ is the Breslow estimator. Note that this definition of martingale residuals is slightly different from the martingale residuals mentioned in the introduction. Martingale residuals have been used as a starting point for model checks by several authors:

- Schoenfeld (1980) proposed a goodness-of-fit test based on partitioning the product space of time and covariates into a finite number of cells $C_1, \dots, C_k$, and comparing the observed and expected number of events in these cells. This amounts to constructing an asymptotically χ^2-distributed test statistic based on the variables

$$D_j := \sum_{i=1}^{n} \int_0^{\mathcal{T}} I\{(t, \boldsymbol{Z}_i(t))^\top \in C_j\}\, \mathrm{d}\widehat{M}_i(t), \quad j = 1, \ldots, k.$$

- Barlow and Prentice (1988) suggested the following weighted martingale residuals:

$$\hat{e}_i(f_i) := \int_0^{\mathcal{T}} f_i(s)\, \mathrm{d}\widehat{M}_i(s)$$

  for some weight functions f_i. They show that several other previously proposed residuals can be written in this form. They suggest to plot $\hat{e}_i(f_i)$ against covariates or against the rank of either failure or censoring time. Instead of considering $\hat{e}_i(f_i)$ directly, they also suggest using $\hat{e}_i(f_i)$ standardized by a heuristic estimator of the variance.

  The resulting plots are hard to interpret, since the distribution of $\hat{e}_i(f_i)$ or its standardized version is not known. Typically, the distribution of $\hat{e}_i(f_i)$ is not close to a normal distribution.

- In Arjas (1988), the individuals are grouped into k different strata defined by a partition $(I_1, \ldots, I_k)$ of $\{1, \ldots, n\}$. The processes $\widehat{\boldsymbol{M}}_{I_j}(t) := \sum_{i \in I_j} \widehat{M}_i(t)$, $j = 1, \ldots, k$, are considered. A graphical procedure is suggested in which, instead of plotting $\widehat{\boldsymbol{M}}_{I_j}$ directly, for each strata, one plots the estimated cumulative hazard against the number of observed failures. If the Cox model holds true, one can expect straight lines with unit slope. Asymptotic results concerning $\widehat{\boldsymbol{M}}_{I_j}$ are derived in Marzec and Marzec (1993). Indeed, it is shown that $\widehat{\boldsymbol{M}}_{I_j}$ converges to a Gaussian process.

- In Therneau et al. (1990), some more discussion on plots based on the martingale residuals is given. Furthermore, the so-called score processes are introduced, which are defined as $\sum_{i=1}^{n} L_{ij}(t)$, where

$$L_{ij}(t) = \int_0^t \left(Z_{ij}(s) - \frac{\sum_{\nu=1}^{n} \rho_\nu(\widehat{\boldsymbol{\beta}}, s) Z_{\nu j}(s)}{\sum_{\nu=1}^{n} \rho_\nu(\widehat{\boldsymbol{\beta}}, s)} \right) \mathrm{d}\widehat{M}_i(s),$$

  where Z_{ij} is the jth covariate of the ith individual. Therneau et al. (1990) state that $\sup_t \sum_{i=1}^{n} L_{ij}(t)$ "should be quite sensitive to alternatives for which covariates have a monotonically increasing or decreasing effect over time, ..." They show that a standardized version of $\sum_{i=1}^{n} L_{ij}(t)$ converges to a Brownian bridge. They suggest plots of standardized versions of $\sum_{i=1}^{n} L_{ij}(t)$, but do not define a test. For a simple proportional hazards model with just one covariate, these residuals have already been considered by Wei (1984). A more recent paper which also considers the score process is Kvaløy and Neef (2004).

- In Lin et al. (1993) the multidimensional stochastic processes

$$W_z(t, \boldsymbol{z}) = \sum_{i=1}^{n} f(Z_i) I\{\boldsymbol{Z}_i \le \boldsymbol{z}\} \widehat{M}_i(t)$$

and

$$W_r(t, r) = \sum_{i=1}^{n} f(Z_i) I\{\boldsymbol{Z}_i \widehat{\boldsymbol{\beta}} \le r\} \widehat{M}_i(t)$$

are considered, where f is some known, real-valued function and the covariates $\boldsymbol{Z}_i$ are assumed to be constant over time. The distributions of $W_z(t, \boldsymbol{z})$ and $W_r(t, r)$ can be approximated by zero mean Gaussian processes. Starting from the above, Lin et al. (1993) suggest several tests. In particular, they suggest an omnibus test based on $\sup_{t,\boldsymbol{z}} |W_z(t, \boldsymbol{z})|$, using the constant weights $f(\boldsymbol{z}) = 1$ for all $\boldsymbol{z}$.

In Spiekerman and Lin (1996), the approach is extended to a setup in which individuals are in n clusters and may be correlated within each cluster, where n is assumed large compared to the number of individuals in each cluster.

- Grønnesby and Borgan (1996) assume time-independent covariates and partition the n individuals according to the risk score $\boldsymbol{Z}_i \widehat{\boldsymbol{\beta}}$ into groups $A_1, \ldots, A_g$ for some fixed g. They consider the vector of processes $\boldsymbol{H} = (H_1, \ldots, H_g)^\top$, where

$$H_j(t) = \sum_{i=1}^{n} I\{i \in A_j\} \widehat{M}_i(t) = \sum_{i=1}^{n} \int_0^t I\{i \in A_j\} \, \mathrm{d}\widehat{M}_i(t).$$

They sketch a proof for the asymptotic distribution of $\boldsymbol{H}$ as $n \to \infty$. They suggest to use plots of $\boldsymbol{H}$ and a formal test based on the asymptotically χ^2_{g-1} distributed random variable

$$(H_1(\tau), \ldots, H_{g-1}(\tau))(\widehat{\boldsymbol{\Sigma}}(\tau))^{-1}(H_1(\tau), \ldots, H_{g-1}(\tau))^\top,$$

where $\widehat{\boldsymbol{\Sigma}}(\tau)$ is an estimator of the covariance of $(H_1(\tau), \ldots, H_{g-1}(\tau))$.

Grønnesby and Borgan (1996) suggest to choose groups of roughly equal size. They also propose some other ad hoc choices for the groups. Some further discussion on this test and on how to choose the groups can be found in May and Hosmer (1998, 2004).

- Marzec and Marzec (1997a) discuss optimal weighting of martingale residuals in the Cox model. In fact, Marzec and Marzec (1997a) use weights in three different places and optimize over the weights in one place. More precisely, they base their considerations on

$$\sum_{j=1}^{k}\int_0^t L_j(\widehat{\boldsymbol{\beta}},s)\sum_{i=1}^{n}\phi_{ij}(s)w(\boldsymbol{Z}_i(s),s)\left(\mathrm{d}N_i(s)-\rho_i(\widehat{\boldsymbol{\beta}},s)\frac{\mathrm{d}\overline{\boldsymbol{N}}(s)}{\overline{\boldsymbol{w\rho}}(\widehat{\boldsymbol{\beta}},s)}\right),$$

  where L_j, ϕ_{ij}, and $w(\boldsymbol{Z}_i(s),s)$ are weights and the estimator $\widehat{\boldsymbol{\beta}}$ in Marzec and Marzec (1997a) is a weighted maximum partial likelihood estimator with weights $w(\boldsymbol{Z}_i(s),s)$.

  They optimize over the weights $L_1(\boldsymbol{\beta},\cdot),\dots,L_k(\boldsymbol{\beta},\cdot)$, where k is a fixed integer. Note that this amounts to choosing optimal time-dependent weights, but not optimal weights for each individual. The optimal choice of the weights ϕ_{ij} and w is not discussed in Marzec and Marzec (1997a), they only make some ad hoc suggestions. In this thesis, we will optimize weights depending on time AND on individual.

- Verweij et al. (1998) suggest to embed the Cox model into a larger model in which there is an additional random effect, i.e. an unobserved covariate. They assume the correlation matrix of the random effect as given and use the null hypothesis that there is no constant effect. They suggest a test statistic based on weighted sums of squared martingale residuals.

Besides the approach based on martingale residuals, there are several other approaches for checking the fit of the Cox model. We list some of them:

- In the paper introducing the Cox model (Cox, 1972), it is suggested to include an additional time-dependent covariate and test whether this covariate has an effect.

- In Andersen (1982), several graphical methods are summarized. Some formal tests for checking the fit are suggested. However, the tests require partitioning of the time axis into several intervals. Furthermore, the formal tests are not for the Cox model itself but for a parametric model with constant hazard in certain intervals.

- In Moreau et al. (1985), the Cox model is embedded into a larger model, where the parameters are allowed to be step-functions with jumps at prespecified time points. The test is then based on a score test with the null hypothesis that the parameters are constant over time.

- Therneau et al. (1990) also introduce the so-called deviance residuals, which are the difference between the log partial likelihood of a saturated Cox model and the Cox model that is to be checked. In the saturated model, each individual has its own parameter vector, i.e. the parameter vector is $nk_{\boldsymbol{\beta}}$-dimensional.

- Lin and Wei (1991) propose a test based on the difference of two different estimators for the inverse of the covariance matrix of the maximum partial likelihood estimators. They show that this difference is asymptotically normally distributed with mean zero. They present extensive simulation studies to compare the test with other goodness-of-fit tests.

- McKeague and Utikal (1991) base a test on the difference between the cumulated hazard rate of a very general model and the cumulated hazard rate of the Cox model. In fact, the general model only assumes that the intensity is a function of time and the covariates. They only consider a one-dimensional covariate that is constant over time. The hazard rate is cumulated with respect to time AND with respect to the covariate to give a multidimensional process. This process is shown to converge to a Gaussian process. Besides using plots to look for model discrepancies, they also partition the product space of time and covariates and construct an asymptotically χ^2-distributed test statistic. As an extension, they consider using the difference between the cumulated hazard rate in a general relative risk model and the cumulated hazard rate in the Cox model. The general relative risk model is given by $\lambda_i(t) = \lambda_0(t)r(Z_i)R_i(t)$, where Z_i is the one-dimensional covariate, $\lambda_0(t)$ is an unknown baseline, r is an unknown function, and R_i is the at-risk indicator.

  In McKeague and Sun (1996), a transformation of the test statistic of McKeague and Utikal (1991) is used to simplify the asymptotic distribution. In fact, after the transformation, the limiting process is a Brownian sheet, i.e. a two-dimensional Gaussian process W with covariance given by $\mathrm{Cov}(W(t,z), W(t',z')) = (t \wedge t')(z \wedge z')$, where t, t' index time and z, z' index the value of the covariates.

- Grambsch and Therneau (1994) suggest to check the Cox model by embedding it into a larger model, where the effect of one covariate may vary over time. Using heuristic arguments, they propose a plotting technique to determine whether - as the Cox model assumes - the effect is indeed constant over time.

- Burke and Yuen (1995) base a test of fit on the Breslow estimator (2.4) using only covariates up to a certain value $\boldsymbol{z}$, i.e. on

$$\hat{\Lambda}_0(t, \boldsymbol{z}) := \int_0^t \frac{\sum_{i=1}^n I\{\boldsymbol{Z}_i \leq \boldsymbol{z}\}\, \mathrm{d}N_i(s)}{\sum_{j=1}^n I\{\boldsymbol{Z}_j \leq \boldsymbol{z}\}\rho_j(\widehat{\boldsymbol{\beta}}, s)}.$$

If the Cox model (2.1) is true then, asymptotically, $\hat{\Lambda}_0(t, \boldsymbol{z})$ should not depend on $\boldsymbol{z}$. They suggest to use $\sqrt{n}\left(\hat{\Lambda}_0(t, \boldsymbol{z}) - \hat{\Lambda}_0(t)\right)$ as basis for tests, where $\hat{\Lambda}_0(t)$ is the usual Breslow estimator. They show asymptotic results and suggest a bootstrap method.

- Fine (2002) considers comparing two nonnested Cox models based on the partial likelihood ratio. Fine (2002) does not assume that one of the two models is correct. It is shown that the maximum partial likelihood estimators converge to some least false values even if the model is misspecified. The null hypothesis is that the expected log partial likelihood (with the least false parameters) of the two models is the same. Fine (2002) gives an interpretation of this null hypothesis in terms of the Kullback-Leibler distance.

  Fine (2002) shows that the test statistic has a different limiting behavior depending on whether the least false intensities of the two models are the same or not. In fact, if they are then the asymptotic distribution of the partial likelihood ratio, suitably normed, is converging in distribution to a weighted sum of χ^2 random variables. Otherwise, the partial likelihood ratio, suitably normed, converges in distribution to a normal distribution. A sequential procedure is suggested to distinguish between the two cases.

- León and Tsai (2004) suggest tests based on what they call censoring consistent residuals. They start with the simple right-censoring model in which one observes $X_i = T_i \wedge C_i$ and $\delta_i = I\{T_i = X_i\}$. They suggest to use a transformation ϕ such that $\mathrm{E}[\phi(X_i, \delta_i)|\boldsymbol{Z}_i] = \mathrm{E}[T_i|\boldsymbol{Z}_i]$, where $\boldsymbol{Z}_i$ are time-independent covariates. Their starting point is the difference $\widehat{R}_i = \tilde{\phi}(X_i, \delta_i) - \hat{m}(\boldsymbol{Z}_i)$, where $\tilde{\phi}$ is the transformation they use and $\hat{m}(\boldsymbol{Z}_i)$ is the estimate of $\mathrm{E}[T_i|\boldsymbol{Z}_i]$ under the Cox model (2.1). They call $\widehat{R}_i$ censoring consistent residuals. They propose the overall test statistic $\tilde{D} = \sup_{\boldsymbol{z}\in\mathbb{R}^{k_\beta}} \left|\sum_{i=1}^n I\{\boldsymbol{Z}_i \leq \boldsymbol{z}\}\widehat{R}_i\right|$ and the following statistic for "testing the functional form of covariate Z_k": $\tilde{D}_k = \sup_{z\in\mathbb{R}} \left|\sum_{i=1}^n I\{Z_{ik} \leq z\}\widehat{R}_i\right|$. To approximate the asymptotic distribution of $\tilde{D}$ and $\tilde{D}_k$ they suggest a bootstrap method.

- Other approaches can be found in Nagelkerke et al. (1984), Lin (1991), Marzec and Marzec (1998) and Parzen and Lipsitz (1999).

In Ng'andu (1997) and in Song and Lee (2000), several of the above tests for the proportional hazards assumption of the Cox model are compared by means of simulation studies. Not surprisingly, the simulations do not yield an overall best test.

2.5.2 Checks for Additive Models

For the additive model (2.6), Aalen (1993) suggested several techniques for model checking based on martingale residuals, among them grouping residuals based on values of the covariates.

Besides checking the Cox model, McKeague and Utikal (1991) also consider using the Aalen model (2.6) as null hypothesis. Simulation studies in McKeague and Utikal (1991) indicate that even for large sample sizes their test exceeds the asymptotic level severely. For example, in their simulation with nominal level 0.05 and $n = 300$ individuals the observed rejection rate was 0.212. This only converges slowly to the asymptotic level, as can be seen by the rejection rate of 0.106 for $n = 1200$ individuals and of 0.066 for $n = 3600$ individuals.

For a subclass of models of the semiparametric additive model (2.7) considered by Lin and Ying (1994), some goodness-of-fit approaches have been suggested by Song et al. (1996), Yuen and Burke (1997), and Kim et al. (1998). The model is given as follows:

$$\lambda_i(t) = (\lambda_0(t) + \boldsymbol{Z}_i\boldsymbol{\beta})R_i(t), \tag{2.15}$$

where $\boldsymbol{Z}_i$ are row vectors of covariates, λ_0 is an unknown baseline, $\boldsymbol{\beta}$ is an unknown finite-dimensional parameter, and R_i is the at-risk indicator. The idea in Song et al. (1996) is similar to the one in Lin et al. (1993) for the Cox model, i.e. they suggest to use partial sums of martingale residuals with respect to time and covariate values. Yuen and Burke (1997) transfer the idea of Burke and Yuen (1995) to (2.15). Indeed they use the difference between two estimators of $\int_0^t \lambda_0(s)\,ds$ as starting point, one estimator uses all individuals, the other only those individuals whose covariates are less than a given value. After that they define Kolmogorov-Smirnov type test statistics and Cramér-von Mises type test statistics and suggest to use a bootstrap approximation of the distribution of the test statistics. Kim et al. (1998) define a function similar to the partial likelihood score function for the Cox model introduced by Therneau et al. (1990) and suggest to use the sup over the absolute value of this score function as test statistic. They prove the asymptotics and the validity of a certain simulation procedure.

Grønnesby and Borgan (1996) also discuss a check for the Aalen model as well as for (2.15). Similar to their approach for the Cox model, they suggest using sums of martingale residuals grouped by the risk score.

To check the semiparametric restriction of the Aalen model given by (2.7), Scheike (2002) proposes to use tests to check whether in the models the parameters $\boldsymbol{\alpha}^c$ are actually constant over time. To this end, he makes several suggestions based on functionals of the difference $\widehat{A}_j^c(t) - \widehat{\alpha_j^c}t$, where $\widehat{A}_j^c(t)$ is an estimator of $\int_0^t \alpha_j^c(s)\,\mathrm{d}s$ that does not assume α_j^c to be constant over time. One test statistic is $\sup_{t\in[0,\tau]} |\widehat{A}_j^c(t) - \widehat{\alpha_j^c}t|$. Similar suggestions can be found in Grund and Polzehl (2001), where only a bootstrap scheme is suggested to approximate the distribution of the test statistic.

2.5.3 Checks for Parametric Models

Section 6 of Hjort (1990) discusses a goodness-of-fit test for the parametric Cox model (2.10). The test statistic is

$$H_n(t) := \sqrt{n} \int_0^t K_n(s)\,\mathrm{d}\left(\hat{A}(s) - A(s,\widehat{\boldsymbol{\theta}})\right),$$

where $\hat{A}(t) = \int_0^t \left(\frac{1}{n}\sum_{i=1}^n \exp(\boldsymbol{Z}_i\widehat{\boldsymbol{\beta}})R_i(s)\right)^{-1} \mathrm{d}\overline{\boldsymbol{N}}(s)$ is a nonparametric estimator for $A(t,\boldsymbol{\theta}) := \int_0^t a(s,\boldsymbol{\theta})\,\mathrm{d}s$, the weights $K_n(s)$ are predictable stochastic processes that converge to a deterministic process, and $\widehat{\boldsymbol{\beta}}$, $\widehat{\boldsymbol{\theta}}$ are the maximum partial likelihood estimators for $\boldsymbol{\beta}$, $\boldsymbol{\theta}$ from the parametric Cox model. Note that $\hat{A}(t)$ is almost identical to the Breslow estimator (2.4) - the only difference is that $\hat{A}(t)$ uses the maximum partial likelihood estimator from the parametric Cox model (2.10) and not the maximum partial likelihood estimator from the semiparametric Cox model (2.1). Hjort (1990) shows convergence of $H_n(t)$ to a mean zero Gaussian process. Based on this, a plotting procedure of a normalized version of $H_n(t)$ as well as χ^2-tests based on a partitioning of the time-axis are suggested. Three different choices of weights are given and some discussion on the consistency of the tests is included. Hjort (1990) relaxes the assumption of predictability of the weights K_n as follows: K_n may depend on a finite-dimensional parameter, say $\boldsymbol{\gamma}$, for which a consistent estimator is plugged in. For fixed $\boldsymbol{\gamma}$, the weights $K_n(\boldsymbol{\gamma}, t)$ have to be predictable. The asymptotic distribution of $H_n(t)$ is not a Gaussian martingale - the covariance structure is more complicated. In Andersen et al. (1993), it is noted that the approach of Khmaladze (1981) could be used to transform $H_n(\cdot)$ to a stochastic process $\widetilde{H}_n(\cdot)$ such that the asymptotic distribution is a mean zero Gaussian martingale.

Lin and Spiekerman (1996) discuss several model checks for parametric regression models with censored data. For the parametric Cox model (2.10), using a similar approach to the one in Hjort (1990), they present Kolmogorov-Smirnov-type tests. Furthermore, they also describe a test for the following accelerated failure time model

$$\lambda_i(t) = \eta(t \exp(-\boldsymbol{Z}_i\boldsymbol{\beta}), \boldsymbol{\theta}) \exp(\boldsymbol{Z}_i\boldsymbol{\beta}) R_i(t),$$

where η is a known function, $\boldsymbol{\beta}$, $\boldsymbol{\theta}$ are finite-dimensional parameters, $\boldsymbol{Z}_i$ are row vectors of covariates , and R_i are the at-risk indicators. The idea is to consider

$$\sqrt{n}\left(\int_0^t \eta(s, \widehat{\boldsymbol{\theta}})\,\mathrm{d}s - \widehat{E}(t, \widehat{\boldsymbol{\beta}})\right),$$

where $\widehat{\boldsymbol{\theta}}, \widehat{\boldsymbol{\beta}}$ are estimators for $\boldsymbol{\theta}, \boldsymbol{\beta}$ and $\widehat{E}(t, \boldsymbol{\beta})$ is an estimator for $\int_0^t \rho(s)\,\mathrm{d}s$ in the semiparametric model

$$\lambda_i(t) = \rho(t \exp(-\boldsymbol{Z}_i\boldsymbol{\beta})) \exp(\boldsymbol{Z}_i\boldsymbol{\beta}) R_i(t),$$

where ρ is an unknown function. Furthermore, for a general model of type

$$\lambda_i(t) = \eta_i(\boldsymbol{\theta}, t) R_i(t),$$

Lin and Spiekerman (1996) describe a similar approach as the one in Lin et al. (1993) based on the martingale residuals

$$\widehat{M_i}(t) = N_i(t) - \int_0^t \eta_i(\widehat{\boldsymbol{\theta}}, s) R_i(s)\,\mathrm{d}s.$$

Using cumulative sums of $\widehat{M_i}(t)$ with respect to failure time or/and covariates, they construct omnibus tests.

In Stute et al. (2000), the following model is considered:

$$Y = m_{\boldsymbol{\theta}}(\boldsymbol{X}) + \epsilon, \quad \boldsymbol{\theta} \in \boldsymbol{\Theta},$$

where $\boldsymbol{\Theta} \subset \mathbb{R}^d$ is finite-dimensional, $\boldsymbol{\theta}$ is unknown, m is a known function, Y is a lifetime, $\boldsymbol{X}$ are finite-dimensional covariates and ϵ is an error term satisfying $\mathrm{E}[\epsilon|\boldsymbol{X}] = 0$. Furthermore, it is assumed that Y is censored by an independent censoring variable C and that $Z = Y \wedge C$ and $\delta = I\{Y \leq Z\}$. The observation consists of n i.i.d. replicates of $(Z, \boldsymbol{X}, \delta)$ denoted by $(Z_i, \boldsymbol{X}_i, \delta_i)$, $i = 1, \ldots, n$.

Let $Z_{1:n} \leq \cdots \leq Z_{n:n}$ be the order statistic of $Z_1, \ldots, Z_n$, and let $\delta_{[i:n]}$ and $\boldsymbol{X}_{[i:n]}$ be the concomitants associated with $Z_{i:n}$. Stute et al. (2000) base their considerations on the "empirical process marked by the weighted residuals," which they define as follows:

$$R_n^1(\boldsymbol{x}) = \sqrt{n}\sum_{i=1}^{n} W_{ni}\left[Z_{i:n} - m_{\widehat{\boldsymbol{\theta}}}(\boldsymbol{X}_{[i:n]})\right] I\{\boldsymbol{X}_{[i:n]} \leq \boldsymbol{x}\},$$

where $\widehat{\boldsymbol{\theta}}$ is the weighted least squares estimator for $\boldsymbol{\theta}$ defined as minimizer of

$$\sum_{i=1}^{n} W_{ni} \left[Z_{i:n} - m_{\boldsymbol{\theta}}(\boldsymbol{X}_{[i:n]}) \right]^2$$

and where W_{ni} are the so-called Kaplan Meier weights given by

$$W_{ni} = \frac{\delta_{[i:n]}}{n-i+1} \prod_{j=1}^{i-1} \left[\frac{n-j}{n-j+1} \right]^{\delta_{[j:n]}}.$$

Stute et al. (2000) show that R_n^1 converges to a d-variate centered Gaussian process R_∞^1. Furthermore, they suggest a bootstrap approximation of the limiting process and show that this approximation converges to R_∞^1. The actual test statistics used in Stute et al. (2000) are $\sup_{\boldsymbol{x}} |R_1(\boldsymbol{x})|$ and $\int (R_n^1(\boldsymbol{x}))^2 F_{1n}(\mathrm{d}\boldsymbol{x})$, where F_{1n} is the empirical distribution function of $\boldsymbol{X}_1, \ldots, \boldsymbol{X}_n$.

Some other ideas for checks of parametric regression models can be found in Andersen et al. (1993).

2.5.4 Checks for Further Models

The approach by Arjas (1988) based on sums of martingale residuals in one strata has been extended to other models as well. Indeed, Marzec and Marzec (1997b) extend the approach to the Cox model with time-dependent coefficients (2.13) and Kraus (2004) extends the approach to the model (2.12), whose intensity is a product of an Aalen and a Cox model.

Scheike and Zhang (2003) suggest a method of checking the model (2.12), whose intensity is the product of a Cox and an Aalen type intensity. Indeed, they suggest to use a certain score process derived from the score function of the time-constant parameters to check the "goodness of fit of the covariates included in the multiplicative part".

In a general Cox model, where some coefficients are time-varying and others are constant over time, Scheike and Martinussen (2004) propose tests whether the coefficients assumed to be constant over time are indeed constant over time. In Martinussen et al. (2002), a similar idea has been proposed for the semiparametric extension of the Cox model given in (2.14).

Chapter 3

Checking Aalen's Additive Risk Model

In this chapter, we consider model checks for Aalen's additive risk model (2.6). Recall that in this model the intensity is given by

$$\lambda_i(t) = \boldsymbol{Y}_i(t)\boldsymbol{\alpha}(t) \text{ for some } \boldsymbol{\alpha} \in \mathrm{bm}(\mathbb{R}^{k_\alpha}), \tag{3.1}$$

where $\boldsymbol{Y}_i$ are $k_{\boldsymbol{\alpha}}$-variate row vectors of predictable, locally bounded processes and where the unknown regression parameter $\boldsymbol{\alpha}$ is a deterministic element of $\mathrm{bm}(\mathbb{R}^{k_\alpha})$, the set of all bounded, measurable mappings from $[0, \tau]$ into $\mathbb{R}^{k_\alpha}$. We want to test whether (3.1) holds true. Writing the hypothesis in terms of the intensity is a convenient shorthand for the following precise formulation of the null hypothesis:

H_0: There exists $\boldsymbol{\alpha} \in \mathrm{bm}(\mathbb{R}^{k_\alpha})$ such that for $i = 1, \ldots, n$, the stochastic process $N_i(t) - \int_0^t \boldsymbol{Y}_i(s)\boldsymbol{\alpha}(s)\,ds$ is a local martingale with mean zero.

When we talk about hypotheses in the upcoming chapters, we will only write down the intensity.

3.1 Setup

We use the setup of Section 2.2. In this chapter, we make the simplifying assumption that the covariates $\boldsymbol{Y}_i, i = 1, \ldots, n$, are bounded, adapted, càglàd stochastic processes. Hereby, càglàd means that the paths of the processes are left-continuous and have right-hand limits. The term càglàd is an abbreviation for the French term 'continue *à g*auche, avec des *l*imites *à d*roite'.

For our test, we need weights given by a predictable process $\boldsymbol{c} = (c_1, \ldots, c_n)^\top$ that satisfies certain orthogonality conditions motivated in the introduction.

Since the Aalen model only has time-dependent parameters, the condition for $\boldsymbol{c}$ is (1.6), which reduces to

$$\sum_{i=1}^{n} c_i(t)\mathbf{Y}_i(t) = \boldsymbol{c}(t)^\top \boldsymbol{Y}(t) = \mathbf{0}, \quad \forall t \in [0,\tau]. \tag{3.2}$$

Using this, we can simplify our test statistic:

$$T(\boldsymbol{\alpha},t) = n^{-\frac{1}{2}} \sum_{i=1}^{n} \int_0^t c_i(s) \left(\mathrm{d}N_i(s) - \mathbf{Y}_i(s)\boldsymbol{\alpha}(s)\,\mathrm{d}s\right) = n^{-\frac{1}{2}} \sum_{i=1}^{n} \int_0^t c_i(s)\,\mathrm{d}N_i(s).$$

Thus $T(\boldsymbol{\alpha},t)$ does not depend on $\boldsymbol{\alpha}$ and hence, we do not need an estimator for $\boldsymbol{\alpha}$ to evaluate $T(\boldsymbol{\alpha},t)$. Therefore we only write $T(t)$.

3.2 Orthogonal Projections

We will ensure (3.2) by using orthogonal projections. Orthogonal projections will be used frequently, therefore we introduce them in some generality.

Let $(A,\mathcal{A},\mu)$ be a measure space. $L_2(A,\mathcal{A},\mu) = L_2(\mu)$ is defined as the vector space of measurable functions $f : A \to \mathbb{R}$ that satisfy

$$\|f\|_\mu := \left(\int_A |f(x)|^2\,\mathrm{d}\mu(x)\right)^{\frac{1}{2}} < \infty.$$

We use the usual identification of functions being equal μ-almost everywhere. Equipped with the scalar product

$$<f,g>_\mu = \int_A f(x)g(x)\,\mathrm{d}\mu(x),$$

$L_2(\mu)$ is a Hilbert space. If X is a subset of $L_2(\mu)$ then $\boldsymbol{Q}_\mu^X$ denotes the orthogonal projection operator onto $X^\perp$, i.e. $\boldsymbol{Q}_\mu^X$ is a linear, continuous operator such that the image of $L_2(\mu)$ under $\boldsymbol{Q}_\mu^X$ is $X^\perp$, and for $a,b \in L_2(\mu)$:

$$\boldsymbol{Q}_\mu^X(\boldsymbol{Q}_\mu^X a) = \boldsymbol{Q}_\mu^X a \quad \text{and} \quad <\boldsymbol{Q}_\mu^X a, b>_\mu = <a, \boldsymbol{Q}_\mu^X b>_\mu .$$

If X is a closed subspace of $L_2(\mu)$ then $\boldsymbol{Q}_\mu^X a = a - \boldsymbol{P}_\mu^X a$, where $\boldsymbol{P}_\mu^X$ denotes the orthogonal projection operator onto X. For $a \in L_2(\mu)$ the following elementary properties hold:

$$\begin{aligned} <\boldsymbol{Q}_\mu^X a, a>_\mu &= \left\|\boldsymbol{Q}_\mu^X a\right\|_\mu^2, \\ \left\|\boldsymbol{Q}_\mu^X a\right\|_\mu &\le \|a\|_\mu, \\ \boldsymbol{Q}_\mu^X a &= a \text{ if } a \in X^\perp, \\ \boldsymbol{Q}_\mu^X a &= \mathbf{0} \text{ if } a \in X. \end{aligned} \tag{3.3}$$

If X is spanned by finitely many vectors $x_1, \dots, x_k$, that are not linearly dependent, then by Lemma B.1, for $a \in L_2(\mu)$,

$$\boldsymbol{Q}_\mu^X a = a - (x_i)_{i=1,\dots,k}^\top \, (<x_i, x_j>_\mu)_{i,j=1,\dots,k}^{-1} \, (<x_i, a>_\mu)_{i=1,\dots,k}. \tag{3.4}$$

The ordinary scalar product and the orthogonal projection in $\mathbb{R}^n$ fits into this framework as follows: Let $\mathbb{P}_n$ be the counting measure on $\{1, \dots, n\}$, i.e. $\mathbb{P}_n(i) = 1$ for $i \in \{1, \dots, n\}$. Then we identify $\mathbb{R}^n$ with $L_2(\mathbb{P}_n)$. Indeed, any vector $\boldsymbol{x} \in \mathbb{R}^n$ corresponds to a function $\tilde{x} \in L_2(\mathbb{P}_n)$ by setting $\tilde{x}(i) = x_i, i = 1, \dots, n$. Using this identification, we clearly have $\boldsymbol{x}^\top \boldsymbol{y} = <\boldsymbol{x}, \boldsymbol{y}>_{\mathbb{P}_n}$ for $\boldsymbol{x}, \boldsymbol{y} \in \mathbb{R}^n$. Furthermore, if X is spanned by the columns of a matrix $\boldsymbol{Y} \in \mathbb{R}^{n \times k}$ and $\boldsymbol{Y}^\top \boldsymbol{Y}$ is invertible then (3.4) is the following:

$$\boldsymbol{Q}_{\mathbb{P}_n}^X \boldsymbol{a} = \boldsymbol{a} - \boldsymbol{Y}(\boldsymbol{Y}^\top \boldsymbol{Y})^{-1} \boldsymbol{Y}^\top \boldsymbol{a}.$$

We often use $\boldsymbol{Q}_{\mathbb{P}_n}^{\boldsymbol{Y}} := \boldsymbol{Q}_{\mathbb{P}_n}^X$. Moreover, we identify functions $\boldsymbol{x} : [0, \tau] \to \mathbb{R}^n$ with elements f of $L_2(\mathbb{P}_n \otimes \lambda\!\!\!\lambda)$ by $f(i, t) = x_i(t)$, where $\lambda\!\!\!\lambda$ is Lebesgue measure on $[0, \tau]$. Hereby, the underlying space is $\{1, \dots, n\} \times [0, \tau]$ equipped with the σ-algebra $\mathcal{P}\{1, \dots, n\} \otimes \mathcal{B}[0, \tau]$, where $\mathcal{P}$ denotes the power set and $\mathcal{B}$ denotes the Borel-σ-algebra. If appropriate, we shall write $\boldsymbol{x} \in L_2(\mathbb{P}_n \otimes \lambda\!\!\!\lambda)$ and use the notation of this section. For example, for functions $\boldsymbol{x}, \boldsymbol{y} \in L_2(\mathbb{P}_n \otimes \lambda\!\!\!\lambda)$:

$$<\boldsymbol{x}, \boldsymbol{y}>_{\mathbb{P}_n \otimes \lambda\!\!\!\lambda} = \int_0^\tau \boldsymbol{x}(s)^\top \boldsymbol{y}(s) \, \mathrm{d}s.$$

For more flexibility, we use the following standard notation: If $w : A \to [0, \infty)$ is measurable, then $w \cdot \mu$ denotes the measure defined by $(w \cdot \mu)(B) = \int_B w(x) \, \mathrm{d}\mu(x)$. We use $w \cdot \mu$ instead of μ in the notation of this section and then w can be interpreted as weights. For example, for a nonnegative vector $\boldsymbol{w} \in \mathbb{R}^n$ and a matrix $\boldsymbol{Y} \in \mathbb{R}^k$ we have

$$\boldsymbol{Q}_{\boldsymbol{w} \cdot \mathbb{P}_n}^{\boldsymbol{Y}} \boldsymbol{a} = \boldsymbol{a} - \boldsymbol{Y}(\boldsymbol{Y}^\top \operatorname{diag}(\boldsymbol{w}) \boldsymbol{Y})^{-1} \boldsymbol{Y}^\top \operatorname{diag}(\boldsymbol{w}) \boldsymbol{a}$$

if $\boldsymbol{Y}^\top \operatorname{diag}(\boldsymbol{w}) \boldsymbol{Y}$ is invertible. Moreover, for nonnegative $\boldsymbol{w} : [0, \tau] \to \mathbb{R}^n$, we shall also use the notation $L_2(\boldsymbol{w} \cdot (\mathbb{P}_n \otimes \lambda\!\!\!\lambda))$.

3.3 Least Squares Weights

Starting from arbitrary weights $\boldsymbol{d}(t) = (d_1(t), \dots, d_n(t))^\top$, one can ensure the orthogonality conditions (3.2) by using

$$\boldsymbol{c}(t) = \boldsymbol{Q}_{\mathbb{P}_n}^{\boldsymbol{Y}(t)} \boldsymbol{d}(t).$$

To emphasize which weights $\boldsymbol{d}$ we are using, we shall write $T(\boldsymbol{d}, \cdot)$. For the asymptotic results we assume the following condition:

(A1) (properties of $\mathrm{E}[\mathbf{Y}_1(t)^\top \mathbf{Y}_1(t)]$)
$\mathrm{E}[\mathbf{Y}_1(t)^\top \mathbf{Y}_1(t)]$ is continuous in t and invertible for all $t \in [0, \tau]$.

If the Aalen model holds true then we have the following convergence result in the space $D[0, \tau]$ of càdlàg functions on $[0, \tau]$ (i.e. right continuous real valued functions with left-hand limits) equipped with the Skorohod topology (Billingsley, 1999).

Theorem 3.1. *Suppose that $(\mathbf{Y}_i, d_i), i = 1, \dots, n$, are i.i.d. and $\mathbf{Y}_i$ and d_i are càglàd, bounded, adapted stochastic processes. If (A1) and the Aalen model* (2.6) *hold true then in $D[0, \tau]$,*

$$T(\boldsymbol{d}, t) = n^{-\frac{1}{2}} \int_0^t (\boldsymbol{Q}_{\mathbb{P}_n}^{\mathbf{Y}(s)} \boldsymbol{d}(s))^\top \mathrm{d}\boldsymbol{N}(s) \xrightarrow{d} m(t),$$

where m is a mean zero Gaussian process with covariance $\mathrm{Cov}(m(s), m(t)) = \sigma^2(\boldsymbol{d}, s \wedge t)$*, where*

$$\sigma^2(\boldsymbol{d}, t) = \int_0^t \mathrm{E}\left[\left(\boldsymbol{Q}_{\mathrm{P}}^{\mathbf{Y}_1(s)} d_1(s)\right)^2 \lambda_1(s)\right] \mathrm{d}s.$$

Furthermore

$$\hat{\sigma}^2(\boldsymbol{d}, t) = \frac{1}{n} \int_0^t (\boldsymbol{Q}_{\mathbb{P}_n}^{\mathbf{Y}(s)} \boldsymbol{d}(s))^\top \mathrm{diag}(\mathrm{d}\boldsymbol{N}(s)) (\boldsymbol{Q}_{\mathbb{P}_n}^{\mathbf{Y}(s)} \boldsymbol{d}(s)) \xrightarrow{\mathrm{P}} \sigma^2(\boldsymbol{d}, t)$$

uniformly in $t \in [0, \tau]$.

Most proofs of this chapter are collected in Section 3.7.
Under the fixed alternative $\lambda_i(t) = h_i(t)$, for some $h_i(t)$, the following asymptotic result holds true:

Theorem 3.2. *Suppose that $(h_i, \mathbf{Y}_i, d_i)$ are i.i.d. and $\mathbf{Y}_i$, h_i, and d_i are càglàd, bounded, adapted stochastic processes. If (A1) is satisfied and $\lambda_i(t) = h_i(t)$ then*

$$n^{-\frac{1}{2}} T(\boldsymbol{d}, t) \xrightarrow{\mathrm{P}} H(t) := \int_0^t <\boldsymbol{Q}_{\mathrm{P}}^{\mathbf{Y}_1(s)} d_1(s), h_1(s)>_{\mathrm{P}} \mathrm{d}s$$

and

$$\hat{\sigma}^2(\boldsymbol{d}, t) \xrightarrow{\mathrm{P}} \int_0^t \left\| \boldsymbol{Q}_{\mathrm{P}}^{\mathbf{Y}_1(s)} d_1(s) \right\|^2_{h_1(s) \cdot \mathrm{P}} \mathrm{d}s.$$

Note that if $\boldsymbol{d} = \boldsymbol{h}$ then by properties of projections,

$$H(t) := \int_0^t \left\| \boldsymbol{Q}_{\mathrm{P}}^{\mathbf{Y}_1(s)} h_1(s) \right\|^2_{\mathrm{P}} \mathrm{d}s \geq 0.$$

With this choice, one-sided tests rejecting for large values seem reasonable.

Furthermore, the following are equivalent:

(i) $H(t) = 0 \quad \forall\, t \in [0,\tau]$

(ii) $h_1(t) \in \mathrm{span}_{\mathrm{P}}(Y_{11}(t), \dots, Y_{1k_\alpha}(t))$ for almost all $t \in [0,\tau]$.

(iii) h_1 is in Aalen's model (2.6).

In (ii), $\mathrm{span}_{\mathrm{P}}$ denotes the span in the space $L_2(\mathrm{P})$ of square integrable random variables. Hence, using $\boldsymbol{d} = \boldsymbol{h}$ we can construct consistent tests.

3.4 Construction of Tests

Under the conditions of Theorem 3.1, dropping the dependence on $\boldsymbol{d}$,

$$T(\cdot) \xrightarrow{\mathrm{d}} m(\cdot),$$

where m is a mean zero Gaussian process with $\mathrm{Cov}(m(s), m(t)) = \sigma^2(s \wedge t)$ given in Theorem 3.1. Furthermore, we have a uniformly consistent estimator $\hat{\sigma}^2(\cdot)$ of $\sigma^2(\cdot)$. There are several ways how to construct one or two-sided tests from this starting point. All of them require

$$\sigma^2(\tau) > 0.$$

The easiest approach is to consider

$$V^{(1)} := \frac{T(\tau)}{\sqrt{\hat{\sigma}^2(\tau)}},$$

which follows asymptotically a normal distribution. Another approach is to consider

$$V^{(2)} := \sup_{t \in [0,\tau]} \left| \frac{\sqrt{\hat{\sigma}^2(\tau)}}{\hat{\sigma}^2(\tau) + \hat{\sigma}^2(t)} T(t) \right| \xrightarrow{\mathrm{d}} \sup_{t \in [0,\frac{1}{2}]} |W^0(t)|,$$

where $W^0(t)$ is a Brownian bridge. This transformation can for example be found in Hall and Wellner (1980). A third approach is to start from

$$V^{(3)} := \sup_{t \in [0,\tau]} \left| \frac{T(t)}{\sqrt{\hat{\sigma}^2(\tau)}} \right| \xrightarrow{\mathrm{d}} \sup_{t \in [0,1]} |W(t)|,$$

where $W(t)$ is a Brownian motion. The convergence is based on the fact that $m(\cdot) \stackrel{\mathrm{d}}{=} W(\sigma^2(\cdot))$, where m and S are as in Theorem 3.1.

An explicit formula for the asymptotic distribution of $V^{(2)}$ can be found in Hall and Wellner (1980). Formulas for the asymptotic distribution of $V^{(3)}$ can be derived from Borodin and Salminen (2002). For the test statistics $V^{(2)}$ and $V^{(3)}$ we always reject at the upper tail. For the test statistic $V^{(1)}$ we will indicate whether we use a two-sided test or a one-sided test (rejecting at the upper tail).

3.5 Choosing the Weights

3.5.1 General Comments

A main restriction on the weights $\boldsymbol{d}$ is that $\boldsymbol{d}$ needs to be predictable. An easy example for predictable weights, is to use a transformations of the covariates as weights, say a weight can be the indicator that a certain covariate exceeds a given threshold.

In the coming chapters, we will often relax considerably the condition that the weights need to be predictable. We will show that one can plug in estimators of finite-dimensional parameters or even an estimator, possibly not predictable, of a nonparametric baseline. However, one main condition remains: To use the asymptotics of Theorem 3.1, we need that the asymptotic variance $\sigma^2(\boldsymbol{d}, \tau)$ is positive. For this to hold true, we need that $d_1(t)$ is not in the model space, i.e. that it cannot be written as $Y_{11}(t)\alpha_1(t) + \cdots + Y_{1k_{\boldsymbol{\alpha}}}(t)\alpha_{k_{\boldsymbol{\alpha}}}(t)$.

If the weights $\boldsymbol{d}$ depend on a parameter $\boldsymbol{\beta}$ for which we have an estimator $\hat{\boldsymbol{\beta}}$ that converges stochastically to some $\boldsymbol{\beta}_0$ then we shall show later that under some regularity conditions $T(\boldsymbol{d}(\hat{\boldsymbol{\beta}}), \cdot)$ and $T(\boldsymbol{d}(\boldsymbol{\beta}_0), \cdot)$ as well as $\sigma^2(\boldsymbol{d}(\hat{\boldsymbol{\beta}}), t)$ and $\sigma^2(\boldsymbol{d}(\boldsymbol{\beta}_0), t)$ are asymptotically equivalent. Hence, we need that under the null hypothesis $d_1(\boldsymbol{\beta}_0, \cdot)$ is not in the model space.

3.5.2 Nested Models

Suppose we want to be powerful against a model in which the Aalen model (2.6) is nested. Say,

$$\lambda_i(t) = \boldsymbol{Y}_i(t)\boldsymbol{\alpha}(t) + Z_i(t)\beta, \tag{3.5}$$

where Z_i is some other covariate which is assumed to be a predictable process. One could try to use an estimate $\boldsymbol{d}(\widehat{\beta})$ of the intensity of the larger model (3.5). As already noted, under some conditions, $T(\boldsymbol{d}(\widehat{\beta}), \cdot)$ and $T(\boldsymbol{d}(\beta_0), \cdot)$ as well as $\sigma^2(\boldsymbol{d}(\widehat{\beta}), \cdot)$ and $\sigma^2(\boldsymbol{d}(\beta_0), \cdot)$ are asymptotically equivalent if (3.5) holds with $\beta = \beta_0$. However, if the smaller model is true, i.e. $\beta_0 = 0$ then $\boldsymbol{d}(\beta_0)$ is in the model space and hence $\sigma^2(\boldsymbol{d}(\beta_0), \tau) = 0$.

To avoid this problem, we suggest to use $d_i(t) = Z_i(t)$ instead. Indeed, if the additional covariate $\boldsymbol{Z} = (Z_1, \ldots, Z_n)^\top$ is not in the model space then $\sigma^2(\boldsymbol{Z}, \tau) > 0$.

For the remainder of this chapter and the next chapters, we will mostly ignore the problem of how to ensure $\sigma^2(\boldsymbol{d}, \tau) > 0$, i.e. that the asymptotic variance of our test does not converge to zero. We will come back to this question in Chapter 7.

3.5.3 Using the Cox Model as Competing Model

In this section, we consider directing our test against a competing Cox model given by

$$\lambda_i(t) = \lambda_0(t) \exp(\boldsymbol{Z}_i(t)\boldsymbol{\beta}) R_i(t), \tag{3.6}$$

where λ_0 is an unknown baseline, $\boldsymbol{\beta}$ is a finite-dimensional regression parameter, R_i is the at-risk indicator, i.e. a stochastic process taking only the values 0 and 1, and $\boldsymbol{Z}_i$ is a $k_{\boldsymbol{\beta}}$-variate row vector of observable processes containing the covariates. The vector $\boldsymbol{Z}_i$ may contain some of the components of $\mathbf{Y}_i$ or transformations of them, but this need not be the case. Using

$$d_i(t) = \exp(\boldsymbol{Z}_i(t)\boldsymbol{\beta}) R_i(t)$$

leads to consistent tests. Indeed, if (3.6) holds true then by Theorem 3.2,

$$n^{-\frac{1}{2}} T(\boldsymbol{d}, t) \xrightarrow{\mathrm{P}} \int_0^t \lambda_0(s) \left\| \boldsymbol{Q}_{\mathrm{P}}^{\mathbf{Y}_1(s)} \exp(\boldsymbol{Z}_1(s)\boldsymbol{\beta}) \right\|_{\mathrm{P}}^2 \mathrm{d}s.$$

As $\boldsymbol{\beta}$ is not known, we shall plug in an estimator $\widehat{\boldsymbol{\beta}}$. Note that we do not need to estimate λ_0. We shall use the maximum partial likelihood estimator $\widehat{\boldsymbol{\beta}}$ given in Subsection 2.4.1. For asymptotic results, we need to know the rate of convergence of $\widehat{\boldsymbol{\beta}}$. If the Cox model (3.6) holds true then this is covered by the classical results, see also Subsection 2.4.1. However, we also need the rates if the Aalen model (2.6) holds true. Lin and Wei (1989), Hjort (1992), Sasieni (1993), and Fine (2002) have shown that even if (3.6) does not hold then $\widehat{\boldsymbol{\beta}}$ still converges to some 'least false' value at a rate of $\sqrt{n}$. They showed this only for the case where each individual has at most one event. In Subsection 4.1.2, we generalize this to multiple events per individual. So the following condition is usually satisfied:

(A2) (convergence of $\widehat{\boldsymbol{\beta}}$) There exists $\boldsymbol{\beta}_0$ such that $\widehat{\boldsymbol{\beta}} - \boldsymbol{\beta}_0 = O_{\mathrm{P}}(n^{-1/2})$.

If the Cox model (3.6) holds true then $\boldsymbol{\beta} = \boldsymbol{\beta}_0$ is the 'true' value. We shall prove that estimating the parameter $\boldsymbol{\beta}$ does not change the asymptotics. For the following, let $\boldsymbol{d}(\boldsymbol{\beta}, t) = \exp(\boldsymbol{Z}_i(t)\boldsymbol{\beta}) R_i(t)$. The first theorem covers the asymptotics under the Aalen model.

Theorem 3.3. *Suppose* $(R_i, \mathbf{Y}_i, \boldsymbol{Z}_i)$ *are i.i.d. and* R_i, $\mathbf{Y}_i$, $\boldsymbol{Z}_i$ *are càglàd, bounded, adapted stochastic processes. If (A1), (A2) are satisfied and the Aalen model* (2.6) *holds true then in* $D[0, \tau]$,

$$T(\boldsymbol{d}(\widehat{\boldsymbol{\beta}}, \cdot), t) \xrightarrow{d} m(t),$$

where m is a mean zero Gaussian process with covariance $\operatorname{Cov}(m(s), m(t)) = \sigma^2(s \wedge t)$, *where*

$$\sigma^2(t) = \int_0^t \mathrm{E}\left[\left(\boldsymbol{Q}_{\mathrm{P}}^{\mathbf{Y}_1(s)} d_1(\boldsymbol{\beta}_0, s)\right)^2 \lambda_1(s)\right] \mathrm{d}s.$$

Furthermore, uniformly in $t \in [0, \tau]$,

$$\hat{\sigma}^2(\boldsymbol{d}(\widehat{\boldsymbol{\beta}}, \cdot), t) \xrightarrow{\mathrm{P}} \sigma^2(t).$$

The next theorem is concerned with the asymptotics if the Cox model holds.

Theorem 3.4. *Suppose* $(R_i, \mathbf{Y}_i, \boldsymbol{Z}_i)$ *are i.i.d. and* R_i, $\mathbf{Y}_i$, $\boldsymbol{Z}_i$ *are càglàd, bounded, adapted stochastic processes. If (A1), (A2) hold true and if*

$$\lambda_i(t) = \lambda_0(t) \exp(\boldsymbol{Z}_i(t)\boldsymbol{\beta}_0) R_i(t) = \lambda_0(t) d_i(\boldsymbol{\beta}_0, t)$$

then uniformly in $t \in [0, \tau]$,

$$n^{-\frac{1}{2}} T(\boldsymbol{d}(\widehat{\boldsymbol{\beta}}, \cdot), t) \xrightarrow{\mathrm{P}} \int_0^t \left\| \boldsymbol{Q}_{\mathrm{P}}^{\mathbf{Y}_1(s)} d_1(\boldsymbol{\beta}_0, s) \right\|_{\mathrm{P}}^2 \lambda_0(s) \, \mathrm{d}s$$

and

$$\hat{\sigma}^2(\boldsymbol{d}(\widehat{\boldsymbol{\beta}}, \cdot), t) \xrightarrow{\mathrm{P}} \int_0^t \left\| \boldsymbol{Q}_{\mathrm{P}}^{\mathbf{Y}_1(s)} d_1(\boldsymbol{\beta}_0, s) \right\|_{d_1(\boldsymbol{\beta}_0, s)\cdot \mathrm{P}}^2 \lambda_0(s) \, \mathrm{d}s.$$

With the above choice of $\boldsymbol{d}$, one-sided tests that reject for large values of $T(\boldsymbol{d}(\widehat{\boldsymbol{\beta}}, \cdot), \tau) \Big/ \sqrt{\hat{\sigma}^2(\boldsymbol{d}(\widehat{\boldsymbol{\beta}}, \cdot), \tau)}$ seem reasonable.

3.6 Optimal Weights

In this section, we derive optimal weights against fixed and local alternatives. Against fixed alternatives, the weights are optimal in the sense of approximate Bahadur efficiency, and against local alternatives, they are optimal in the sense of Pitman efficiency. However, the optimal weights will depend on some unknown quantities. Plugging in estimators for them gives weights for which the asymptotic behavior of T is not covered by the results for the least squares estimators in Section 3.3. We will not prove the necessary extensions here since in Chapter 5, we prove more general results that cover these estimated optimal weights.

3.6.1 Fixed Alternatives

We want to choose optimal weights against a fixed alternative given by $\lambda_i(t) = h_i(t)$, $i = 1, \dots, n$, where h_i is assumed to be known. As optimality criterion, we use the concept of approximate Bahadur efficiency, see Bahadur (1960) and Nikitin (1995, p. 10f). We shall motivate the approximate Bahadur efficiency briefly for our special case. We want to consider tests based on

$$V(\boldsymbol{d}) := V^{(1)}(\boldsymbol{d}) = T(\boldsymbol{d},\tau)/\sqrt{\hat{\sigma}^2(\boldsymbol{d},\tau)}$$

that reject for large values of $V(\boldsymbol{d})$. If the null hypothesis holds true then under the conditions of Theorem 3.1,

$$V(\boldsymbol{d}) \xrightarrow{\mathrm{d}} N(0,1).$$

If $\lambda_i(t) = h_i(t)$ then under the conditions of Theorem 3.2,

$$n^{-\frac{1}{2}}V(\boldsymbol{d}) \xrightarrow{\mathrm{P}} b(\boldsymbol{d}) := \frac{\int_0^\tau < \boldsymbol{Q}_{\mathrm{P}}^{\mathbf{Y}_1(s)} d_1(s), h_1(s) >_{\mathrm{P}} \mathrm{d}s}{\left(\int_0^\tau \left\| \boldsymbol{Q}_{\mathrm{P}}^{\mathbf{Y}_1(s)} d_1(s)\right\|_{h_1(s)\cdot\mathrm{P}}^2 \mathrm{d}s\right)^{\frac{1}{2}}}.$$

The approximated p-value is in our case given by

$$L(\boldsymbol{d}) = 1 - \Phi(V(\boldsymbol{d})),$$

where Φ denotes the cumulative standard normal distribution function. $L(\boldsymbol{d})$ is the (approximate) probability that the test statistic exceeds the observed value $V(\boldsymbol{d})$. One rejects the null hypothesis if $L(\boldsymbol{d})$ is less than the asymptotic level α of the test.

Assume we are given two different choices of weights $\boldsymbol{d}^{(1)}$ and $\boldsymbol{d}^{(2)}$. Then, using the words of Bahadur (1960), it would be fair to say that the test based on $V(\boldsymbol{d}^{(1)})$ is less successful than the test based on $V(\boldsymbol{d}^{(2)})$ if, under the alternative, $L(\boldsymbol{d}^{(1)}) > L(\boldsymbol{d}^{(2)})$. By Bahadur (1960),

$$-\frac{1}{n}\log\left(L(\boldsymbol{d})\right) = \frac{1}{2}b(\boldsymbol{d})^2 + o_{\mathrm{P}}(1).$$

The quantity $b(\boldsymbol{d})^2$ is called approximate Bahadur slope. Hence, for the two choices of weights $\boldsymbol{d}^{(1)}$ and $\boldsymbol{d}^{(2)}$,

$$\frac{\log(L(\boldsymbol{d}^{(1)}))}{\log(L(\boldsymbol{d}^{(2)}))} \xrightarrow{\mathrm{P}} \left(\frac{b(\boldsymbol{d}^{(1)})}{b(\boldsymbol{d}^{(2)})}\right)^2.$$

As $\log(L(\boldsymbol{d})) < 0$, one can conclude that if $b(\boldsymbol{d}^{(1)}) < b(\boldsymbol{d}^{(2)})$ then the test based on $V(\boldsymbol{d}^{(1)})$ is less successful than the test based on $V(\boldsymbol{d}^{(2)})$. Hence, it seems reasonable to use weights that maximize $b(\boldsymbol{d})$. This is precisely what we are going to do. So the optimization problem we consider is maximizing $b(\boldsymbol{d})$. We can give the following upper bound for $b(\boldsymbol{d})$ (recall that we use the convention $0/0 = 0$):

Lemma 3.1. *Suppose that the conditions of Theorem 3.2 are satisfied, that $h_i(t) = 0$ implies $\mathbf{Y}_i(t) = \mathbf{0}$, and that there exists $K > 0$ such that $h_i(t) > 0$ implies $h_i(t) > K$. Let*

$$d_i^*(t) := \begin{cases} \boldsymbol{Q}_{h_i(t)\cdot \mathrm{P}}^{B_i(t)} 1, & h_i(t) > 0 \\ 0, & h_i(t) = 0 \end{cases}$$

where $B_i(t) = \left\{ \frac{Y_{ij}(t)}{h_i(t)} : j = 1, \dots, k_{\boldsymbol{\alpha}} \right\}$. Then

$$b(\boldsymbol{d}) \le b(\boldsymbol{d}^*).$$

Proof. The conditions on h_i ensure that $\frac{Y_{ij}(t)}{h_i(t)} \in L_2(h_i(t) \cdot \mathrm{P})$ for $j = 1, \dots, k_{\boldsymbol{\alpha}}$. The key idea is to rewrite $b(\boldsymbol{d})$ by using the scalar products $< \cdot, \cdot >_{h_1(t)\cdot \mathrm{P}}$ and applying the Cauchy-Schwarz inequality. Let $\tilde{d}(t) := \boldsymbol{Q}_{\mathrm{P}}^{\mathbf{Y}_1(t)} d_1(t)$. As $< \cdot, Y_{1j}(t) >_{\mathrm{P}} = < \cdot, Y_{1j}(t)/h_1(t) >_{h_1(t)\cdot \mathrm{P}}$ we have $\tilde{d}(t) = \boldsymbol{Q}_{h_1(t)\cdot \mathrm{P}}^{B_1(t)} \tilde{d}(t)$. Using some of the properties of projection matrices, for $s \in [0, \tau]$,

$$\begin{aligned} <\tilde{d}(s), h_1(s)>_{\mathrm{P}} = <\tilde{d}(s), 1>_{h_1(s)\cdot \mathrm{P}} &= <\boldsymbol{Q}_{h_1(s)\cdot \mathrm{P}}^{B_1(s)} \tilde{d}(s), 1>_{h_1(s)\cdot \mathrm{P}} \\ &= <\tilde{d}(s), \boldsymbol{Q}_{h_1(s)\cdot \mathrm{P}}^{B_1(s)} 1>_{h_1(s)\cdot \mathrm{P}} . \end{aligned}$$

Thus, by using twice the Cauchy-Schwarz inequality,

$$\begin{aligned} \int_0^\tau <\tilde{d}(s), h_1(s)>_{\mathrm{P}} \,\mathrm{d}s &\le \int_0^\tau \left\| \tilde{d}(s) \right\|_{h_1(s)\cdot \mathrm{P}} \left\| \boldsymbol{Q}_{h_1(s)\cdot \mathrm{P}}^{B_1(s)} 1 \right\|_{h_1(s)\cdot \mathrm{P}} \mathrm{d}s \\ &\le \left(\int_0^\tau \left\| \tilde{d}(s) \right\|_{h_1(s)\cdot \mathrm{P}}^2 \mathrm{d}s \right)^{\frac{1}{2}} \left(\int_0^\tau \left\| \boldsymbol{Q}_{h_1(s)\cdot \mathrm{P}}^{B_1(s)} 1 \right\|_{h_1(s)\cdot \mathrm{P}}^2 \mathrm{d}s \right)^{\frac{1}{2}}. \end{aligned}$$

Therefore, $b(\boldsymbol{d}) \le \left(\int_0^\tau \|\boldsymbol{d}^*(s)\|_{h_1(s)\cdot \mathrm{P}}^2 \,\mathrm{d}s \right)^{1/2}$. By the definition of $B_1(s)$,

$$<d_1^*(s), Y_{1j}(s)>_{\mathrm{P}} = <\boldsymbol{Q}_{h_1(s)\cdot \mathrm{P}}^{B_1(s)} 1, \frac{Y_{1j}(s)}{h_1(s)}>_{h_1(s)\cdot \mathrm{P}} = 0, \quad \text{for } j = 1, \dots, k_{\boldsymbol{\alpha}},$$

and hence, $\boldsymbol{Q}_{\mathrm{P}}^{\mathbf{Y}_1(s)} d_1^*(s) = d_1^*(s)$. Thus,

$$b(\boldsymbol{d}^*) = \int_0^\tau <\boldsymbol{d}^*(s), 1>_{h_1(s)\cdot \mathrm{P}} \,\mathrm{d}s \left(\int_0^\tau \|\boldsymbol{d}^*(s)\|_{h_1(s)\cdot \mathrm{P}}^2 \,\mathrm{d}s \right)^{-\frac{1}{2}}$$

Hence, by properties of projections (3.3), $b(\boldsymbol{d}^*) = \left(\int_0^\tau \|\boldsymbol{d}^*(s)\|_{h_1(s)\cdot \mathrm{P}}^2 \,\mathrm{d}s \right)^{1/2}$. □

Hence, using the weights $\boldsymbol{d}^*$ is optimal - if we can show the asymptotics for $\boldsymbol{d}^*$. If the matrix

$$\boldsymbol{A}(s) = \left(< \frac{Y_{1j}(s)}{h_1(s)}, \frac{Y_{1\nu}(s)}{h_1(s)} >_{h_1(s)\cdot P} \right)_{j,\nu=1,\dots,k_{\boldsymbol{\alpha}}}$$

is invertible for all $s \in [0, \tau]$, $\|\boldsymbol{A}^{-1}(\cdot)\|$ is bounded, and h_1 is bounded away from zero on the event $\{h_1(s) > 0\}$ then

$$d_1^*(s) = \left(\mathbf{1} - \boldsymbol{Y}_1(s)\frac{1}{h_1(s)}\left(\boldsymbol{A}(s)\right)^{-1}\mathrm{E}(\boldsymbol{Y}_1(s))^\top\right) I\{h_1(s) > 0\}.$$

Unfortunately, usually $\boldsymbol{A}(s)$ and $\mathrm{E}(\boldsymbol{Y}_1(s))$ are not known, so one needs to plug in estimators, e.g. the consistent estimators $\overline{\boldsymbol{Y}\boldsymbol{h}^{-1}\boldsymbol{Y}}(s)$ and $\overline{\boldsymbol{Y}}(s)$. Doing so, we get the weights

$$\widehat{\boldsymbol{d}^*}(s) = \mathrm{diag}(\boldsymbol{R}(s))\boldsymbol{Q}^{\boldsymbol{B}(s)}_{\boldsymbol{h}(s)\cdot\mathbb{P}_n}\mathbf{1},$$

where $\boldsymbol{B}(s) = \{\boldsymbol{y}(s)/\boldsymbol{h}(s) : \boldsymbol{y}(s) \text{ column of } \boldsymbol{Y}(s)\}$ and $R_i(s) = I\{h_i(s) > 0\}$. Plugging this into our test statistic one sees that $\boldsymbol{Q}^{\boldsymbol{Y}(s)}_{\mathbb{P}_n}\widehat{\boldsymbol{d}^*}(s) = \widehat{\boldsymbol{d}^*}(s)$. Hence,

$$T(\widehat{\boldsymbol{d}^*}, t) = n^{-\frac{1}{2}}\int_0^t \widehat{\boldsymbol{d}^*}(s)^\top \mathrm{d}\boldsymbol{N}(s),$$

meaning that $\widehat{\boldsymbol{d}^*}(s)$ need not be projected any more.

Note that we cannot use Theorem 3.1 and Theorem 3.2, since the components of $\widehat{\boldsymbol{d}^*}(s)$ are not i.i.d. So we need to relax this assumption. One possibility would be to extend Theorem 3.1 and Theorem 3.2 to cover weighted projections. Instead, we refer to Chapter 5, where we will prove the needed results in a more general setting.

Furthermore, in applications, one could be interested in not just a fixed alternative, but another model as alternative. For this we could plug in estimators into $\boldsymbol{h}$ to get an estimator $\widehat{\boldsymbol{h}}$. In general, $\widehat{\boldsymbol{h}}$ will not be predictable, and hence, similar to the situation described in Subsection 3.5.3, we need to relax the assumption of predictability of the weights as well. The asymptotic results in Chapter 5 cover this for many competing models.

3.6.2 Local Alternatives

The idea of local alternatives is that their distance to the null hypothesis goes to zero at a rate which ensures that distinguishing between the null and the alternative does not become trivial. We shall consider the following set of local alternatives to the Aalen model (2.6):

$$\lambda_i(t) = \boldsymbol{Y}_i(t)\boldsymbol{\alpha}(t) + n^{-\frac{1}{2}}g_i(t), \tag{3.7}$$

where $\boldsymbol{Y}_i$ is as in (2.6) and g_i are some predictable processes. We want to derive optimal choices against this local alternative using the so-called Pitman-efficiency. For this, we first need to derive the asymptotic distribution of $T(\boldsymbol{d}, \cdot)$ under (3.7).

Theorem 3.5. *Suppose that $(g_i, \mathbf{Y}_i, d_i)$ are i.i.d. and g_i, $\mathbf{Y}_i$, d_i, are càglàd, adapted, bounded stochastic processes. If (A1) and the local alternative (3.7) hold true then in $D[0,\tau]$,*

$$T(\boldsymbol{d},\cdot) = n^{-\frac{1}{2}} \int_0^{\cdot} (\boldsymbol{Q}_{\mathbb{P}_n}^{\boldsymbol{Y}(t)} \boldsymbol{d}(t))^\top \mathrm{d}\boldsymbol{N}(t) \xrightarrow{d} m(\cdot),$$

where m is a Gaussian process with mean

$$\mathrm{E}[m(t)] = \mu(\boldsymbol{d},t) := \int_0^t \mathrm{E}[(\boldsymbol{Q}_{\mathrm{P}}^{\mathbf{Y}_1(s)} d_1(s)) g_1(s)]\,\mathrm{d}s$$

and covariance $\mathrm{Cov}(m(s), m(t)) = \sigma^2(\boldsymbol{d}, s \wedge t)$, where

$$\sigma^2(\boldsymbol{d},t) = \int_0^t \mathrm{E}\left[\left(\boldsymbol{Q}_{\mathrm{P}}^{\mathbf{Y}_1(s)} d_1(s)\right)^2 \mathbf{Y}_1(s)\boldsymbol{\alpha}(s)\right] \mathrm{d}s.$$

Furthermore,

$$\hat{\sigma}^2(\boldsymbol{d},t) = \frac{1}{n}\int_0^t (\boldsymbol{Q}_{\mathbb{P}_n}^{\boldsymbol{Y}(s)} \boldsymbol{d}(s))^\top \mathrm{diag}(\mathrm{d}\boldsymbol{N}(s))(\boldsymbol{Q}_{\mathbb{P}_n}^{\boldsymbol{Y}(s)} \boldsymbol{d}(s)) \xrightarrow{\mathrm{P}} \sigma^2(\boldsymbol{d},t)$$

uniformly in $t \in [0,\tau]$.

The previous theorem can be used to derive optimal tests under local alternatives of type (3.7) within the class of one-sided tests based on $T(\boldsymbol{d},\tau)/\sqrt{\hat{\sigma}^2(\boldsymbol{d},\tau)}$. If $\sigma^2(\boldsymbol{d},\tau) \neq 0$ then, by a Slutsky argument, the previous theorem implies $T(\boldsymbol{d},\tau)/\sqrt{\hat{\sigma}^2(\boldsymbol{d},\tau)} \xrightarrow{\mathrm{P}} N(\zeta(\boldsymbol{d}),1)$, where $\zeta(\boldsymbol{d}) = \mu(\boldsymbol{d},\tau)/\sqrt{\sigma^2(\boldsymbol{d},\tau)}$. Thus the power against (3.7) converges to $1 - \Phi(\Phi^{-1}(1-\alpha) - \zeta(\boldsymbol{d}))$, where α is the asymptotic level of the test and Φ is the cumulative distribution function of the standard normal distribution. Therefore, an asymptotically most powerful sequence of tests is achieved when $\zeta(\boldsymbol{d})$ is maximized. This can be done similarly to the previous section, as the following lemma shows.

Lemma 3.2. *Suppose that the conditions of Theorem 3.5 hold true and that with $\eta_i(t) := \mathbf{Y}_i(t)\boldsymbol{\alpha}(t)$, there exists $K > 0$ such that $\eta_i(t) > 0$ implies $\eta_i(t) > K$ and $\eta_i(t) = 0$ implies ($g_i(t) = 0$ and $\mathbf{Y}_i(t) = \mathbf{0}$). Let*

$$d_i^*(t) := \begin{cases} \boldsymbol{Q}_{\eta_i(t)\cdot \mathrm{P}}^{B_i(t)} \frac{g_i(t)}{\eta_i(t)}, & \eta_i(t) > 0 \\ 0, & \eta_i(t) = 0 \end{cases}$$

where $B_i(t) = \left\{\frac{Y_{ij}(t)}{\eta_i(t)}, j = 1, \ldots, k_{\boldsymbol{\alpha}}\right\}$. Then

$$\zeta(\boldsymbol{d}^*) \geq \zeta(\boldsymbol{d}).$$

Proof. Again, the idea is to rewrite $\zeta(\boldsymbol{d})$ by using the scalar products $<\cdot,\cdot>_{\eta_1(t)\cdot \mathrm{P}}$ and applying the Cauchy-Schwarz inequality. Let $\tilde{d}(t) := \boldsymbol{Q}_{\mathrm{P}}^{\mathbf{Y}_1(t)} d_1(t)$. Using some of the properties of projection matrices given in (3.3),

$$\begin{aligned}
\mu(\boldsymbol{d},\tau) &= \int_0^\tau <\tilde{d}(s), g_1(s)>_{\mathrm{P}} \,\mathrm{d}s = \int_0^\tau <\tilde{d}(s), \frac{g_1(s)}{\eta_1(s)}>_{\eta_1(s)\cdot\mathrm{P}} \,\mathrm{d}s \\
&= \int_0^\tau <\boldsymbol{Q}_{\eta_1(s)\cdot\mathrm{P}}^{B_1(s)}\tilde{d}(s), \frac{g_1(s)}{\eta_1(s)}>_{\eta_1(s)\cdot\mathrm{P}} \,\mathrm{d}s \\
&= \int_0^\tau <\tilde{d}(s), \boldsymbol{Q}_{\eta_1(s)\cdot\mathrm{P}}^{B_1(s)}\frac{g_1(s)}{\eta_1(s)}>_{\eta_1(s)\cdot\mathrm{P}} \,\mathrm{d}s.
\end{aligned}$$

Hence, using twice the Cauchy-Schwarz inequality,

$$\begin{aligned}
\mu(\boldsymbol{d},\tau) &\leq \int_0^\tau \left\|\tilde{d}(s)\right\|_{\eta_1(s)\cdot\mathrm{P}} \left\|\boldsymbol{Q}_{\eta_1(s)\cdot\mathrm{P}}^{B_1(s)}\frac{g_1(s)}{\eta_1(s)}\right\|_{\eta_1(s)\cdot\mathrm{P}} \mathrm{d}s \\
&\leq \left(\int_0^\tau \left\|\tilde{d}(s)\right\|^2_{\eta_1(s)\cdot\mathrm{P}} \mathrm{d}s\right)^{\frac{1}{2}} \left(\int_0^\tau \left\|\boldsymbol{Q}_{\eta_1(s)\cdot\mathrm{P}}^{B_1(s)}\frac{g_1(s)}{\eta_1(s)}\right\|^2_{\eta_1(s)\cdot\mathrm{P}} \mathrm{d}s\right)^{\frac{1}{2}} \\
&= \sqrt{\sigma^2(\boldsymbol{d},\tau)} \left(\int_0^\tau \|d_1^*(s)\|^2_{\eta_1(s)\cdot\mathrm{P}} \,\mathrm{d}s\right)^{\frac{1}{2}}.
\end{aligned}$$

Thus, $\zeta(\boldsymbol{d}) \leq \left(\int_0^\tau \|d_1^*(s)\|^2_{\eta_1(s)\cdot\mathrm{P}} \,\mathrm{d}s\right)^{\frac{1}{2}}$. By the definition of $B_1(s)$,

$$<d_1^*(s), Y_{1j}(s)>_{\mathrm{P}} = <\boldsymbol{Q}_{\eta_1(s)\cdot\mathrm{P}}^{B_1(s)} 1, \frac{Y_{1j}(s)}{\eta_1(s)}>_{\eta_1(s)\cdot\mathrm{P}} = 0, \quad \text{for } j=1,\dots,k_{\boldsymbol{\alpha}},$$

and hence, $\boldsymbol{Q}_{\mathrm{P}}^{\mathbf{Y}_1(s)} d_1^*(s) = d_1^*(s)$. Thus, by properties of projections (3.3),

$$\mu(\boldsymbol{d}^*,\tau) = \int_0^\tau <d_1^*(s), 1>_{h_1(s)\cdot\mathrm{P}} \,\mathrm{d}s = \int_0^\tau \|d_1^*(s)\|^2_{h_1(s)\cdot\mathrm{P}} \,\mathrm{d}s.$$

Therefore, $\zeta(\boldsymbol{d}^*) = \left(\int_0^\tau \|d_1^*(s)\|^2_{\eta_1(s)\cdot\mathrm{P}} \,\mathrm{d}s\right)^{1/2}$. □

3.7 Proofs

Before we prove the results, we need some preparations.

3.7.1 Convergence of Inverted Matrices

This section shows how the uniform stochastic convergence of time-dependent random matrices can be carried over to their inverses. We use this to get uniform stochastic convergence of $(\overline{\boldsymbol{Y}\boldsymbol{Y}}(t))^{-1}$ from that of $\overline{\boldsymbol{Y}\boldsymbol{Y}}(t)$. A lemma similar to Lemma 3.3 was already stated in McKeague (1988). Lemma 3.3 and Lemma 3.4 are slight extensions of a similar lemma in Gandy (2002). Lemma 3.5 is directly from Gandy (2002).

Lemma 3.3. *Suppose X is some compact subset of a Euclidean space and $\boldsymbol{A}^{(n)}(x), x \in X$, $n \in \mathbb{N}$, are $k \times k$ matrices of random processes. If there exists a continuous function $\boldsymbol{a} : X \to \mathbb{R}^{k\times k}$ such that $\boldsymbol{a}(x)$ is invertible for all $x \in X$ and $\sup_{x\in X} \|\boldsymbol{A}^{(n)}(x) - \boldsymbol{a}(x)\| \xrightarrow{\mathrm{P}} 0$ then*

i) $\mathrm{P}(\boldsymbol{A}^{(n)}(x)$ *is invertible* $\forall x \in X) \to 1$,

ii) $\exists K > 0$ *s.t.* $\mathrm{P}(\| \left(\boldsymbol{A}^{(n)}(x)\right)^{-1} \| < K \,\forall x \in X) \to 1$, *and*

iii) $\sup_{x\in X} \| \left(\boldsymbol{A}^{(n)}(x)\right)^{-1} - \boldsymbol{a}^{-1}(x)\| = O_{\mathrm{P}}(\sup_{x\in X} \|\boldsymbol{A}^{(n)}(x) - \boldsymbol{a}(x)\|)$.

We prepare the proof of Lemma 3.3 by giving two lemmas.

Lemma 3.4. *Let X be a compact topological space and $k \in \mathbb{N}$. If $\boldsymbol{a} : X \to \mathbb{R}^{k\times k}$ is a continuous mapping such that $\boldsymbol{a}(x)$ is invertible for all $x \in X$ then there exists an $\epsilon > 0$ such that for all $\boldsymbol{B} : X \to \mathbb{R}^{k\times k}$, $\sup_{x\in X} \|\boldsymbol{a}(x) - \boldsymbol{B}(x)\| < \epsilon$ implies $\boldsymbol{B}(x)$ invertible $\forall x \in X$.*

Proof. Since $\boldsymbol{a}(X)$ is compact, the set of invertible matrices $\mathrm{GL} \subset \mathbb{R}^{k\times k}$ is open, and $\boldsymbol{a}(X) \subset \mathrm{GL}$ we can find $x_1, \ldots, x_\nu \in X$ and $\delta_1, \ldots, \delta_\nu > 0$ such that $\boldsymbol{a}(X) \subset \bigcup_{i=1}^{\nu} U(\boldsymbol{a}(x_i), \delta_i)$ and $U(\boldsymbol{a}(x_i), 2\delta_i) \subset \mathrm{GL}$, where $U(\boldsymbol{C}, \xi) := \{\boldsymbol{D} \in \mathbb{R}^{k\times k} : \|\boldsymbol{C} - \boldsymbol{D}\| < \xi\}$. It can be verified that $\epsilon := \min\{\delta_1, \ldots, \delta_\nu\}$ satisfies the claim. □

Lemma 3.5. *Let $k \in \mathbb{N}$. If $\boldsymbol{C}, \boldsymbol{D} \in \mathbb{R}^{k\times k}$ are invertible and $\|\boldsymbol{D}^{-1}\|\|\boldsymbol{D} - \boldsymbol{C}\| < 1$ then*

$$\|\boldsymbol{C}^{-1} - \boldsymbol{D}^{-1}\| \leq \frac{\|\boldsymbol{D}^{-1}\|^2\|\boldsymbol{C} - \boldsymbol{D}\|}{1 - \|\boldsymbol{D}^{-1}\|\|\boldsymbol{C} - \boldsymbol{D}\|}. \tag{3.8}$$

Proof. Let $\boldsymbol{I}$ denote the identity matrix in $\mathbb{R}^{k\times k}$.

$$\begin{aligned}\|\boldsymbol{C}^{-1} - \boldsymbol{D}^{-1}\| &= \|\boldsymbol{C}^{-1}(\boldsymbol{D} - \boldsymbol{C})\boldsymbol{D}^{-1}\| \leq \|\boldsymbol{C}^{-1}\|\|\boldsymbol{C} - \boldsymbol{D}\|\|\boldsymbol{D}^{-1}\| \\ &= \|\boldsymbol{C}^{-1}\boldsymbol{D}\boldsymbol{D}^{-1}\|\|\boldsymbol{C} - \boldsymbol{D}\|\|\boldsymbol{D}^{-1}\| \leq \|\boldsymbol{C}^{-1}\boldsymbol{D}\|\|\boldsymbol{D}^{-1}\|^2\|\boldsymbol{C} - \boldsymbol{D}\|\end{aligned}$$

By the assumption, $\|\boldsymbol{D}^{-1}(\boldsymbol{D} - \boldsymbol{C})\| \leq \|\boldsymbol{D}^{-1}\|\|\boldsymbol{D} - \boldsymbol{C}\| < 1$. Hence,

$$\begin{aligned}\|\boldsymbol{C}^{-1}\boldsymbol{D}\| &= \|(\boldsymbol{D}^{-1}\boldsymbol{C})^{-1}\| = \|(\boldsymbol{I} - (\boldsymbol{D}^{-1}(\boldsymbol{D} - \boldsymbol{C}))^{-1}\| = \|\sum_{n=0}^{\infty}(\boldsymbol{D}^{-1}(\boldsymbol{D} - \boldsymbol{C}))^n\| \\ &\leq \sum_{n=0}^{\infty} \|\boldsymbol{D}^{-1}(\boldsymbol{D} - \boldsymbol{C})\|^n = \left(1 - \|\boldsymbol{D}^{-1}(\boldsymbol{D} - \boldsymbol{C})\|\right)^{-1} \\ &\leq \left(1 - \|\boldsymbol{D}^{-1}\|\|\boldsymbol{C} - \boldsymbol{D}\|\right)^{-1}.\end{aligned}$$

□

Proof of Lemma 3.3. Choose $\epsilon > 0$ as in Lemma 3.4. Then

$$\mathrm{P}(\exists x \in X \text{ s.t. } \boldsymbol{A}^{(n)}(x) \text{ is singular}) \leq \mathrm{P}(\sup_{x \in X} \|\boldsymbol{a}(x) - \boldsymbol{A}^{(n)}(x)\| \geq \epsilon) \to 0.$$

Since $\boldsymbol{a}$, taking inverses, and $\|\cdot\|$ are continuous mappings, the compactness of X implies that $\{\|\boldsymbol{a}^{-1}(x)\| : x \in X\}$ is compact. Hence, there exists a constant $L > 0$ such that $\sup_{x \in X} \|\boldsymbol{a}^{-1}(x)\| \leq L$. On the event

$$D_n := \{\|\boldsymbol{a}^{-1}(x)\| \|\boldsymbol{a}(x) - \boldsymbol{A}^{(n)}(x)\| < 1/2, \boldsymbol{A}^{(n)}(x) \text{ invertible } \forall x \in X\},$$

we can use Lemma 3.5 to see that $\forall x \in X$,

$$\|\boldsymbol{a}^{-1}(x) - \left(\boldsymbol{A}^{(n)}(x)\right)^{-1}\| \leq 2L^2 \|\boldsymbol{a}(x) - \boldsymbol{A}^{(n)}(x)\|.$$

Since

$$D_n \supset \{\|\boldsymbol{a}(x) - \boldsymbol{A}^{(n)}(x)\| < (2L)^{-1}, \boldsymbol{A}^{(n)}(x) \text{ invertible } \forall x \in X\},$$

we have $\mathrm{P}(D_n) \to 1$. Hence,

$$\sup_{x \in X} \| \left(\boldsymbol{A}^{(n)}(x)\right)^{-1} - \boldsymbol{a}^{-1}(x)\| = O_{\mathrm{P}}(\sup_{x \in X} \|\boldsymbol{A}^{(n)}(x) - \boldsymbol{a}(x)\|).$$

Let $K := L + 1$. Since $\| \left(\boldsymbol{A}^{(n)}(x)\right)^{-1} \| \leq L + \| \left(\boldsymbol{A}^{(n)}(x)\right)^{-1} - \boldsymbol{a}^{-1}(x)\|$, we get iii). □

3.7.2 Properties of Projected Vectors of Stochastic Processes

Lemma 3.6. *Suppose $(Y_{i1}, \dots, Y_{ik_\alpha}, a_i, b_i, c_i), i \in \mathbb{N}$, are i.i.d. vectors with elements that are bounded processes with càglàd paths. If (A1) holds true then uniformly in $t \in [0, \tau]$,*

$$\overline{\boldsymbol{a}(\boldsymbol{Q}^{\boldsymbol{Y}}_{\mathbb{P}_n} \boldsymbol{b})}(t) \xrightarrow{\mathrm{P}} \mathrm{E}[a_1(t) \boldsymbol{Q}_{\mathrm{P}}^{\boldsymbol{Y}_1(t)} b_1(t)]$$

and

$$\begin{aligned}\overline{(\boldsymbol{Q}^{\boldsymbol{Y}}_{\mathbb{P}_n} \boldsymbol{a}) \boldsymbol{c} (\boldsymbol{Q}^{\boldsymbol{Y}}_{\mathbb{P}_n} \boldsymbol{b})}(t) &\xrightarrow{\mathrm{P}} \mathrm{E}\left[\left(\boldsymbol{Q}_{\mathrm{P}}^{\boldsymbol{Y}_1(t)} a_1(t)\right) c_1(t) \left(\boldsymbol{Q}_{\mathrm{P}}^{\boldsymbol{Y}_1(t)} b_1(t)\right)\right] \\ &= < \boldsymbol{Q}_{\mathrm{P}}^{\boldsymbol{Y}_1(t)} a_1(t), \boldsymbol{Q}_{\mathrm{P}}^{\boldsymbol{Y}_1(t)} b_1(t) >_{c_1(t) \cdot \mathrm{P}} .\end{aligned}$$

Proof. By a strong law of large numbers (Theorem A.4),

$$\frac{1}{n} \boldsymbol{Y}(t)^\top \boldsymbol{Y}(t) = \overline{\boldsymbol{Y}\boldsymbol{Y}}(t) \xrightarrow{\mathrm{P}} \mathrm{E}[\boldsymbol{Y}_1(t)^\top \boldsymbol{Y}_1(t)]$$

uniformly in t. Hence, by Lemma 3.3 and the assumptions,

$$(\overline{\boldsymbol{Y}\boldsymbol{Y}}(t))^{-1} \xrightarrow{\mathrm{P}} (\mathrm{E}[\boldsymbol{Y}_1(t)^\top \boldsymbol{Y}_1(t)])^{-1}$$

and the event $A_n := \{\overline{\boldsymbol{YY}}(t) \text{ invertible for all } t \in [0,\tau]\}$ satisfies $\mathrm{P}(A_n) \to 1$. On A_n, we have

$$\overline{\boldsymbol{a}(\boldsymbol{Q}^{\boldsymbol{Y}}_{\mathbb{P}_n}\boldsymbol{b})}(t) = \overline{\boldsymbol{ab}}(t) - \overline{\boldsymbol{aY}}(t)(\overline{\boldsymbol{YY}}(t))^{-1}\overline{\boldsymbol{Yb}}(t).$$

Dropping the dependence on t, we use the strong law of large numbers again to get that uniformly on $[0,\tau]$,

$$\begin{aligned}\overline{\boldsymbol{a}(\boldsymbol{Q}^{\boldsymbol{Y}}_{\mathbb{P}_n}\boldsymbol{b})} &\xrightarrow{\mathrm{P}} \mathrm{E}[a_1b_1] - \mathrm{E}[a_1\mathbf{Y}_1](\mathrm{E}[\mathbf{Y}_1\mathbf{Y}_1^\top])^{-1}\,\mathrm{E}[\mathbf{Y}_1b_1]^\top \\ &= \mathrm{E}\left[a_1\left(b_1 - \mathbf{Y}_1((<Y_{1j},Y_{1l}>_{\mathrm{P}})_{j,l=1,\dots,k_\alpha})^{-1}(<Y_{1j},b_1>_{\mathrm{P}})_{j=1,\dots,k_\alpha}\right)\right] \\ &= \mathrm{E}[a_1(\boldsymbol{Q}^{\mathbf{Y}_1}_{\mathrm{P}}b_1)].\end{aligned}$$

Similarly, on A_n we have for $x := \overline{(\boldsymbol{Q}^{\boldsymbol{Y}}_{\mathbb{P}_n}\boldsymbol{a})\boldsymbol{c}(\boldsymbol{Q}^{\boldsymbol{Y}}_{\mathbb{P}_n}\boldsymbol{b})}(t)$ that

$$\begin{aligned}x =& \frac{1}{n}(\boldsymbol{a} - \boldsymbol{Y}(\boldsymbol{Y}^\top\boldsymbol{Y})^{-1}\boldsymbol{Y}^\top\boldsymbol{a})^\top \mathrm{diag}(\boldsymbol{c})(\boldsymbol{b} - \boldsymbol{Y}(\boldsymbol{Y}^\top\boldsymbol{Y})^{-1}\boldsymbol{Y}^\top\boldsymbol{b}) \\ =& \overline{\boldsymbol{acb}} - \overline{\boldsymbol{aY}}(\overline{\boldsymbol{YY}})^{-1}\overline{\boldsymbol{Ycb}} - \overline{\boldsymbol{acY}}(\overline{\boldsymbol{YY}})^{-1}\overline{\boldsymbol{Yb}} \\ &+ \overline{\boldsymbol{aY}}(\overline{\boldsymbol{YY}})^{-1}\overline{\boldsymbol{YcY}}(\overline{\boldsymbol{YY}})^{-1}\overline{\boldsymbol{Yb}}.\end{aligned}$$

Using the law of large numbers again, we get

$$\begin{aligned}x \xrightarrow{\mathrm{P}} & \mathrm{E}[a_1c_1b_1] - \mathrm{E}[a_1\mathbf{Y}_1]\left(\mathrm{E}[\mathbf{Y}_1^\top\mathbf{Y}_1]\right)^{-1}\mathrm{E}[\mathbf{Y}_1c_1b_1]^\top \\ & - \mathrm{E}[a_1c_1\mathbf{Y}_1]\left(\mathrm{E}[\mathbf{Y}_1^\top\mathbf{Y}_1]\right)^{-1}\mathrm{E}[\mathbf{Y}_1b_1]^\top \\ & + \mathrm{E}[a_1\mathbf{Y}_1]\left(\mathrm{E}[\mathbf{Y}_1^\top\mathbf{Y}_1]\right)^{-1}\left(\mathrm{E}[\mathbf{Y}_1^\top c_1\mathbf{Y}_1]\right)\left(\mathrm{E}[\mathbf{Y}_1^\top\mathbf{Y}_1]\right)^{-1}\mathrm{E}[\mathbf{Y}_1b_1]^\top \\ &= \mathrm{E}\left[\left(a_1 - \mathbf{Y}_1\left(\mathrm{E}[\mathbf{Y}_1^\top\mathbf{Y}_1]\right)^{-1}\mathrm{E}[\mathbf{Y}_1a_1]^\top\right)c_1\left(b_1 - \mathbf{Y}_1\left(\mathrm{E}[\mathbf{Y}_1^\top\mathbf{Y}_1]\right)^{-1}\mathrm{E}[\mathbf{Y}_1b_1]^\top\right)\right] \\ &= \mathrm{E}\left[\left(\boldsymbol{Q}^{\mathbf{Y}_1}_{\mathrm{P}}a_1\right)c_1\left(\boldsymbol{Q}^{\mathbf{Y}_1}_{\mathrm{P}}b_1\right)\right].\end{aligned}$$

□

The following lemma is formulated in a slightly more general way than necessary for this chapter; the weights $\boldsymbol{w}$ are only needed for the following chapters.

Lemma 3.7. *Suppose all elements of the $n \times \nu$-dimensional matrix $\boldsymbol{Y}$ and the n-variate vectors $\boldsymbol{w}$ and $\mathbf{e}$ are predictable, locally bounded processes and suppose that $\boldsymbol{w} \geq 0$ and for each $i = 1,\dots,n$, $w_i(t)^{-1}$ is locally bounded.*

Then $\boldsymbol{Q}^{\boldsymbol{Y}(t)}_{\boldsymbol{w}(t)\cdot\mathbb{P}_n}\mathbf{e}(t)$ is an n-variate vector of locally bounded predictable processes, where we set $(\boldsymbol{Q}^{\boldsymbol{Y}(t)}_{\boldsymbol{w}(t)\cdot\mathbb{P}_n}\mathbf{e}(t))_i = 0$ if $w_i(t) = 0$.

Proof. Recall that a stochastic process is predictable if it is measurable with respect to the so-called predictable σ-algebra when considered as a mapping from $\Omega \times [0, \tau]$ to $\mathbb{R}$. One can choose a bijective mapping $f : \{1, \ldots, 2^\nu\} \to \mathcal{P}(\{1, \ldots, \nu\})$, where $\mathcal{P}$ stands for power set, such that $i \leq j$ implies $\#f(i) \geq \#f(j)$, where $\#$ stands for the cardinality of the set. Then $\boldsymbol{Q}^{\boldsymbol{Y}(t)}_{\boldsymbol{w}(t)\cdot\mathbb{P}_n} \boldsymbol{e}(t) = \boldsymbol{e}(t) - \boldsymbol{P}^{\boldsymbol{Y}(t)}_{\boldsymbol{w}(t)\cdot\mathbb{P}_n} \boldsymbol{e}(t)$, where

$$\boldsymbol{P}^{\boldsymbol{Y}(t)}_{\boldsymbol{w}(t)\cdot\mathbb{P}_n} \boldsymbol{e}(t) = \sum_{j=1}^{2^\nu} \Bigg[\left(\prod_{i<j} I\{\det(\boldsymbol{Z}_i(t)^\top \boldsymbol{Z}_i(t)) = 0\} \right) I\{\det(\boldsymbol{Z}_j(t)^\top \boldsymbol{Z}_j(t)) \neq 0\} \\ \times \boldsymbol{Z}_j(t) \left(\overline{\boldsymbol{Z}_j \boldsymbol{w} \boldsymbol{Z}_j}(t)\right)^{-1} \overline{\boldsymbol{Z}_j \boldsymbol{w} \boldsymbol{e}}(t) \Bigg], \tag{3.9}$$

where, deviating from our usual notation, $\boldsymbol{Z}_i(t)$ is the matrix consisting of the columns of $\boldsymbol{Y}(t)$ indexed by $f(i)$. Since the determinant can be written as polynomial, $\det(\boldsymbol{Z}_i(t)^\top \boldsymbol{Z}_i(t))$ is predictable for all $i = 1, \ldots, 2^\nu$. Since $g : \mathbb{R} \to \mathbb{R}, g(x) = I\{x \neq 0\}$ is measurable, the indicators in (3.9) are predictable processes. Setting the inverse in (3.9) to the matrix containing only zeros if $\boldsymbol{Z}_i(t)^\top \boldsymbol{Z}_i(t)$ is not invertible one can see that the remaining terms of (3.9) are predictable as well. Hence, $\boldsymbol{Q}^{\boldsymbol{Y}(t)}_{\boldsymbol{w}(t)\cdot\mathbb{P}_n} \boldsymbol{e}(t)$ is predictable as a product/sum of predictable processes.

Next, we show that the projection is locally bounded. If $w_j(t) = 0$ then $(\boldsymbol{Q}^{\boldsymbol{Y}(t)}_{\boldsymbol{w}(t)\cdot\mathbb{P}_n} \boldsymbol{e}(t))_j = 0$. Otherwise, using the fact that projections have an operator norm of 1,

$$\begin{aligned} |(\boldsymbol{Q}^{\boldsymbol{Y}(t)}_{\boldsymbol{w}(t)\cdot\mathbb{P}_n} \boldsymbol{e}(t))_j| &\leq w_j(t)^{-\frac{1}{2}} \left\| \boldsymbol{Q}^{\boldsymbol{Y}(t)}_{\boldsymbol{w}(t)\cdot\mathbb{P}_n} \boldsymbol{e}(t) \right\|_{\boldsymbol{w}(t)\cdot\mathbb{P}_n} \leq w_j(t)^{-\frac{1}{2}} \left\| \boldsymbol{e}(t) \right\|_{\boldsymbol{w}(t)\cdot\mathbb{P}_n} \\ &\leq w_j(t)^{-\frac{1}{2}} \sqrt{n} \max_{i=1,\ldots,n} |e_i(t)| \sqrt{w_i(t)}. \end{aligned}$$

Hence, since $w_i^{-1}(\cdot)$, $e_i(\cdot), w_i(\cdot)$, $i = 1, \ldots, n$, are locally bounded, $(\boldsymbol{Q}^{\boldsymbol{Y}(t)}_{\boldsymbol{w}(t)\cdot\mathbb{P}_n} \boldsymbol{e}(t))_j$ is locally bounded. □

3.7.3 Proofs of the Asymptotic Results

Proof of Theorem 3.1. Since, $\boldsymbol{c}(s)^\top \boldsymbol{\lambda}(s) = (\boldsymbol{Q}^{\boldsymbol{Y}(s)}_{\mathbb{P}_n} \boldsymbol{d}(s))^\top \boldsymbol{Y}(s) \boldsymbol{\alpha}(s) = 0$,

$$T(t) = n^{-\frac{1}{2}} \int_0^t (\boldsymbol{Q}^{\boldsymbol{Y}(s)}_{\mathbb{P}_n} \boldsymbol{d}(s))^\top (\mathrm{d}\boldsymbol{N}(s) - \boldsymbol{\lambda}(s)\,\mathrm{d}s). \tag{3.10}$$

By Lemma 3.7, $\boldsymbol{Q}^{\boldsymbol{Y}(s)}_{\mathbb{P}_n} \boldsymbol{d}(s)$ is predictable and locally bounded. Hence, $T(t)$ is a mean zero locally square integrable martingale. We want to apply a central limit theorem for T for which we need convergence of the predictable variation process:

$$\langle T\rangle(t)=\frac{1}{n}\int_0^t(\boldsymbol{Q}_{\mathbb{P}_n}^{\boldsymbol{Y}(s)}\boldsymbol{d}(s))^\top\operatorname{diag}(\boldsymbol{\lambda}(s))(\boldsymbol{Q}_{\mathbb{P}_n}^{\boldsymbol{Y}(s)}\boldsymbol{d}(s))\,\mathrm{d}s.$$

By Lemma 3.6, $\langle T\rangle(t)\xrightarrow{\mathrm{P}}\sigma^2(t)$. To apply the central limit theorem we also need some condition on the jumps of $T(t)$. We show

$$n^{-\frac{1}{2}}\sup_{\substack{i=1\dots n\\ t\in[0,\tau]}}\left|\left(\boldsymbol{Q}_{\mathbb{P}_n}^{\boldsymbol{Y}(t)}\boldsymbol{d}(t)\right)_i\right|\xrightarrow{\mathrm{P}}0. \tag{3.11}$$

Indeed, on A_n,

$$\begin{aligned}n^{-\frac{1}{2}}\left|\left(\boldsymbol{Q}_{\mathbb{P}_n}^{\boldsymbol{Y}(t)}\boldsymbol{d}(t)\right)_i\right|=&n^{-\frac{1}{2}}\left|d_i(t)-\boldsymbol{Y}_i(t)(\overline{\boldsymbol{Y}\boldsymbol{Y}}(t))^{-1}\overline{\boldsymbol{Y}\boldsymbol{d}}(t)\right|\\ \leq&n^{-\frac{1}{2}}|d_i(t)|+n^{-\frac{1}{2}}\|\boldsymbol{Y}_i(t)\|\left\|\left(\overline{\boldsymbol{Y}\boldsymbol{Y}}(t)\right)^{-1}\right\|\left\|\overline{\boldsymbol{Y}\boldsymbol{d}}(t)\right\|.\end{aligned}$$

$\overline{\boldsymbol{Y}\boldsymbol{d}}(t)$ converges uniformly to $\mathrm{E}[\boldsymbol{Y}_1(t)^\top d_1(t)]$ which is bounded. Lemma 3.3 implies that $\left(\overline{\boldsymbol{Y}\boldsymbol{Y}}(t)\right)^{-1}$ is stochastically bounded uniformly in t. Hence, as $\boldsymbol{Y}_i$ and d_i are bounded, (3.11) holds.

Rebolledo's central limit (Theorem A.3) finishes the proof. □

Proof of Theorem 3.2.

$$n^{-\frac{1}{2}}T(t)=n^{-\frac{1}{2}}\boldsymbol{a}(t)+\frac{1}{n}\int_0^t(\boldsymbol{Q}_{\mathbb{P}_n}^{\boldsymbol{Y}(s)}\boldsymbol{d}(s))^\top\boldsymbol{h}(s)\,\mathrm{d}s, \tag{3.12}$$

where $\boldsymbol{a}(t)=n^{-\frac{1}{2}}\int_0^t(\boldsymbol{Q}_{\mathbb{P}_n}^{\boldsymbol{Y}(s)}\boldsymbol{d}(s))^\top(\mathrm{d}\boldsymbol{N}(s)-\boldsymbol{\lambda}(s)\,\mathrm{d}s)$. Since $\boldsymbol{a}(t)$ is identical to the right hand side in (3.10), we can repeat the steps of the proof of Theorem 3.1 to see that $\boldsymbol{a}$ converges in distribution and that $\hat\sigma^2(\boldsymbol{d},t)$ converges in probability. Hence, $n^{-1/2}\boldsymbol{a}(t)$ vanishes uniformly in t. By Lemma 3.6,

$$\int_0^t\overline{(\boldsymbol{Q}_{\mathbb{P}_n}^{\boldsymbol{Y}}\boldsymbol{d})\boldsymbol{h}}(s)\,\mathrm{d}s\xrightarrow{\mathrm{P}}H(t).$$

□

Proof of Theorem 3.3. By Theorem 3.1,

$$T(\boldsymbol{d}(\boldsymbol{\beta}_0,\cdot),t)\xrightarrow{\mathrm{d}}m(t)\quad\text{and}\quad\hat\sigma^2(\boldsymbol{d}(\boldsymbol{\beta}_0,\cdot),t)\xrightarrow{\mathrm{P}}\sigma^2(t).$$

We will show that uniformly in $t\in[0,\tau]$,

$$T(\boldsymbol{d}(\widehat{\boldsymbol{\beta}},\cdot),t)-T(\boldsymbol{d}(\boldsymbol{\beta}_0,\cdot),t)\xrightarrow{\mathrm{P}}0\quad\text{and}\quad\hat\sigma^2(\boldsymbol{d}(\widehat{\boldsymbol{\beta}},\cdot),t)-\hat\sigma^2(\boldsymbol{d}(\boldsymbol{\beta}_0,\cdot),t)\xrightarrow{\mathrm{P}}0.$$

Let $\nabla_j = \frac{\partial}{\partial \beta_j}$ and $\nabla = (\nabla_1, \dots, \nabla_{k_\beta})$. Note that $\nabla d_i(\boldsymbol{\beta}, t) = \boldsymbol{Z}_i(t) d_i(\boldsymbol{\beta}, t)$ and $\nabla^\top \nabla d_i(\boldsymbol{\beta}, t) = \boldsymbol{Z}_i(t)^\top \boldsymbol{Z}_i(t) d_i(\boldsymbol{\beta}, t)$. By Taylor's theorem,

$$
\begin{aligned}
&T(\boldsymbol{d}(\widehat{\boldsymbol{\beta}}, \cdot), t) - T(\boldsymbol{d}(\boldsymbol{\beta}_0, \cdot), t) = \\
&= n^{\frac{1}{2}} (\widehat{\boldsymbol{\beta}} - \boldsymbol{\beta}_0)^\top \frac{1}{n} \int_0^t \left(\boldsymbol{Q}_{\mathbb{P}_n}^{\boldsymbol{Y}(s)} (\nabla \boldsymbol{d})(\boldsymbol{\beta}_0, s) \right)^\top \mathrm{d}\boldsymbol{M}(s) + \\
&\quad + \frac{1}{2} n^{\frac{1}{2}} (\widehat{\boldsymbol{\beta}} - \boldsymbol{\beta}_0)^\top \left(\frac{1}{n} \int_0^t n^{-\frac{1}{2}} \boldsymbol{x}_{\mu\nu}(s)^\top \mathrm{d}\boldsymbol{M}(s) \right)_{\mu,\nu=1,\dots,k_\beta} n^{\frac{1}{2}} (\widehat{\boldsymbol{\beta}} - \boldsymbol{\beta}_0),
\end{aligned}
\tag{3.13}
$$

where

$$
\boldsymbol{x}_{\mu\nu}(s) := (x_{i\mu\nu}(s))_{i=1,\dots,n} := \boldsymbol{Q}_{\mathbb{P}_n}^{\boldsymbol{Y}(s)} (\nabla_\mu \nabla_\nu \boldsymbol{d})(\widetilde{\boldsymbol{\beta}}, s),
$$

for some $\widetilde{\boldsymbol{\beta}}$ between $\widehat{\boldsymbol{\beta}}$ and $\boldsymbol{\beta}_0$. To justify the interchange of differentiation and integration, first note that the part of the integral with respect to $\boldsymbol{N}$ poses no problem, since this part is just a finite sum. For the other part of the integral, one can use e.g. Bauer (1992, p. 102), since the derivative is bounded uniformly in $\boldsymbol{\beta}$ in a neighborhood of $\boldsymbol{\beta}_0$. We show that the right hand side of (3.13) vanishes. We start with the first term.

$$
\begin{aligned}
&\left\langle \frac{1}{n} \int_0^\cdot \left(\boldsymbol{Q}_{\mathbb{P}_n}^{\boldsymbol{Y}(s)} (\nabla \boldsymbol{d})(\boldsymbol{\beta}_0, s) \right)^\top \mathrm{d}\boldsymbol{M}(s) \right\rangle (t) = \\
&\qquad = n^{-2} \int_0^t \left(\boldsymbol{Q}_{\mathbb{P}_n}^{\boldsymbol{Y}(s)} (\nabla \boldsymbol{d})(\boldsymbol{\beta}_0, s) \right)^\top \operatorname{diag}(\boldsymbol{\lambda}(s)) \left(\boldsymbol{Q}_{\mathbb{P}_n}^{\boldsymbol{Y}(s)} (\nabla \boldsymbol{d})(\boldsymbol{\beta}_0, s) \right) \mathrm{d}s \\
&\qquad = \frac{1}{n} \int_0^t \overline{(\boldsymbol{Q}_{\mathbb{P}_n}^{\boldsymbol{Y}} (\nabla \boldsymbol{d})) \boldsymbol{\lambda} (\boldsymbol{Q}_{\mathbb{P}_n}^{\boldsymbol{Y}} (\nabla \boldsymbol{d}))} (\boldsymbol{\beta}_0, s) \, \mathrm{d}s
\end{aligned}
$$

which converges uniformly to $\mathbf{0}$ by Lemma 3.6. Hence, by Lenglart's inequality,

$$
\frac{1}{n} \int_0^t \left(\boldsymbol{Q}_{\mathbb{P}_n}^{\boldsymbol{Y}(s)} (\nabla \boldsymbol{d})(\boldsymbol{\beta}_0, s) \right)^\top \mathrm{d}\boldsymbol{M}(s) \xrightarrow{\mathrm{P}} \mathbf{0}
$$

uniformly in $t \in [0, \tau]$.

Next, we show that the second term on the right hand side of (3.13) vanishes. Let $\mathcal{C} \subset \mathbb{R}^{k_\beta}$ be a compact set containing $\boldsymbol{\beta}_0$ in its interior. Consider the event

$$
A_n := \left\{ \overline{\boldsymbol{Y}\boldsymbol{Y}}(t) \text{ invertible } \forall t \in [0, \tau], \widehat{\boldsymbol{\beta}} \in \mathcal{C} \right\}.
$$

By (A1), Lemma 3.3, and (A2), we have $\mathrm{P}(A_n) \to 1$. On A_n,

$$
x_{i\mu\nu}(s) = Z_{i\mu}(s) Z_{i\nu}(s) d_i(\widetilde{\boldsymbol{\beta}}, s) + \boldsymbol{Y}_i(s) (\overline{\boldsymbol{Y}\boldsymbol{Y}}(s))^{-1} \overline{\boldsymbol{Y}(\nabla_\mu \nabla_\nu \boldsymbol{d})}(\widetilde{\boldsymbol{\beta}}, s) \tag{3.14}
$$

We show that the right hand side of (3.14) is stochastically bounded, uniformly in s and i. Since $\mathcal{C}$ is compact, $d_i(\widetilde{\boldsymbol{\beta}}, s) \le \sup_{\substack{s \in [0,\tau] \\ \boldsymbol{\beta} \in \mathcal{C}}} |d_i(\boldsymbol{\beta}, s)|$ is bounded. By Lemma 3.3, $\|(\overline{\boldsymbol{Y}\boldsymbol{Y}}(s))^{-1}\|$ is stochastically bounded, uniformly in s. By a strong law of large numbers (Theorem A.5), $\overline{\boldsymbol{Y}(\nabla_\mu\nabla_\nu \boldsymbol{d})}(\boldsymbol{\beta}, s)$ converges uniformly on $\mathcal{C} \times [0,\tau]$ in probability to the finite limit $\mathrm{E}[\boldsymbol{Y}_1(s)^\top Z_{1\mu}(s) Z_{1\nu}(s) d_1(\boldsymbol{\beta}, s)]$. Hence,

$$\|\overline{\boldsymbol{Y}(\nabla_\mu\nabla_\nu \boldsymbol{d})}(\widetilde{\boldsymbol{\beta}}, s)\| \leq \sup_{\substack{t\in[0,\tau]\\ \boldsymbol{\beta}\in\mathcal{C}}} \|\overline{\boldsymbol{Y}(\nabla_\mu\nabla_\nu \boldsymbol{d})}(\boldsymbol{\beta}, t)\|,$$

which is stochastically bounded. Since all other terms on the right hand side of (3.14) are uniformly bounded, $x_{i\mu\nu}(s) = O_{\mathrm{P}}(1)$ uniformly in i and s. By Lemma A.3 and (A2), this implies that the second term on the right hand side of (3.13) vanishes.

Next, we show $\hat{\sigma}^2(\boldsymbol{d}(\widehat{\boldsymbol{\beta}}, \cdot), t) - \hat{\sigma}^2(\boldsymbol{d}(\boldsymbol{\beta}_0, \cdot), t) \xrightarrow{\mathrm{P}} 0$. By a Taylor expansion,

$$\hat{\sigma}^2(\boldsymbol{d}(\widehat{\boldsymbol{\beta}}, \cdot), t) - \hat{\sigma}^2(\boldsymbol{d}(\boldsymbol{\beta}_0, \cdot), t) = 2\boldsymbol{A}(t) n^{\frac{1}{2}}(\widehat{\boldsymbol{\beta}} - \boldsymbol{\beta}_0), \tag{3.15}$$

where

$$\boldsymbol{A}(t) = \frac{1}{n}\sum_{i=1}^{n}\int_0^t n^{-\frac{1}{2}}(\boldsymbol{Q}_{\mathbb{P}_n}^{\boldsymbol{Y}(t)}(\nabla \boldsymbol{d})(\widetilde{\boldsymbol{\beta}}, t))_i (\boldsymbol{Q}_{\mathbb{P}_n}^{\boldsymbol{Y}(t)} \boldsymbol{d}(\widetilde{\boldsymbol{\beta}}, t))_i \,\mathrm{d}N_i(t),$$

for some $\widetilde{\boldsymbol{\beta}}$ between $\widehat{\boldsymbol{\beta}}$ and $\boldsymbol{\beta}$. Similarly to the considerations for the second term of (3.13), one can show that

$$(\boldsymbol{Q}_{\mathbb{P}_n}^{\boldsymbol{Y}(t)}(\nabla \boldsymbol{d})(\widetilde{\boldsymbol{\beta}}, t))_i (\boldsymbol{Q}_{\mathbb{P}_n}^{\boldsymbol{Y}(t)} \boldsymbol{d}(\widetilde{\boldsymbol{\beta}}, t))_i$$

is stochastically bounded, uniformly in i, t. Hence, we may use Lemma A.3 and (A2) to see that (3.15) vanishes uniformly in $t \in [0,\tau]$. □

Proof of Theorem 3.4. We will show that uniformly in $t \in [0,\tau]$,

$$n^{-\frac{1}{2}} T(\boldsymbol{d}(\widehat{\boldsymbol{\beta}}, \cdot), t) - n^{-\frac{1}{2}} T(\boldsymbol{d}(\boldsymbol{\beta}_0, \cdot), t) \xrightarrow{\mathrm{P}} 0$$

and

$$\hat{\sigma}^2(\boldsymbol{d}(\widehat{\boldsymbol{\beta}}, \cdot), t) - \hat{\sigma}^2(\boldsymbol{d}(\boldsymbol{\beta}_0, \cdot), t) \xrightarrow{\mathrm{P}} 0.$$

For this we use a Taylor expansion and argue similarly to the previous proof. Indeed,

$$\begin{aligned} n^{-\frac{1}{2}} T(\boldsymbol{d}(\widehat{\boldsymbol{\beta}}, \cdot), t) - n^{-\frac{1}{2}} T(\boldsymbol{d}(\boldsymbol{\beta}_0, \cdot), t) &= \\ = \frac{1}{n}\sum_{i=1}^{n}\int_0^t n^{-\frac{1}{2}}(\boldsymbol{Q}_{\mathbb{P}_n}^{\boldsymbol{Y}(t)}(\nabla \boldsymbol{d})(\widetilde{\boldsymbol{\beta}}, t))_i \,\mathrm{d}N_i(s) n^{\frac{1}{2}}(\widehat{\boldsymbol{\beta}} - \boldsymbol{\beta}_0) \end{aligned} \tag{3.16}$$

Similar to the proof of Theorem 3.3, one shows that $(\boldsymbol{Q}_{\mathbb{P}_n}^{\boldsymbol{Y}(t)}(\nabla \boldsymbol{d})(\widetilde{\boldsymbol{\beta}}, t))_i$ is stochastically bounded. After that one can use Lemma A.3 to show that the right hand side of (3.16) vanishes.

$\hat{\sigma}^2(\boldsymbol{d}(\widehat{\boldsymbol{\beta}}, \cdot), t) - \hat{\sigma}^2(\boldsymbol{d}(\boldsymbol{\beta}_0, \cdot), t) \xrightarrow{\mathrm{P}} 0$ can be shown as in the proof of Theorem 3.3.

□

Proof of Theorem 3.5. We can rewrite the test statistic as follows:

$$T(t) = U(t) + \int_0^t \overline{(\boldsymbol{Q}_{\mathbb{P}_n}^{\boldsymbol{Y}} \boldsymbol{d}) \boldsymbol{g}}(s)\,\mathrm{d}s, \tag{3.17}$$

where

$$U(t) = n^{-\frac{1}{2}} \int_0^t (\boldsymbol{Q}_{\mathbb{P}_n}^{\boldsymbol{Y}(s)} \boldsymbol{d}(s))^\top (\mathrm{d}\boldsymbol{N}(s) - \boldsymbol{\lambda}(s)\,\mathrm{d}s).$$

By Lemma 3.7, $\boldsymbol{Q}_{\mathbb{P}_n}^{\boldsymbol{Y}(s)} \boldsymbol{d}(s)$ is predictable and locally bounded. Hence, $U(t)$ is a mean zero locally square integrable martingale. For applying a central limit theorem for U we need convergence of the predictable variation process:

$$\begin{aligned} \langle U \rangle (t) &= \frac{1}{n} \int_0^t (\boldsymbol{Q}_{\mathbb{P}_n}^{\boldsymbol{Y}(s)} \boldsymbol{d}(s))^\top \operatorname{diag}(\boldsymbol{\lambda}(s)) (\boldsymbol{Q}_{\mathbb{P}_n}^{\boldsymbol{Y}(s)} \boldsymbol{d}(s))\,\mathrm{d}s \\ &= \int_0^t \overline{(\boldsymbol{Q}_{\mathbb{P}_n}^{\boldsymbol{Y}} \boldsymbol{d})(\boldsymbol{Y}\boldsymbol{\alpha})(\boldsymbol{Q}_{\mathbb{P}_n}^{\boldsymbol{Y}} \boldsymbol{d})}(s)\,\mathrm{d}s + n^{-\frac{1}{2}} \int_0^t \overline{(\boldsymbol{Q}_{\mathbb{P}_n}^{\boldsymbol{Y}} \boldsymbol{d})\boldsymbol{g}(\boldsymbol{Q}_{\mathbb{P}_n}^{\boldsymbol{Y}} \boldsymbol{d})}(s)\,\mathrm{d}s. \end{aligned}$$

By Lemma 3.6, $\langle U \rangle (t) \xrightarrow{\mathrm{P}} \sigma^2(t)$. Since the jumps of $T(t)$ are bounded (see (3.11) in the proof of Theorem 3.1) we can use Rebolledo's central limit (Theorem A.3) to see that U converges in distribution to a mean zero martingale V with covariance $\mathrm{Cov}(V(s), V(t)) = \sigma^2(s \wedge t)$. By Lemma 3.6, the second term on the right hand side of (3.17) converges to $\mu(t)$ uniformly on $[0, \tau]$. Hence, $T \xrightarrow{\mathrm{d}} m$.

□

Chapter 4

Misspecified Models and Smoothing of Estimated Functions

Whenever a statistical model is proposed in the literature, usually estimators of the parameters are given and their (asymptotic) properties are studied. Most asymptotic results are based on the assumption, that the model is true. For our discussion, we need to know the asymptotic behavior of the estimators if the model is misspecified. For example in Subsection 3.5.3, we have seen that we need to know the rate of convergence of the maximum partial likelihood estimator of a Cox model if an Aalen model holds true. The goal of Section 4.1 is to show for the Aalen model (2.6) and the Cox model (2.1) that, even if the model is not true, the estimated parameters converge to some limit at the same rate as if the model was true, i.e. they are $\sqrt{n}$-consistent.

Many models contain parameters that are functions over time, for example the baseline $\lambda_0(t)$ in the Cox model. As already mentioned, estimators usually estimate the integrated functions, for example in the Cox model, the standard Breslow estimator $\hat{\Lambda}_0(t)$ estimates $\Lambda_0(t) := \int_0^t \lambda_0(s)\, ds$. For the following development, we need estimates of the functions themselves and we need their asymptotic convergence rates. We also need rates of convergence of the total variation of the difference between the estimated function and its limit. In Section 4.2, we describe a modified kernel smoothing approach that starts from the estimates of the integrated functions and yields rates of convergence of the estimated function and its derivative. For example, starting from a $\sqrt{n}$-consistent estimator for the integrated function, one can get an estimator of the function itself which converges at a rate of $n^{1/3}$.

4.1 Convergence Rates of Estimators under Misspecified Models

In this section, we only show that the standard estimators in the Aalen model and in the Cox model are $\sqrt{n}$-consistent even if the model does not hold true.

For other models, it is reasonable to expect that under certain regularity conditions, estimated parameters or functions converge at a rate of $\sqrt{n}$ as well. For example, in Hjort (1992), the maximum likelihood estimator of the parametric Cox model (2.10) is studied and it is shown that it is $\sqrt{n}$-consistent for a 'least false' value.

4.1.1 Misspecified Aalen Model

Recall that the least squares estimator for $\int_0^t \boldsymbol{\alpha}(s)\,\mathrm{d}s$ in the Aalen model 2.6 is given by

$$\widehat{\boldsymbol{A}}(t) = \int_0^t \boldsymbol{Y}^-(s)\,\mathrm{d}\boldsymbol{N}(s),$$

where $\boldsymbol{Y}^-(s)$ is a generalized inverse of $\boldsymbol{Y}(s)$. We show that $\widehat{\boldsymbol{A}}(t)$ converges uniformly in probability to

$$\boldsymbol{A}_0(t) = \int_0^t (\mathrm{E}[\boldsymbol{Y}_1(s)^\top \boldsymbol{Y}_1(s)])^{-1}\,\mathrm{E}[\boldsymbol{Y}_1(s)\lambda_1(s)]^\top\,\mathrm{d}s,$$

even if the Aalen model (2.6) does not hold. Of course, if the Aalen model (2.6) holds true then $\boldsymbol{A}_0(t) = \int_0^t \boldsymbol{\alpha}(s)\,\mathrm{d}s$.

Lemma 4.1. *Suppose that $(\boldsymbol{Y}_i, \lambda_i)$ are i.i.d. and that $\boldsymbol{Y}_i$, λ_i are càglàd, bounded, adapted stochastic processes. If (A1) holds true then*

$$\sup_{t\in[0,\tau]} \|\widehat{\boldsymbol{A}}(t) - \boldsymbol{A}_0(t)\| = O_{\mathrm{P}}(n^{-\frac{1}{2}}).$$

Proof. Let $K := \inf\{\det(\overrightarrow{\boldsymbol{YY}}(t)), t \in [0,\tau]\}$. Define the event

$$B_n = \{\det(\overline{\boldsymbol{YY}}(t)) > K/2 \text{ for all } \ t \in [0,\tau]\}$$

and the predictable process $J(t) = I\{\det(\overline{\boldsymbol{YY}}(t)) > K/2\}$. Consider the decomposition

$$\|\widehat{\boldsymbol{A}}(t) - \boldsymbol{A}_0(t)\| \le \|\widehat{\boldsymbol{A}}(t) - \widetilde{\boldsymbol{A}}(t)\| + \|\widetilde{\boldsymbol{A}}(t) - \boldsymbol{A}^*(t)\| + \|\boldsymbol{A}^*(t) - \boldsymbol{A}_0(t)\|,$$

where

$$\widetilde{\boldsymbol{A}}(t) = \frac{1}{n}\int_0^t J(s)(\overline{\boldsymbol{YY}}(s))^{-1}\boldsymbol{Y}(s)^\top\mathrm{d}\boldsymbol{N}(s)$$

and

$$\boldsymbol{A}^*(t) = \int_0^t J(s)(\overline{\boldsymbol{Y}\boldsymbol{Y}}(s))^{-1}\overline{\boldsymbol{Y}\boldsymbol{\lambda}}(s)\,\mathrm{d}s.$$

By the strong law of large numbers (Theorem A.4) and Lemma 3.3, $\mathrm{P}(B_n) \to 1$. Hence, $n^{1/2}\sup_{t\in[0,\tau]} \|\widehat{\boldsymbol{A}}(t) - \widetilde{\boldsymbol{A}}(t)\| \leq I(B_n^c) \xrightarrow{\mathrm{P}} 0$, where $I(B_n^c)$ is the indicator of the complement of B_n. Since

$$n^{\frac{1}{2}}(\widetilde{\boldsymbol{A}}(t) - \boldsymbol{A}^*(t)) = n^{-\frac{1}{2}} \int_0^t J(s)(\overline{\boldsymbol{Y}\boldsymbol{Y}}(s))^{-1}\boldsymbol{Y}(s)^\top \mathrm{d}\boldsymbol{M}(s)$$

is a martingale, whose predictable quadratic variation converges and whose jumps are bounded, we can use Rebolledo's theorem (Theorem A.3) to show that

$$\sup_{t\in[0,\tau]} \|\widetilde{\boldsymbol{A}}(t) - \boldsymbol{A}^*(t)\| = O_{\mathrm{P}}(n^{-\frac{1}{2}}).$$

By the assumptions, we may consider the components of $\boldsymbol{Y}_i^\top\boldsymbol{Y}_i$ and $\boldsymbol{Y}_i\lambda_i$ as elements of $L_2(\lambda\!\!\lambda)$. Using a central limit theorem for random elements in Hilbert spaces (Theorem A.6), we get that the components of $n^{-1/2}\sum_{i=1}^n(\boldsymbol{Y}_i(\cdot)^\top\boldsymbol{Y}_i(\cdot) - \mathrm{E}[\boldsymbol{Y}_1(\cdot)^\top\boldsymbol{Y}_1(\cdot)])$ and $n^{-1/2}\sum_{i=1}^n(\boldsymbol{Y}_i(\cdot)\lambda_i(\cdot) - \mathrm{E}[\boldsymbol{Y}_1(\cdot)\lambda_1(\cdot)])$ converge weakly to mean zero Gaussian random elements in $L_2(\lambda\!\!\lambda)$. Since the norm in $L_2(\lambda\!\!\lambda)$ is continuous, we may conclude

$$\left(\int_0^\tau \left\|\overline{\boldsymbol{Y}\boldsymbol{Y}}(s) - \mathrm{E}[\boldsymbol{Y}_1(s)^\top\boldsymbol{Y}_1(s)]\right\|^2 \mathrm{d}s\right)^{\frac{1}{2}} = O_{\mathrm{P}}(n^{-\frac{1}{2}})$$

and

$$\left(\int_0^\tau \left\|\overline{\boldsymbol{Y}\boldsymbol{\lambda}}(s) - \mathrm{E}[\boldsymbol{Y}_1(s)^\top\lambda_1(s)]\right\|^2 \mathrm{d}s\right)^{\frac{1}{2}} = O_{\mathrm{P}}(n^{-\frac{1}{2}}).$$

Using Lemma 3.3 and Lemma 3.5, we get

$$\left(\int_0^\tau \left\|(\overline{\boldsymbol{Y}\boldsymbol{Y}}(s))^{-1} - \boldsymbol{B}(s)^{-1}\right\|^2 \mathrm{d}s\right)^{\frac{1}{2}} = O_{\mathrm{P}}(n^{-\frac{1}{2}}),$$

where $\boldsymbol{B}(s) := \mathrm{E}[\boldsymbol{Y}_1(s)^\top\boldsymbol{Y}_1(s)]$. Using triangular inequalities,

$$\begin{aligned}\sup_{t\in[0,\tau]} \|\boldsymbol{A}^*(t) - \boldsymbol{A}_0(t)\| &\leq \int_0^\tau \left\|\overline{\boldsymbol{Y}\boldsymbol{Y}}(s)^{-1}\overline{\boldsymbol{Y}\boldsymbol{\lambda}}(s) - \boldsymbol{B}(s)^{-1}\,\mathrm{E}[\boldsymbol{Y}_1(s)^\top\lambda_1(s)]\right\| \mathrm{d}s \\ &\leq \int_0^\tau \left\|\overline{\boldsymbol{Y}\boldsymbol{Y}}(s)^{-1}\left(\overline{\boldsymbol{Y}\boldsymbol{\lambda}}(s) - \mathrm{E}[\boldsymbol{Y}_1(s)^\top\lambda_1(s)]\right)\right\| \mathrm{d}s+ \\ &\quad + \int_0^\tau \left\|\left(\overline{\boldsymbol{Y}\boldsymbol{Y}}(s)^{-1} - \boldsymbol{B}(s)^{-1}\right)\mathrm{E}[\boldsymbol{Y}_1(s)^\top\lambda_1(s)]\right\| \mathrm{d}s.\end{aligned}$$

Thus, by the Cauchy-Schwarz inequality,

$$\sup_{t\in[0,\tau]} \|\boldsymbol{A}^*(t) - \boldsymbol{A}_0(t)\| \leq$$

$$\leq \left(\int_0^\tau \left\|\overline{\boldsymbol{Y}\boldsymbol{Y}}(s)^{-1}\right\|^2 \mathrm{d}s \int_0^\tau \left\|\overline{\boldsymbol{Y}\boldsymbol{\lambda}}(s) - \mathrm{E}[\boldsymbol{Y}_1(s)^\top \lambda_1(s)]\right\|^2 \mathrm{d}s\right)^{\frac{1}{2}}$$

$$+ \left(\int_0^\tau \left\|\overline{\boldsymbol{Y}\boldsymbol{Y}}(s)^{-1} - \boldsymbol{B}(s)^{-1}\right\|^2 \mathrm{d}s \int_0^\tau \left\|\mathrm{E}[\boldsymbol{Y}_1(s)^\top \lambda_1(s)]\right\|^2 \mathrm{d}s\right)^{\frac{1}{2}}.$$

Hence, $\sup_{t\in[0,\tau]} \|\boldsymbol{A}^*(t) - \boldsymbol{A}_0(t)\| = O_{\mathrm{P}}(n^{-1/2})$. □

4.1.2 Misspecified Cox Model

If Cox's model (2.1) holds true, asymptotic normality of $n^{1/2}(\hat{\boldsymbol{\beta}} - \boldsymbol{\beta}_0)$ is shown in Andersen and Gill (1982, p. 1105), where $\hat{\boldsymbol{\beta}}$ is the partial maximum likelihood estimator and $\boldsymbol{\beta}_0$ is the 'true' parameter. Of course, this implies $\hat{\boldsymbol{\beta}} - \boldsymbol{\beta}_0 = O_{\mathrm{P}}(n^{-1/2})$. For our purposes we need to know the asymptotic behavior of $\hat{\boldsymbol{\beta}}$ if the Cox model does not hold true. For the Cox model, some results are available on the behavior of estimators under misspecified models. Asymptotic normality of $n^{1/2}(\hat{\boldsymbol{\beta}} - \boldsymbol{\beta}_0)$ for some $\boldsymbol{\beta}_0$ under misspecified models has been considered previously by Lin and Wei (1989), Hjort (1992), Sasieni (1993), and Fine (2002) under an i.i.d. setup allowing only one event per individual. We relax these requirements but confine ourselves to showing $\hat{\boldsymbol{\beta}} - \boldsymbol{\beta}_0 = O_{\mathrm{P}}(n^{-1/2})$ if the Cox model is misspecified. In fact, $\boldsymbol{\beta}_0$ is defined as follows:

$$\boldsymbol{\beta}_0 := \operatorname{argmax}_{\boldsymbol{\beta}\in\boldsymbol{\mathcal{X}}_{\boldsymbol{\beta}}} a(\boldsymbol{\beta}, \tau),$$

where $\boldsymbol{\mathcal{X}}_{\boldsymbol{\beta}}$ is the parameter space for the Cox model,

$$a(\boldsymbol{\beta}, t) := \int_0^t \left(\boldsymbol{\beta}^\top \overrightarrow{\boldsymbol{Z}\boldsymbol{\lambda}}(s) - \log(\overrightarrow{\boldsymbol{\rho}}(\boldsymbol{\beta}, s))\overrightarrow{\boldsymbol{\lambda}}(s)\right) \mathrm{d}s,$$

and

$$\rho_i(\boldsymbol{\beta}, s) = \exp(\boldsymbol{Z}_i(s)\boldsymbol{\beta})R_i(s).$$

In the above, we have used the notation introduced in Section 2.3. For simplicity, we only consider the classical Cox model with the link function exp. One can generalize the results to other link functions i.e. the model given in (2.2), or even to Cox-type models (2.5) with intensity $\lambda_i(t) = \lambda_0(t)\rho_i(\boldsymbol{\beta}, t)$.

The main idea for the proof of $\hat{\boldsymbol{\beta}} - \boldsymbol{\beta}_0 = O_{\mathrm{P}}(n^{-1/2})$ under an arbitrary intensity is similar to the proof under the Cox model (2.1) given in Andersen and Gill (1982). First, uniform stochastic convergence of $X(\boldsymbol{\beta}, t) := \frac{1}{n}C(\boldsymbol{\beta}, t) + (\log n)\overline{\boldsymbol{N}}(t)$ to $a(\boldsymbol{\beta}, t)$ is shown, where

$$C(\boldsymbol{\beta}, t) := \sum_{i=1}^n \int_0^t \log(\rho_i(\boldsymbol{\beta}, s))\, \mathrm{d}N_i(s) - \int_0^t \log(n\overline{\boldsymbol{\rho}}(\boldsymbol{\beta}, s))n\, \mathrm{d}\overline{\boldsymbol{N}}(s).$$

$C(\boldsymbol{\beta},\tau)$ is the log partial likelihood function given in (2.3). In Lemma 4.3, it will be shown that $C(\boldsymbol{\beta},t)$ is concave in $\boldsymbol{\beta}$. Hence, $X(\boldsymbol{\beta},\tau)$ is concave in $\boldsymbol{\beta}$, and thus convex analysis can be used to transfer the convergence of $X(\boldsymbol{\beta},\tau)$ to its maximizer $\hat{\boldsymbol{\beta}}$. A suitable Taylor expansion of $X(\boldsymbol{\beta},\tau)$ around $\boldsymbol{\beta}_0$ allows to show $\hat{\boldsymbol{\beta}}-\boldsymbol{\beta}_0 = O_{\mathrm{P}}(n^{-1/2})$. The following lemma is formulated in a more general setup than the i.i.d. setup, similar in spirit to the setup of Andersen and Gill (1982).

Lemma 4.2. *Suppose that*

i) for each $\boldsymbol{\beta} \in \boldsymbol{\mathcal{X}}_{\boldsymbol{\beta}}$, $\overline{\boldsymbol{\rho}}(\boldsymbol{\beta},\cdot)$ *converges uip on* $[0,\tau]$ *and the mapping* $\overrightarrow{\boldsymbol{\rho}}(\boldsymbol{\beta},\cdot)$ *is bounded away from* 0,

ii) $\overline{\boldsymbol{\lambda}}$, $\overline{\boldsymbol{\lambda Z}}$ *and* $\overline{\boldsymbol{ZZ\lambda}}$ *converge uip on* $[0,\tau]$,

iii) $\boldsymbol{\beta}_0$ *exists and is unique.*

Then $\hat{\boldsymbol{\beta}} \xrightarrow{\mathrm{P}} \boldsymbol{\beta}_0$.
If furthermore there exists an open set $\mathcal{C} \subset \boldsymbol{\mathcal{X}}_{\boldsymbol{\beta}}$ *with* $\boldsymbol{\beta}_0 \in \mathcal{C}$ *such that*

iv) $\overline{\boldsymbol{\rho}}$, $\overline{\boldsymbol{Z\rho}}$, $\overline{\boldsymbol{Z\rho Z}}$ *converge uip on* $\mathcal{C}\times[0,\tau]$,

v) $\int_0^\tau \left(\overrightarrow{\boldsymbol{\rho}}(\boldsymbol{\beta},s)^{-2}\overrightarrow{\boldsymbol{Z\rho}}^{\otimes 2}(\boldsymbol{\beta},s) - \overrightarrow{\boldsymbol{\rho}}(\boldsymbol{\beta},s)^{-1}\overrightarrow{\boldsymbol{Z\rho Z}}(\boldsymbol{\beta},s)\right)\overrightarrow{\boldsymbol{\lambda}}(s)\,\mathrm{d}s$ *is invertible for* $\boldsymbol{\beta}=\boldsymbol{\beta}_0$ *and continuous at* $\boldsymbol{\beta}=\boldsymbol{\beta}_0$,

vi) $\int_0^\tau |\overline{\boldsymbol{\rho}}(\boldsymbol{\beta}_0,s)-\overrightarrow{\boldsymbol{\rho}}(\boldsymbol{\beta}_0,s)|\,\mathrm{d}s = O_{\mathrm{P}}(n^{-\frac{1}{2}})$, $\int_0^\tau \|\overline{\boldsymbol{Z\lambda}}(s)-\overrightarrow{\boldsymbol{Z\lambda}}(s)\|\,\mathrm{d}s = O_{\mathrm{P}}(n^{-\frac{1}{2}})$, $\int_0^\tau \|\overline{\boldsymbol{Z\rho}}(\boldsymbol{\beta}_0,s)-\overrightarrow{\boldsymbol{Z\rho}}(\boldsymbol{\beta}_0,s)\|\,\mathrm{d}s = O_{\mathrm{P}}(n^{-\frac{1}{2}})$, $\int_0^\tau |\overline{\boldsymbol{\lambda}}(s)-\overrightarrow{\boldsymbol{\lambda}}(s)|\,\mathrm{d}s = O_{\mathrm{P}}(n^{-\frac{1}{2}})$,

then $\hat{\boldsymbol{\beta}}-\boldsymbol{\beta}_0 = O_{\mathrm{P}}(n^{-1/2})$.

Above, we used the notation $\boldsymbol{a}^{\otimes 2} = \boldsymbol{a}^\top\boldsymbol{a}$ for column vectors $\boldsymbol{a}$. We prepare the proof by the following lemma, the result of which has been used previously in the literature, see e.g. Andersen and Gill (1982), but no explicit proof was given.

Lemma 4.3. $C(\boldsymbol{\beta},t)$ *is concave in* $\boldsymbol{\beta} \in \boldsymbol{\mathcal{X}}_{\boldsymbol{\beta}}$ *for each* $t\in[0,\tau]$.

Proof. $\frac{\partial}{\partial\boldsymbol{\beta}}\overline{\boldsymbol{\rho}}(\boldsymbol{\beta},t) = \overline{\boldsymbol{\rho Z}}(\boldsymbol{\beta},t)$ and $\left(\frac{\partial}{\partial\boldsymbol{\beta}}\right)^2\overline{\boldsymbol{\rho}}(\boldsymbol{\beta},t) = \overline{\boldsymbol{Z\rho Z}}(\boldsymbol{\beta},t)$. Hence,

$$\frac{\partial}{\partial\boldsymbol{\beta}}C(\boldsymbol{\beta},t) = \int_0^t \boldsymbol{Z}(s)^\top \mathrm{d}\boldsymbol{N}(s) - \int_0^t (\overline{\boldsymbol{\rho}}(\boldsymbol{\beta},s))^{-1}\overline{\boldsymbol{\rho Z}}(\boldsymbol{\beta},s) n\,\mathrm{d}\overline{\boldsymbol{N}}(s)$$

and $\left(\frac{\partial}{\partial\boldsymbol{\beta}}\right)^2 C(\boldsymbol{\beta},t) = -\int_0^t \boldsymbol{A}(\boldsymbol{\beta},s) n d\overline{\boldsymbol{N}}(s)$, where

$$\begin{aligned}\boldsymbol{A}(\boldsymbol{\beta},s) &= (\overline{\boldsymbol{\rho}}(\boldsymbol{\beta},s))^{-1}\,\overline{\boldsymbol{Z\rho Z}}(\boldsymbol{\beta},s) - (\overline{\boldsymbol{\rho}}(\boldsymbol{\beta},s))^{-2}\,\overline{\boldsymbol{\rho Z}}(\boldsymbol{\beta},s)\overline{\boldsymbol{Z\rho}}(\boldsymbol{\beta},s)\\ &= (\overline{\boldsymbol{\rho}}(\boldsymbol{\beta},s))^{-2}\left(\overline{\boldsymbol{Z\rho Z}}(\boldsymbol{\beta},s)\overline{\boldsymbol{\rho}}(\boldsymbol{\beta},s) - \overline{\boldsymbol{\rho Z}}(\boldsymbol{\beta},s)\overline{\boldsymbol{Z\rho}}(\boldsymbol{\beta},s)\right).\end{aligned}$$

Dropping the dependence on $\boldsymbol{\beta}$ and s,

$$\begin{aligned}\boldsymbol{A} &= (n\overline{\boldsymbol{\rho}})^{-2}\left(\sum_{i,j=1}^{n}\rho_i\rho_j\boldsymbol{Z}_j^\top\boldsymbol{Z}_j - \sum_{i,j=1}^{n}\rho_i\rho_j\boldsymbol{Z}_i^\top\boldsymbol{Z}_j\right)\\ &= (n\overline{\boldsymbol{\rho}})^{-2}\sum_{i=1}^{n-1}\sum_{j=i+1}^{n}\rho_i\rho_j\left(\boldsymbol{Z}_j^\top\boldsymbol{Z}_j + \boldsymbol{Z}_i^\top\boldsymbol{Z}_i - \boldsymbol{Z}_i^\top\boldsymbol{Z}_j - \boldsymbol{Z}_j^\top\boldsymbol{Z}_i\right)\\ &= (n\overline{\boldsymbol{\rho}})^{-2}\sum_{i=1}^{n-1}\sum_{j=i+1}^{n}\rho_i\rho_j(\boldsymbol{Z}_i - \boldsymbol{Z}_j)^\top(\boldsymbol{Z}_i - \boldsymbol{Z}_j)\,.\end{aligned}$$

As $\rho_i(\boldsymbol{\beta}, s) \geq 0$ for all $i = 1,\dots,n$, the above implies that $\boldsymbol{A}(\boldsymbol{\beta}, s)$ is positive semidefinite. Hence, $\left(\frac{\partial}{\partial\boldsymbol{\beta}}\right)^2 C(\boldsymbol{\beta}, t)$ is negative semidefinite. □

Proof of Lemma 4.2. Let $\boldsymbol{\beta} \in \boldsymbol{\mathcal{X}}_{\boldsymbol{\beta}}$ and $X(\boldsymbol{\beta}, t) := \frac{1}{n}C(\boldsymbol{\beta}, t) + (\log n)\overline{\boldsymbol{N}}(t)$. By Lemma 4.3, $X(\boldsymbol{\beta}, t)$ is concave in $\boldsymbol{\beta} \in \boldsymbol{\mathcal{X}}_{\boldsymbol{\beta}}$ for each $t \in [0, \tau]$. Furthermore,

$$\begin{aligned}X(\boldsymbol{\beta}, t) &= \frac{1}{n}\int_0^t \left(\boldsymbol{\beta}^\top\boldsymbol{Z}(s)^\top - \log(\overline{\boldsymbol{\rho}}(\boldsymbol{\beta}, s))\boldsymbol{1}^\top\right)\mathrm{d}\boldsymbol{N}(s)\\ &= \frac{1}{n}\int_0^t \left(\boldsymbol{\beta}^\top\boldsymbol{Z}(s)^\top - I\{\overline{\boldsymbol{\rho}}(\boldsymbol{\beta}, s) > 0\}\log(\overline{\boldsymbol{\rho}}(\boldsymbol{\beta}, s))\boldsymbol{1}^\top\right)\mathrm{d}\boldsymbol{N}(s)\end{aligned}$$

and hence for each $\boldsymbol{\beta} \in \boldsymbol{\mathcal{X}}_{\boldsymbol{\beta}}$, $X(\boldsymbol{\beta}, t) - A(\boldsymbol{\beta}, t)$ is a local square integrable martingale, where

$$A(\boldsymbol{\beta}, t) := \frac{1}{n}\int_0^t \left(\boldsymbol{\beta}^\top\boldsymbol{Z}(s)^\top - I\{\overline{\boldsymbol{\rho}}(\boldsymbol{\beta}, s) > 0\}\log(\overline{\boldsymbol{\rho}}(\boldsymbol{\beta}, s))\boldsymbol{1}^\top\right)\boldsymbol{\lambda}(s)\,\mathrm{d}s.$$

For $\boldsymbol{\beta} \in \boldsymbol{\mathcal{X}}_{\boldsymbol{\beta}}$,

$$\begin{aligned}&n\left\langle X(\boldsymbol{\beta}, \cdot) - A(\boldsymbol{\beta}, \cdot)\right\rangle(t) =\\ &= \frac{1}{n}\int_0^t \left(\boldsymbol{\beta}^\top\boldsymbol{Z}(s)^\top - I\{\overline{\boldsymbol{\rho}}(\boldsymbol{\beta}, s) > 0\}\log(\overline{\boldsymbol{\rho}}(\boldsymbol{\beta}, s))\boldsymbol{1}^\top\right)\operatorname{diag}(\boldsymbol{\lambda}(s))\\ &\qquad (\boldsymbol{Z}(s)\boldsymbol{\beta} - I\{\overline{\boldsymbol{\rho}}(\boldsymbol{\beta}, s) > 0\}\log(\overline{\boldsymbol{\rho}}(\boldsymbol{\beta}, s))\boldsymbol{1})\,\mathrm{d}s\\ &= \int_0^t \Big(\boldsymbol{\beta}^\top\overline{\boldsymbol{Z\lambda Z}}(s)\boldsymbol{\beta} - 2\log(\overline{\boldsymbol{\rho}}(\boldsymbol{\beta}, s))I\{\overline{\boldsymbol{\rho}}(\boldsymbol{\beta}, s) > 0\}\overline{\boldsymbol{\lambda Z}}(s)\boldsymbol{\beta}\\ &\qquad + I\{\overline{\boldsymbol{\rho}}(\boldsymbol{\beta}, s) > 0\}\log^2(\overline{\boldsymbol{\rho}}(\boldsymbol{\beta}, s))\overline{\boldsymbol{\lambda}}(s)\Big)\,\mathrm{d}s.\end{aligned}$$

The convergence of $\overline{\boldsymbol{Z\lambda Z}}$ is implied by that of $\overline{\boldsymbol{ZZ\lambda}}$. Since $\overrightarrow{\boldsymbol{\rho}}$ is bounded away from 0, the convergence of $\overline{\boldsymbol{\rho}}$ implies the convergence of $I\{\overline{\boldsymbol{\rho}}(\boldsymbol{\beta}, s) > 0\}$ and $\log(\overline{\boldsymbol{\rho}}(\boldsymbol{\beta}, s))$. Hence, $n\left\langle X(\boldsymbol{\beta}, \cdot) - A(\boldsymbol{\beta}, \cdot)\right\rangle(t)$ converges in probability, to

$$\int_0^t \left(\boldsymbol{\beta}^\top\overrightarrow{\boldsymbol{Z\lambda Z}}(s)\boldsymbol{\beta} - 2\log(\overrightarrow{\boldsymbol{\rho}}(\boldsymbol{\beta}, s))\overrightarrow{\boldsymbol{\lambda Z}}(s)\boldsymbol{\beta} + \log^2(\overrightarrow{\boldsymbol{\rho}}(\boldsymbol{\beta}, s))\overrightarrow{\boldsymbol{\lambda}}(s)\right)\mathrm{d}s,$$

uniformly in $t \in [0,\tau]$. By the above and Lenglart's inequality (Lemma A.2) $\sup_{t\in[0,\tau]} |X(\boldsymbol{\beta},t) - A(\boldsymbol{\beta},t)| \xrightarrow{\mathrm{P}} 0$ for all $\boldsymbol{\beta} \in \boldsymbol{\mathcal{X}}_{\boldsymbol{\beta}}$. Furthermore, since

$$A(\boldsymbol{\beta},t) = \int_0^t \left(\boldsymbol{\beta}^\top \overline{\boldsymbol{Z\lambda}}(s) - I\{\overline{\boldsymbol{\rho}}(\boldsymbol{\beta},s) > 0\} \log(\overline{\boldsymbol{\rho}}(\boldsymbol{\beta},s))\overline{\boldsymbol{\lambda}}(s)\right) \mathrm{d}s,$$

we can use assumptions i) and ii) to get $\sup_{t\in[0,\tau]} |A(\boldsymbol{\beta},t) - a(\boldsymbol{\beta},t)| \xrightarrow{\mathrm{P}} 0$. Hence, $\sup_{t\in[0,\tau]} |X(\boldsymbol{\beta},t) - a(\boldsymbol{\beta},t)| \xrightarrow{\mathrm{P}} 0$. By Andersen and Gill (1982, Cor.II.2) this implies $\hat{\boldsymbol{\beta}} - \boldsymbol{\beta}_0 \xrightarrow{\mathrm{P}} 0$.

To show $\hat{\boldsymbol{\beta}} - \boldsymbol{\beta}_0 = O_{\mathrm{P}}(n^{-1/2})$ we proceed as follows. Let

$$\begin{aligned}
U(\boldsymbol{\beta},t) &:= \frac{\partial}{\partial\boldsymbol{\beta}} X(\boldsymbol{\beta},t) = \frac{1}{n}\int_0^t \boldsymbol{Z}(s)^\top \mathrm{d}\boldsymbol{N}(s) - \int_0^t \overline{\boldsymbol{\rho}}(\boldsymbol{\beta},s)^{-1}\overline{\boldsymbol{Z\rho}}(\boldsymbol{\beta},s)\,\mathrm{d}\overline{\boldsymbol{N}}(s),\\
J(\boldsymbol{\beta},t) &:= \left(\frac{\partial}{\partial\boldsymbol{\beta}}\right)^2 X(\boldsymbol{\beta},t)\\
&= \int_0^t \left(\overline{\boldsymbol{\rho}}(\boldsymbol{\beta},s)^{-2}\overline{\boldsymbol{Z\rho}}(\boldsymbol{\beta},s)^{\otimes 2} - \overline{\boldsymbol{\rho}}(\boldsymbol{\beta},s)^{-1}\overline{\boldsymbol{Z\rho Z}}(\boldsymbol{\beta},s)\right) \mathrm{d}\overline{\boldsymbol{N}}(s).
\end{aligned}$$

A Taylor expansion of U around $\boldsymbol{\beta}_0$ yields

$$U(\boldsymbol{\beta},t) - U(\boldsymbol{\beta}_0,t) = J(\tilde{\boldsymbol{\beta}},t)(\boldsymbol{\beta} - \boldsymbol{\beta}_0)$$

for some $\tilde{\boldsymbol{\beta}}$ between $\boldsymbol{\beta}$ and $\boldsymbol{\beta}_0$. By definition of $\hat{\boldsymbol{\beta}}$, we have $U(\hat{\boldsymbol{\beta}},\tau) = 0$ and hence

$$-n^{\frac{1}{2}}U(\boldsymbol{\beta}_0,\tau) = J(\tilde{\boldsymbol{\beta}},\tau)n^{\frac{1}{2}}(\hat{\boldsymbol{\beta}} - \boldsymbol{\beta}_0).$$

We will show $U(\boldsymbol{\beta}_0,\tau) = O_{\mathrm{P}}(n^{-1/2})$ and $J(\tilde{\boldsymbol{\beta}},\tau) \xrightarrow{\mathrm{P}} \overrightarrow{J}(\boldsymbol{\beta}_0,\tau)$ where

$$\overrightarrow{J}(\boldsymbol{\beta},t) := \int_0^t \left(\overrightarrow{\boldsymbol{\rho}}(\boldsymbol{\beta},s)^{-2}\overrightarrow{\boldsymbol{Z\rho}}(\boldsymbol{\beta},s)^{\otimes 2} - \overrightarrow{\boldsymbol{\rho}}(\boldsymbol{\beta},s)^{-1}\overrightarrow{\boldsymbol{Z\rho Z}}(\boldsymbol{\beta},s)\right)\overrightarrow{\boldsymbol{\lambda}}(s)\,\mathrm{d}s.$$

Since we assumed $\overrightarrow{J}(\boldsymbol{\beta}_0,\tau)$ to be invertible we will get $\hat{\boldsymbol{\beta}} - \boldsymbol{\beta}_0 = O_{\mathrm{P}}(n^{-1/2})$.

The convergence of $J(\hat{\boldsymbol{\beta}},t)$ is immediate from $\hat{\boldsymbol{\beta}} \xrightarrow{\mathrm{P}} \boldsymbol{\beta}_0$, Lemma A.5, and the assumptions. The boundedness of $U(\hat{\boldsymbol{\beta}},\tau)$ needs some more work. Let

$$V(t) := \int_0^t \left(\overline{\boldsymbol{Z\lambda}}(s) - \overline{\boldsymbol{\rho}}(\boldsymbol{\beta}_0,s)^{-1}\overline{\boldsymbol{Z\rho}}(\boldsymbol{\beta}_0,s)\overline{\boldsymbol{\lambda}}(s)\right) \mathrm{d}s$$

and

$$\overrightarrow{V}(t) := \int_0^t \left(\overrightarrow{\boldsymbol{Z\lambda}}(s) - \overrightarrow{\boldsymbol{\rho}}(\boldsymbol{\beta}_0,s)^{-1}\overrightarrow{\boldsymbol{Z\rho}}(\boldsymbol{\beta}_0,s)\overrightarrow{\boldsymbol{\lambda}}(s)\right) \mathrm{d}s.$$

By definition of $\boldsymbol{\beta}_0$ we have $\overrightarrow{V}(\tau) = 0$. Hence,

$$n^{\frac{1}{2}}U(\boldsymbol{\beta}_0,\tau) = n^{\frac{1}{2}}(U(\boldsymbol{\beta}_0,\tau) - V(\tau)) + n^{\frac{1}{2}}(V(\tau) - \overrightarrow{V}(\tau)). \tag{4.1}$$

We show that both terms on the right hand side are stochastically bounded. Since

$$\left\langle n^{\frac{1}{2}}\left(U(\boldsymbol{\beta}_0,\cdot)-V(\cdot)\right)\right\rangle(\tau)=\int_0^\tau\left(\overline{\boldsymbol{Z}\boldsymbol{\lambda}\boldsymbol{Z}}(s)-\overline{\boldsymbol{\rho}}(\boldsymbol{\beta}_0,s)^{-2}\overline{\boldsymbol{Z}\boldsymbol{\rho}}(\boldsymbol{\beta}_0,s)^{\otimes 2}\overline{\boldsymbol{\lambda}}(s)\right)\mathrm{d}s$$

converges in probability, $\langle n^{1/2}\left(U(\boldsymbol{\beta}_0,\cdot)-V(\cdot)\right)\rangle(\tau)$ is stochastically bounded. Hence, using Lenglart's inequality, we can conclude that $n^{1/2}\left(U(\boldsymbol{\beta}_0,t)-V(t)\right)$ is stochastically bounded uniformly in $t\in[0,\tau]$. The second term on the right hand side of (4.1) can be dealt with as follows. Since

$$\begin{aligned}V(\tau)-\vec{V}(\tau)=&\int_0^\tau\left(\overline{\boldsymbol{Z}\boldsymbol{\lambda}}(s)-\overrightarrow{\boldsymbol{Z}\boldsymbol{\lambda}}(s)\right)\mathrm{d}s\\&+\int_0^\tau\left(\overline{\boldsymbol{\rho}}(\boldsymbol{\beta}_0,s)^{-1}-\vec{\boldsymbol{\rho}}(\boldsymbol{\beta}_0,s)^{-1}\right)\overline{\boldsymbol{Z}\boldsymbol{\rho}}(\boldsymbol{\beta}_0,s)\overline{\boldsymbol{\lambda}}(s)\,\mathrm{d}s\\&+\int_0^\tau\vec{\boldsymbol{\rho}}(\boldsymbol{\beta}_0,s)^{-1}\left(\overline{\boldsymbol{Z}\boldsymbol{\rho}}(\boldsymbol{\beta}_0,s)-\overrightarrow{\boldsymbol{Z}\boldsymbol{\rho}}(\boldsymbol{\beta}_0,s)\right)\overline{\boldsymbol{\lambda}}(s)\,\mathrm{d}s\\&+\int_0^\tau\vec{\boldsymbol{\rho}}(\boldsymbol{\beta}_0,s)^{-1}\overrightarrow{\boldsymbol{Z}\boldsymbol{\rho}}(\boldsymbol{\beta}_0,s)\left(\overline{\boldsymbol{\lambda}}(s)-\vec{\boldsymbol{\lambda}}(s)\right)\mathrm{d}s,\end{aligned}$$

we have

$$\begin{aligned}\left\|V(\tau)-\vec{V}(\tau)\right\|\le&\int_0^\tau\left\|\overline{\boldsymbol{Z}\boldsymbol{\lambda}}(s)-\overrightarrow{\boldsymbol{Z}\boldsymbol{\lambda}}(s)\right\|\mathrm{d}s\\&+\int_0^\tau\left|\overline{\boldsymbol{\rho}}(\boldsymbol{\beta}_0,s)^{-1}-\vec{\boldsymbol{\rho}}(\boldsymbol{\beta}_0,s)^{-1}\right|\mathrm{d}s\sup_{s\in[0,\tau]}\left\|\overline{\boldsymbol{Z}\boldsymbol{\rho}}(\boldsymbol{\beta}_0,s)\right\|\overline{\boldsymbol{\lambda}}(s)\\&+\int_0^\tau\left\|\overline{\boldsymbol{Z}\boldsymbol{\rho}}(\boldsymbol{\beta}_0,s)-\overrightarrow{\boldsymbol{Z}\boldsymbol{\rho}}(\boldsymbol{\beta}_0,s)\right\|\mathrm{d}s\sup_{s\in[0,\tau]}\vec{\boldsymbol{\rho}}(\boldsymbol{\beta}_0,s)^{-1}\overline{\boldsymbol{\lambda}}(s)\\&+\int_0^\tau\left|\overline{\boldsymbol{\lambda}}(s)-\vec{\boldsymbol{\lambda}}(s)\right|\mathrm{d}s\sup_{s\in[0,\tau]}\vec{\boldsymbol{\rho}}(\boldsymbol{\beta}_0,s)^{-1}\left\|\overrightarrow{\boldsymbol{Z}\boldsymbol{\rho}}(\boldsymbol{\beta}_0,s)\right\|.\end{aligned}$$

Hence, by the assumptions and Lemma 3.5, $V(\tau)-\vec{V}(\tau)=O_{\mathrm{P}}(n^{-1/2})$. □

Next, we want to show $\sup_{t\in[0,\tau]}|\hat{\Lambda}_0(t)-\Lambda_0(t)|=O_{\mathrm{P}}(n^{-1/2})$ if the Cox model does not necessarily hold true, where $\Lambda_0(t)=\int_0^t\vec{\boldsymbol{\rho}}(\boldsymbol{\beta}_0,s)^{-1}\vec{\boldsymbol{\lambda}}(s)\,\mathrm{d}s$ and $\hat{\Lambda}_0(t)$ is the Breslow estimator. The Breslow estimator under misspecified models has been considered previously by Lin and Wei (1989), Sasieni (1993), and Fine (2002) under an i.i.d. setup allowing only one event per individual. The next lemma relaxes these assumptions for the Breslow estimator.

Lemma 4.4. *Suppose there exists an open neighborhood $\mathcal{C}\subset\boldsymbol{\mathcal{X}}_{\boldsymbol{\beta}}$ of $\boldsymbol{\beta}_0$ such that the following conditions are satisfied:*

(i) $\hat{\boldsymbol{\beta}}-\boldsymbol{\beta}_0=O_{\mathrm{P}}(n^{-1/2})$,

(ii) $\overline{\boldsymbol{\rho}}$ *and* $\overline{\boldsymbol{\rho}\boldsymbol{Z}}$ *converge uip on* $\mathcal{C}\times[0,\tau]$,

(iii) $\vec{\boldsymbol{\rho}}(\boldsymbol{\beta}_0, \cdot)$ *is bounded away from* 0 *uniformly on* $[0, \tau]$,

(iv) $\overline{\boldsymbol{\lambda}}$ *converges uip on* $[0, \tau]$,

(v) $\int_0^\tau \overline{\boldsymbol{\rho}^{-2}\boldsymbol{\lambda}}(\boldsymbol{\beta}_0, s)\,\mathrm{d}s$ *converges in probability,*

(vi) $\int_0^\tau |\overline{\boldsymbol{\lambda}}(s) - \vec{\boldsymbol{\lambda}}(s)|\,\mathrm{d}s = O_{\mathrm{P}}(n^{-1/2})$, *and*

(vii) $\int_0^\tau |\overline{\boldsymbol{\rho}}(\boldsymbol{\beta}_0, s) - \vec{\boldsymbol{\rho}}(\boldsymbol{\beta}_0, s)|\,\mathrm{d}s = O_{\mathrm{P}}(n^{-1/2})$.

Then $\sup_{t\in[0,\tau]} |\hat{\Lambda}_0(t) - \Lambda_0(t)| = O_{\mathrm{P}}(n^{-1/2})$.

Proof. Consider the decomposition

$$\begin{aligned}\hat{\Lambda}_0(t) - \Lambda_0(t) = &\int_0^t \left(\overline{\boldsymbol{\rho}}(\hat{\boldsymbol{\beta}}, s)^{-1} - \overline{\boldsymbol{\rho}}(\boldsymbol{\beta}_0, s)^{-1}\right)\left(\mathrm{d}\overline{\boldsymbol{N}}(s) - \overline{\boldsymbol{\lambda}}(s)\,\mathrm{d}s\right)\\ &+ \int_0^t \overline{\boldsymbol{\rho}}(\boldsymbol{\beta}_0, s)^{-1}\left(\mathrm{d}\overline{\boldsymbol{N}}(s) - \overline{\boldsymbol{\lambda}}(s)\,\mathrm{d}s\right)\\ &+ \int_0^t \left(\overline{\boldsymbol{\rho}}(\hat{\boldsymbol{\beta}}, s)^{-1}\overline{\boldsymbol{\lambda}}(s) - \vec{\boldsymbol{\rho}}(\boldsymbol{\beta}_0, s)^{-1}\vec{\boldsymbol{\lambda}}(s)\right)\mathrm{d}s.\end{aligned}$$

Denote the terms on the right hand side by $A(t)$, $B(t)$, and $C(t)$. By a Taylor expansion,

$$A(t) = \int_0^t -\overline{\boldsymbol{\rho}}(\tilde{\boldsymbol{\beta}}, s)^{-2}\overline{\boldsymbol{\rho}\boldsymbol{Z}}(\tilde{\boldsymbol{\beta}}, s)\,\mathrm{d}\overline{\boldsymbol{M}}(s)(\hat{\boldsymbol{\beta}} - \boldsymbol{\beta}_0)$$

for some $\tilde{\boldsymbol{\beta}}$ between $\hat{\boldsymbol{\beta}}$ and $\boldsymbol{\beta}_0$. Hence,

$$\sup_{t\in[0,\tau]} |A(t)| \le \|\hat{\boldsymbol{\beta}} - \boldsymbol{\beta}_0\| \sup_{t\in[0,\tau],\boldsymbol{\beta}\in\mathcal{C}} |\overline{\boldsymbol{\rho}}(\boldsymbol{\beta}, t)|^{-2} \sup_{t\in[0,\tau],\boldsymbol{\beta}\in\mathcal{C}} |\overline{\nabla\boldsymbol{\rho}}(\boldsymbol{\beta}, t)| \int_0^\tau |\,\mathrm{d}\overline{\boldsymbol{M}}(s)|.$$

Since $\int_0^\tau |\,\mathrm{d}\overline{\boldsymbol{M}}(s)| = \overline{\boldsymbol{N}}(\tau) + \int_0^\tau \overline{\boldsymbol{\lambda}}(s)\,\mathrm{d}s = \left(\overline{\boldsymbol{N}}(\tau) - \int_0^\tau \overline{\boldsymbol{\lambda}}(s)\,\mathrm{d}s\right) + 2\int_0^\tau \overline{\boldsymbol{\lambda}}(s)\,\mathrm{d}s \xrightarrow{\mathrm{P}} 2\int_0^\tau \vec{\boldsymbol{\lambda}}(s)\,\mathrm{d}s$ by Lenglart's inequality, we have $\sup_{t\in[0,\tau]} |A(t)| = O_{\mathrm{P}}(n^{-1/2})$. As $B(t)$ is a local martingale, Lenglart's inequality (Lemma A.2) and (v) imply $\sup_{t\in[0,\tau]} |B(t)| = O_{\mathrm{P}}(n^{-1/2})$. It remains to consider $C(t)$. By triangular inequalities,

$$\begin{aligned}|C(t)| &\le \int_0^\tau \left|\overline{\boldsymbol{\rho}}(\hat{\boldsymbol{\beta}}, s)^{-1}\overline{\boldsymbol{\lambda}}(s) - \vec{\boldsymbol{\rho}}(\boldsymbol{\beta}_0, s)^{-1}\vec{\boldsymbol{\lambda}}(s)\right|\mathrm{d}s\\ &\le \int_0^\tau \left|\overline{\boldsymbol{\rho}}(\hat{\boldsymbol{\beta}}, s)^{-1} - \overline{\boldsymbol{\rho}}(\boldsymbol{\beta}_0, s)^{-1}\right|\overline{\boldsymbol{\lambda}}(s)\,\mathrm{d}s\\ &\quad + \int_0^\tau \left|\overline{\boldsymbol{\rho}}(\boldsymbol{\beta}_0, s)^{-1} - \vec{\boldsymbol{\rho}}(\boldsymbol{\beta}_0, s)^{-1}\right|\overline{\boldsymbol{\lambda}}(s)\,\mathrm{d}s\\ &\quad + \int_0^\tau \vec{\boldsymbol{\rho}}(\boldsymbol{\beta}_0, s)^{-1}\left|\overline{\boldsymbol{\lambda}}(s) - \vec{\boldsymbol{\lambda}}(s)\right|\mathrm{d}s.\end{aligned}$$

If $\hat{\boldsymbol{\beta}} \in \mathcal{C}$ then by Taylor's theorem, Lemma 3.5 and the assumptions,

$$\begin{aligned}\sup_{t\in[0,\tau]} |C(t)| \leq & \|\hat{\boldsymbol{\beta}} - \boldsymbol{\beta}_0\| \sup_{\substack{i=1\dots n\\ t\in[0,\tau],\boldsymbol{\beta}\in\mathcal{C}}} \frac{|\overline{\boldsymbol{Z}\boldsymbol{\rho}}(\boldsymbol{\beta},t)|}{\overline{\boldsymbol{\rho}}(\boldsymbol{\beta},t)^2}\overline{\boldsymbol{\lambda}}(t) \\ &+ \int_0^\tau \left|\overline{\boldsymbol{\rho}}(\boldsymbol{\beta}_0,s)^{-1} - \vec{\boldsymbol{\rho}}(\boldsymbol{\beta}_0,s)^{-1}\right| \mathrm{d}s \sup_{s\in[0,\tau]} \overline{\boldsymbol{\lambda}}(s) \\ &+ \int_0^\tau \left|\overline{\boldsymbol{\lambda}}(s) - \vec{\boldsymbol{\lambda}}(s)\right| \mathrm{d}s \sup_{s\in[0,\tau]} \vec{\boldsymbol{\rho}}(\boldsymbol{\beta}_0,s)^{-1} \\ =& O_{\mathrm{P}}(n^{-\frac{1}{2}}).\end{aligned}$$

□

4.2 Smoothing of Nonparametric Estimators

In Chapters 5 and 6, we need to plug in estimators of functions into our test statistic. In this context, we are often in a situation, where we want to show that the difference

$$\int_0^t \hat{\alpha}_n(s)\,\mathrm{d}X_n(s) - \int_0^t \alpha(s)\,\mathrm{d}X_n(s) = \int_0^t (\hat{\alpha}_n(s) - \alpha(s))\,\mathrm{d}X_n(s),$$

converges to zero, where $\hat{\alpha}_n(s)$ are estimators for $\alpha(s)$ and X_n is some stochastic process of finite variation whose total variation may go to infinity as $n \to \infty$. If $\hat{\alpha}_n$ is predictable and X_n is a local martingale, we can use martingale theory, e.g. Lenglart's inequality. Otherwise, the following lemma can be used:

Lemma 4.5. *If $X_n(t)$ is a sequence of càdlàg stochastic processes of finite variation and $a_n(t)$ are càglàd stochastic processes then*

$$\sup_{t\in[0,\tau]} \left|\int_0^t a_n(s)\,\mathrm{d}X_n(s)\right| = O_{\mathrm{P}}\left(\left(\sup_{t\in[0,\tau]} |a_n(t)| + \int_0^\tau |\mathrm{d}a_n(s)|\right) \sup_{t\in[0,\tau]} |X_n(t)|\right),$$

where $\int_0^\tau |\mathrm{d}a_n(s)|$ denotes the total variation of a_n.

Proof. One merely has to apply partial integration (see e.g. Fleming and Harrington (1991, Theorem A.1.2)) as follows:

$$\begin{aligned}\left|\int_0^t a_n(s)\,\mathrm{d}X_n(s)\right| &\leq \left|\int_0^t X_n(s)\,\mathrm{d}a_n(s)\right| + |X_n(t)a_n(t)| + |X_n(0)a_n(0)| \\ &\leq \sup_{s\in[0,\tau]} |X_n(s)| \int_0^t |\mathrm{d}a_n(s)| + |X_n(t)a_n(t)| + |X_n(0)a_n(0)| \\ &\leq \sup_{s\in[0,\tau]} |X_n(s)| \left(2\sup_{s\in[0,\tau]} |a_n(s)| + \int_0^\tau |\mathrm{d}a_n(s)|\right).\end{aligned}$$

□

In nonparametric or semiparametric models from survival analysis, usually one first estimates $A(t) := \int_0^t \alpha(s)\,\mathrm{d}s$ with an estimator $\hat{A}(t)$ and derives asymptotic properties of $\hat{A}(t) - A(t)$. Usually $\hat{A}$ is a càdlàg step function. In this section, we shall restrict ourselves to this case.

If an estimator of α itself is needed then the standard approach is to smooth $\hat{A}(t)$ by means of kernel smoothers to get an estimator $\hat{\alpha}(t)$ of $\alpha(t)$. In order to apply the previous lemma, we need to ensure that $\hat{\alpha}(t) - \alpha(t)$ vanishes uniformly and that $\int_0^\tau |\mathrm{d}(\hat{\alpha}(t) - \alpha(t))|$ vanishes. The classical kernel smoothing approach is to use the estimator $\tilde{\alpha}(t) = \int_0^\tau b^{-1} K\left(\frac{t-s}{b}\right) \mathrm{d}\widehat{A}(s)$ with a kernel function K and a bandwidth b. If K is differentiable, then the derivative of $\tilde{\alpha}(t)$ is $\int_0^\tau b^{-2} K'\left(\frac{t-s}{b}\right) \mathrm{d}\widehat{A}(s)$ which, together with a partial integration, can be used to bound the total variation of $\tilde{\alpha}(t) - \alpha(t)$. Since we are working on a finite interval and need uniform properties of $\hat{\alpha}(t) - \alpha(t)$, we need to take care of boundary effects. Using a boundary correction by boundary kernels complicates things, since the kernels one uses close to the boundary depend on t as well.

In this section, we modify the kernel smoothing approach such that for the resulting estimator $\hat{\alpha}(t)$ of $\alpha(t)$ we know properties of $\hat{\alpha}(t) - \alpha(t)$ as well as of $\hat{\alpha}'(t) - \alpha'(t)$, where $'$ denotes differentiation with respect to t. We include a boundary correction based on boundary kernels. The main idea is to estimate the derivative $\alpha'(t)$ using kernel smoothers with boundary correction and define $\hat{\alpha}(t)$ as integral. The estimate of the derivative $\alpha'(t)$ we use is

$$\widehat{(\alpha')}(t) = \int_0^\tau h_n(s,t)\,\mathrm{d}\hat{A}(s),$$

where, for $s, t \in [0, \tau]$,

$$h_n(s,t) := \begin{cases} \frac{1}{b_n^2} K_1'\left(\frac{t-s}{b_n}\right), & t \in [b_n, \tau - b_n] \\ \frac{1}{b_n^2} K_{t/b_n}'\left(\frac{t-s}{b_n}\right), & t < b_n \\ -\frac{1}{b_n^2} K_{(\tau-t)/b_n}'\left(-\frac{t-s}{b_n}\right), & t > \tau - b_n \end{cases} \tag{4.2}$$

and $K_q'(x) : \mathbb{R} \to \mathbb{R}, q \in [0,1]$, is of bounded variation uniformly in q, $K_q'(x) = 0$ for $x \notin [-1, q]$,

$$\int_{-1}^q K_q'(x)\,\mathrm{d}x = 0, \int_{-1}^q K_q'(x)\,x\,\mathrm{d}x = -1, \sup_{q\in[0,1]} \int_{-1}^q |K_q'(x)|\,x^2\,\mathrm{d}x < \infty, \tag{4.3}$$

and the bandwidth $b_n > 0$ are random variables such that $b_n \xrightarrow{\mathrm{P}} 0$. We assume $\tau \geq 2b_n$. Now the idea is to use an estimator of $\alpha(t)$ defined by $\hat{\alpha}(0) + \int_0^t \widehat{(\alpha')}(s)\,\mathrm{d}s$, where $\hat{\alpha}(0)$ is some estimator for $\alpha(0)$. This ensures $\widehat{(\alpha')}(t) = \hat{\alpha}'(t)$. More precisely, the smoothed estimator for $\alpha(t)$ we will consider is

$$\hat{\alpha}(t) := \int_0^\tau H_n(s,t)\,\mathrm{d}\hat{A}(s), \tag{4.4}$$

where

$$H_n(s,t) := \int_0^t h_n(s,u)\,\mathrm{d}u + H_n(s,0),$$

with $H_n(s,0)$ chosen such that

$$\begin{aligned} &\int_0^\tau H_n(s,0)\,\mathrm{d}s = 1, \quad \int_0^\tau |H_n(s,0)|\,\mathrm{d}s < \infty, \\ &b_n^2 \int_0^\tau |H_n(\mathrm{d}s,0)| = O_{\mathrm{P}}(1), \quad \text{and} \quad H_n(s,0) = 0 \text{ for } s \geq 2b_n. \end{aligned} \tag{4.5}$$

Example 4.1. Next, we give one explicit $H_n(s,t)$ using the above construction. Starting with the assumption that $K_q'(x) := a_1(q)x + a_0(q)$ the condition (4.3) implies that $a_1(q) = -12(q+1)^{-3}$ and $a_0(q) = 6(q-1)(q+1)^{-3}$. An illustration can be seen in Figure 4.1. This leaves us with the choice of $H_n(s,0)$. We require $H_n(s,b_n) = \frac{3}{4b_n}\left(1-\left(\frac{b_n-s}{b_n}\right)^2\right)$, since we want that outside of the boundary regions our smoothing approach reduces to a usual kernel smoother with the Epanechnikov kernel. It follows that for $s,t \in [0,\tau]$,

$$H_n(s,t) = \begin{cases} -\frac{3}{4}\frac{(2b_n-s)^2}{{b_n}^3}, & t \leq b_n, t+b_n \leq s \leq 2b_n \\ -\frac{3}{4{b_n}^3}\left(s^2 + \frac{4b_n\left(-t^2-2tb_n+{b_n}^2\right)}{(t+b_n)^2}s + \frac{4{b_n}^2(t^2-{b_n}^2)}{(t+b_n)^2}\right), & t \leq b_n, t+b_n \geq s \\ \frac{3}{4b_n}\left(1-\left(\frac{t-s}{b_n}\right)^2\right), & b_n \leq t \leq \tau - b_n, |t-s| \leq b_n \\ H_n(\tau-s,\tau-t), & t \geq \tau - b_n \\ 0, & \text{otherwise} \end{cases} \tag{4.6}$$

For a sketch of H_n, see Figure 4.2. One can check that indeed $\int_0^\tau H_n(s,0)\,\mathrm{d}s = 1$, $\int_0^\tau |H_n(s,0)|\,\mathrm{d}s = 16\sqrt{2}-21$, and $\int_0^\tau |H_n(\mathrm{d}s,0)| = 9/(2b_n)$.

Lemma 4.6. *Suppose that $\alpha : [0,\tau] \to \mathbb{R}$ is twice continuously differentiable, that $\hat{\alpha}$ is given by* (4.4)*, and that conditions* (4.3) *and* (4.5) *hold true. If*

$$\sup_{t\in[0,\tau]} |\hat{A}(t) - A(t)| = O_{\mathrm{P}}(n^{-\frac{1}{2}})$$

then

$$\sup_{t\in[0,\tau]} |\hat{\alpha}'(t) - \alpha'(t)| = O_{\mathrm{P}}(n^{-\frac{1}{2}}b_n^{-2} + b_n).$$

For an optimal rate we would need that $n^{-1/2}b_n^{-2}$ and b_n are of the same order. Assuming $b_n = n^{-\alpha}$, this leads to the equation $-1/2 + 2\alpha = -\alpha$ and thus $\alpha = 1/6$. With this choice, $\sup_{t\in[0,\tau]} |\hat{\alpha}'(t) - \alpha'(t)| = O_{\mathrm{P}}(n^{-1/6})$.

Before we begin with the proof of Lemma 4.6 we state a lemma concerning h_n given in (4.2).

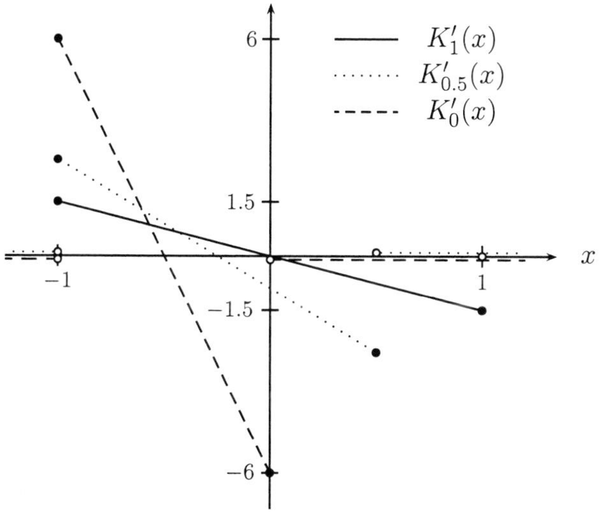

Figure 4.1: Sketch of K_q from Example 4.1.

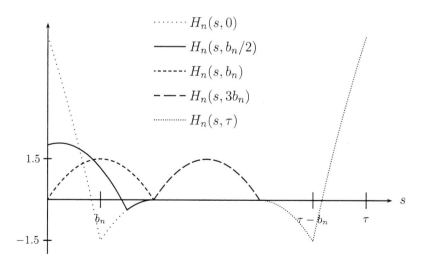

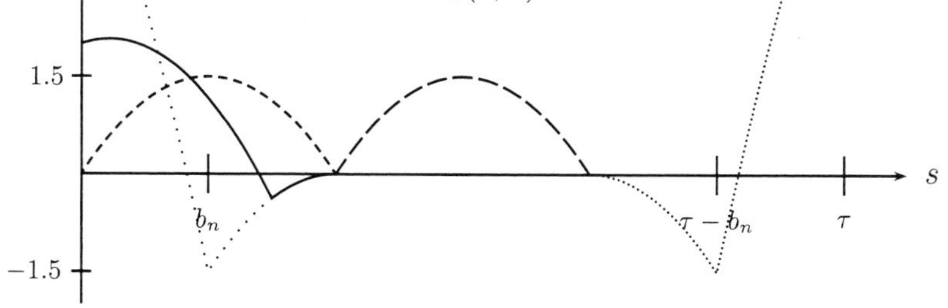

Figure 4.2: Sketch of H_n from Example 4.1 using $\tau = 3$ and $b_n = 0.5$.

Lemma 4.7. *$h_n : [0,\tau]^2 \to \mathbb{R}$ satisfies $\int_0^\tau h_n(s,t)\,\mathrm{d}s = 0$, $\int_0^\tau h_n(s,t)s\,\mathrm{d}s = 1$, and $\int_0^\tau |h_n(s,t)|(s-t)^2\,\mathrm{d}s = O_{\mathrm{P}}(b_n)$ uniformly in $t \in [0,\tau]$.*

Proof. For $t \in [b_n, \tau - b_n]$,

$$\int_0^\tau h_n(s,t)\,\mathrm{d}s = \int_0^\tau \frac{1}{b_n^2} K_1'\left(\frac{t-s}{b_n}\right)\mathrm{d}s = -\frac{1}{b_n}\int_{t/b_n}^{(t-\tau)/b_n} K_1'(u)\,\mathrm{d}u$$
$$= \frac{1}{b_n}\int_{-1}^{1} K_1'(u)\,\mathrm{d}u = 0,$$

$$\int_0^\tau h_n(s,t)s\,\mathrm{d}s = -\int_0^\tau h_n(s,t)(t-s)\,\mathrm{d}s = -\int_0^\tau K_1'\left(\frac{t-s}{b_n}\right)\frac{t-s}{b_n}\frac{\mathrm{d}s}{b_n}$$
$$= \int_{t/b_n}^{(t-\tau)/b_n} K_1'(u)\,u\,\mathrm{d}u = -\int_{-1}^{1} K_1'(u)\,u\,\mathrm{d}u = 1,$$

$$\int_0^\tau |h_n(s,t)|(t-s)^2\,\mathrm{d}s = b_n\int_0^\tau \left|K_1'\left(\frac{t-s}{b_n}\right)\right|\left(\frac{t-s}{b_n}\right)^2\frac{\mathrm{d}s}{b_n}$$
$$= -b_n\int_{t/b_n}^{(t-\tau)/b_n} \left|K_1'(u)\right| u^2\,\mathrm{d}u = b_n\int_{-1}^{1}\left|K_1'(u)\right| u^2\,\mathrm{d}u = O_{\mathrm{P}}(b_n).$$

For $t < b_n$,

$$\int_0^\tau h_n(s,t)\,\mathrm{d}s = \int_0^\tau \frac{1}{b_n^2} K_{t/b_n}'\left(\frac{t-s}{b_n}\right)\mathrm{d}s = -\frac{1}{b_n}\int_{t/b_n}^{(t-\tau)/b_n} K_{t/b_n}'(u)\,\mathrm{d}u$$
$$= \frac{1}{b_n}\int_{-1}^{t/b_n} K_{t/b_n}'(u)\,\mathrm{d}u = 0,$$

$$\int_0^\tau h_n(s,t)s\,\mathrm{d}s = -\int_0^\tau h_n(s,t)(t-s)\,\mathrm{d}s = -\int_0^\tau K_{t/b_n}'\left(\frac{t-s}{b_n}\right)\frac{t-s}{b_n}\frac{\mathrm{d}s}{b_n}$$
$$= \int_{t/b_n}^{(t-\tau)/b_n} K_{t/b_n}'(u)\,u\,\mathrm{d}u = -\int_{-1}^{t/b_n} K_{t/b_n}'(u)\,u\,\mathrm{d}u = 1,$$

$$\int_0^\tau |h_n(s,t)|(t-s)^2\,\mathrm{d}s = b_n\int_0^\tau \left|K_{t/b_n}'\left(\frac{t-s}{b_n}\right)\right|\left(\frac{t-s}{b_n}\right)^2\frac{\mathrm{d}s}{b_n}$$
$$= -b_n\int_{t/b_n}^{(t-\tau)/b_n} \left|K_{t/b_n}'(u)\right| u^2\,\mathrm{d}u = b_n\int_{-1}^{t/b_n}\left|K_{t/b_n}'(u)\right| u^2\,\mathrm{d}u = O_{\mathrm{P}}(b_n).$$

For $t > \tau - b_n$,

$$\int_0^\tau h_n(s,t)\,\mathrm{d}s = -\int_0^\tau \frac{1}{b_n^2} K_{(\tau-t)/b_n}'\left(-\frac{t-s}{b_n}\right)\mathrm{d}s$$
$$= -\frac{1}{b_n}\int_{t/b_n}^{(t-\tau)/b_n} K_{(t-\tau)/b_n}'(u)\,\mathrm{d}u = -\frac{1}{b_n}\int_{-1}^{(t-\tau)/b_n} K_{t/b_n}'(u)\,\mathrm{d}u = 0,$$

$$\int_0^\tau h_n(s,t)s\,\mathrm{d}s = \int_0^\tau h_n(s,t)(s-t)\,\mathrm{d}s = -\int_0^\tau K'_{(\tau-t)/b_n}\left(\frac{s-t}{b_n}\right)\frac{s-t}{b_n}\frac{\mathrm{d}s}{b_n}$$
$$= -\int_{-t/b_n}^{(\tau-t)/b_n} K'_{(\tau-t)/b_n}(u)\,u\,\mathrm{d}u = -\int_{-1}^{(\tau-t)/b_n} K'_{(\tau-t)/b_n}(u)\,u\,\mathrm{d}u = 1,$$

$$\int_0^\tau |h_n(s,t)|(t-s)^2\,\mathrm{d}s = b_n\int_0^\tau \left|K'_{(\tau-t)/b_n}\left(\frac{s-t}{b_n}\right)\right|\left(\frac{s-t}{b_n}\right)^2\frac{\mathrm{d}s}{b_n}$$
$$= -b_n\int_{t/b_n}^{(t-\tau)/b_n}\left|K'_{(\tau-t)/b_n}(u)\right|u^2\,\mathrm{d}u$$
$$= b_n\int_{-1}^{t/b_n}\left|K'_{(\tau-t)/b_n}(u)\right|u^2\,\mathrm{d}u = O_{\mathrm{P}}(b_n).$$

□

Proof of Lemma 4.6. Since the integral is a finite sum, $\hat{\alpha}'(t) = \int_0^\tau h_n(s,t)\,\mathrm{d}\hat{A}(s)$. Let $\alpha^*(t) := \int_0^\tau h_n(s,t)\alpha(s)\,\mathrm{d}s$. By Taylor's theorem,

$$\alpha^*(t) = \int_0^\tau h_n(s,t)\,\mathrm{d}s\alpha(t) + \int_0^\tau h_n(s,t)(s-t)\,\mathrm{d}s\alpha'(t)$$
$$+ \int_0^\tau h_n(s,t)(s-t)^2\frac{1}{2}\alpha''(\tilde{s}(s))\,\mathrm{d}s$$
$$=0 + \alpha'(t) + \int_0^\tau h_n(s,t)(s-t)^2\frac{1}{2}\alpha''(\tilde{s}(s))\,\mathrm{d}s$$

for $\tilde{s}(s) \in [0,\tau]$. Hence,

$$|\alpha^*(t) - \alpha'(t)| \leq \frac{1}{2}\sup_{s\in[0,\tau]}|\alpha''(s)|\int_0^\tau |h_n(s,t)|(s-t)^2\,\mathrm{d}s = O_{\mathrm{P}}(b_n)$$

uniformly in $t \in [0,\tau]$. Using partial integration,

$$|\hat{\alpha}'(t) - \alpha^*(t)| = \left|\int_0^\tau h_n(s,t)\,\mathrm{d}\left(\hat{A}(s) - A(s)\right)\right|$$
$$= \left|\left(\hat{A}(\tau) - A(\tau)\right)h_n(\tau,t) + \int_0^\tau \left(\hat{A}(s) - A(s)\right)h_n(\mathrm{d}s,t)\right|$$
$$\leq \sup_{s\in[0,\tau]}\left|\hat{A}(s) - A(s)\right|\left(|h_n(\tau,t)| + \int_0^\tau |h_n(\mathrm{d}s,t)|\right)$$

Since $|h_n(\tau,t)| = O_{\mathrm{P}}(b_n^{-2})$ and $\int_0^\tau |h_n(\mathrm{d}s,t)| = b_n^{-2}\int_{-1}^q |K_q(\,\mathrm{d}x)| = O_{\mathrm{P}}(b_n^{-2})$,

$$\sup_{t\in[0,\tau]}|\hat{\alpha}'(t) - \alpha'(t)| \leq \sup_{t\in[0,\tau]}|\hat{\alpha}'(t) - \alpha^*(t)| + \sup_{t\in[0,\tau]}|\alpha^*(t) - \alpha'(t)|$$
$$=O_{\mathrm{P}}(b_n + n^{-\frac{1}{2}}b_n^{-2}).$$

□

We also need to know at what rate $\sup_{s\in[0,\tau]} |\hat{\alpha}(s) - \alpha(s)|$ converges to 0. The next theorem gives such a result. Note that H_n given by (4.6) satisfies the assumptions.

Theorem 4.1. *Suppose* $\int_0^\tau H_n(s,0)s\,\mathrm{d}s = 0$ *and uniformly in* t, $H_n(0,t) = O_\mathrm{P}(b_n^{-1})$, $\int_0^\tau |H_n(s,t)|\,(s-t)^2\,\mathrm{d}s = O_\mathrm{P}(b_n^2)$, $\int_0^\tau |H_n(\mathrm{d}s,t)| = O_\mathrm{P}(b_n^{-1})$. *If* α *is twice differentiable then*

$$\sup_{s\in[0,\tau]} |\hat{\alpha}(s) - \alpha(s)| = O_\mathrm{P}(\sup_{t\in[0,\tau]} |\hat{A}(t) - A(t)| b_n^{-1} + b_n^2).$$

For the proof, we need the following elementary lemma.

Lemma 4.8. *For all* $t \in [0,\tau]$, $\int_0^\tau H_n(s,t)\,\mathrm{d}s = 1$ *and* $\int_0^\tau H_n(s,t)(s-t)\,\mathrm{d}s = \int_0^\tau H_n(s,0)s\,\mathrm{d}s$.

Proof.

$$\int_0^\tau H_n(s,t)\,\mathrm{d}s = \int_0^\tau \int_0^t h_n(s,u)\,\mathrm{d}u\,\mathrm{d}s + \int_0^\tau H_n(s,0)\,\mathrm{d}s = 0 + 1 = 1$$

$$\begin{aligned}\int_0^\tau H_n(s,t)(s-t)\,\mathrm{d}s &= \int_0^\tau \int_0^t h_n(s,u)\,\mathrm{d}u(s-t)\,\mathrm{d}s + \int_0^\tau H_n(s,0)(s-t)\,\mathrm{d}s \\ &= \int_0^t \left(\int_0^\tau h_n(s,u)s\,\mathrm{d}s - t\int_0^\tau h_n(s,u)\,\mathrm{d}s\right)\mathrm{d}u \\ &\quad + \int_0^\tau H_n(s,0)s\,\mathrm{d}s - t\int_0^\tau H_n(s,0)\,\mathrm{d}s \\ &= \int_0^t (1-0)\,\mathrm{d}u + \int_0^\tau H_n(s,0)s\,\mathrm{d}s - t = \int_0^\tau H_n(s,0)s\,\mathrm{d}s\end{aligned}$$

□

Proof of Theorem 4.1. Let $\alpha^*(t) = \int_0^\tau H_n(s,t)\alpha(s)\,\mathrm{d}s$. By Lemma 4.8, Taylor's theorem, and the assumptions,

$$\begin{aligned}|\alpha^*(t) - \alpha(t)| &= \left|\int_0^\tau H_n(s,t)(\alpha(s) - \alpha(t))\,\mathrm{d}s\right| \\ &\le \left|\int_0^\tau H_n(s,t)(s-t)\,\mathrm{d}s\alpha'(t)\right| + \frac{1}{2}\sup_{s\in[0,\tau]} |\alpha''(s)| \int_0^\tau \left|H_n(s,t)(s-t)^2\right|\mathrm{d}s \\ &= 0 + \frac{1}{2}\sup_{s\in[0,\tau]} |\alpha''(s)| \int_0^\tau |H_n(s,t)|\,(s-t)^2\,\mathrm{d}s \\ &= O_\mathrm{P}(b_n^2).\end{aligned}$$

Using partial integration,

$$\begin{aligned}
|\hat{\alpha}(t) - \alpha^*(t)| =& \left| \int_0^\tau H_n(s,t)\,\mathrm{d}\left(\hat{A}(s) - A(s)\right) \right| \\
\leq& \left| \int_0^\tau \left(\hat{A}(s) - A(s)\right) H_n(\mathrm{d}s,t) \right| + \left|\hat{A}(\tau) - A(\tau)\right| |H_n(\tau,t)| \\
&+ \left| \left(\hat{A}(0) - A(0)\right) H_n(0,t) \right| \\
\leq& \sup_{s\in[0,\tau]} \left|\hat{A}(s) - A(s)\right| 2 \left(|H_n(0,t)| + \int_0^\tau |H_n(\mathrm{d}s,t)| \right) \\
=& O_{\mathrm{P}}\left(\sup_{s\in[0,\tau]} |\hat{A}(s) - A(s)| b_n^{-1} \right)
\end{aligned}$$

□

The choice of the bandwidth parameter b_n is critical. Assume that

$$\sup_{t\in[0,\tau]} |\hat{A}(t) - A(t)| = O_{\mathrm{P}}(n^{-1/2}).$$

For an optimal rate we need that $n^{-1/2}b_n^{-1}$ is of the same order as b_n^2. Assuming $b_n = n^{-\alpha}$, this leads to the equation $-1/2 + \alpha = -2\alpha$ and thus $\alpha = 1/6$. With this choice, $\sup_{t\in[0,\tau]} |\hat{\alpha}(t) - \alpha(t)| = O_{\mathrm{P}}(n^{-1/3})$. Note that the optimal rate coincides with the optimal rate for the derivatives.

Simulation studies show that if we use the smoothing procedure in our tests of the following chapters and the bandwidth is too small then the tests can be very liberal. The following ad hoc choice for b_n worked reasonably well in a situation, where we smoothed the Breslow estimator $\hat{\Lambda}_0(t)$ in the Cox model:

$$b_n := \begin{cases} \inf A & \text{if } A \neq \emptyset \\ \tau/2 & \text{otherwise} \end{cases}, \tag{4.7}$$

where

$$A := \left\{ \tilde{b} \in [0, \frac{\tau}{2}] : \sum_{i=1}^n (N_i(t + 2\tilde{b}) - N_i(t)) \geq \left(\sum_{i=1}^n N_i(\tau) \right)^{5/6} \quad \forall t \in [0, \tau - 2\tilde{b}] \right\}.$$

The H_n given by (4.6) requires $b_n \geq \tau/2$. Besides this, our choice of b_n ensures that for each $t \in [0, \tau]$, the estimator $\hat{\alpha}(t)$ is based on at least $(\sum_{i=1}^n N_i(\tau))^{5/6}$ events. One can show that under mild conditions the above b_n actually satisfies $b_n \xrightarrow{\mathrm{P}} 0$ and $b_n^4 n \xrightarrow{\mathrm{P}} \infty$.

Chapter 5

Checking a Nonparametric Model

In Chapter 3, we introduced checks for the Aalen model in which the unknown parameters act linearly on the intensity. We want to extend these checks to models in which parameters may act nonlinearly on the intensity. Basically, we want to consider the model (1.4) mentioned in the introduction with only time-dependent parameters, i.e. we want to consider

$$\lambda_i(t) = f(\boldsymbol{X}_i(t), \boldsymbol{\alpha}(t)) \text{ for some } \boldsymbol{\alpha} \in \mathrm{bm}(\boldsymbol{\mathcal{X}}_{\boldsymbol{\alpha}}), \tag{5.1}$$

where $\boldsymbol{\mathcal{X}}_{\boldsymbol{\alpha}} \subset \mathbb{R}^{k_{\boldsymbol{\alpha}}}$ is a convex set, f is a known continuous function, and the observable covariates are given by a vector $\boldsymbol{X}_i$ of locally bounded predictable stochastic processes. By Lemma A.6 and Lemma A.7, $f_i(\boldsymbol{\alpha}(\cdot), \cdot)$ is a locally bounded predictable stochastic process for all $\boldsymbol{\alpha} \in \mathrm{bm}(\boldsymbol{\mathcal{X}}_{\boldsymbol{\alpha}})$. Hence, the following model is more general than (5.1):

$$\lambda_i(t) = f_i(\boldsymbol{\alpha}(t), t) \text{ for some } \boldsymbol{\alpha} \in \mathrm{bm}(\boldsymbol{\mathcal{X}}_{\boldsymbol{\alpha}}), \tag{5.2}$$

where $\boldsymbol{\mathcal{X}}_{\boldsymbol{\alpha}} \subset \mathbb{R}^{k_{\boldsymbol{\alpha}}}$ is a convex set, and for each $\boldsymbol{\alpha} \in \mathrm{bm}(\boldsymbol{\mathcal{X}}_{\boldsymbol{\alpha}})$, the observable stochastic processes $f_i(\boldsymbol{\alpha}(\cdot), \cdot)$ are predictable and locally bounded. In this chapter, we shall be working with the more general model (5.2).

Recall that we want to construct tests that are sensitive against certain competing models. We do so by adjusting a weight process which will be called $\boldsymbol{c}$. The value of the weight process $\boldsymbol{c}$ at time t may depend on $\boldsymbol{\alpha}(t)$ as well as on $\boldsymbol{\beta}(t)$, where $\boldsymbol{\beta} \in \mathrm{bm}(\boldsymbol{\mathcal{X}}_{\boldsymbol{\beta}})$, for some convex set $\boldsymbol{\mathcal{X}}_{\boldsymbol{\beta}} \subset \mathbb{R}^{k_{\boldsymbol{\beta}}}$, is some other parameter. Usually, $\boldsymbol{\beta}$ will be a parameter from a competing model. For ease of notation, we consider the combined parameter $\boldsymbol{\theta}(t) = (\boldsymbol{\alpha}(t)^\top, \boldsymbol{\beta}(t)^\top)^\top$. Thus $\boldsymbol{\theta} \in \mathrm{bm}(\boldsymbol{\mathcal{X}}_{\boldsymbol{\theta}})$, where $\boldsymbol{\mathcal{X}}_{\boldsymbol{\theta}} := \boldsymbol{\mathcal{X}}_{\boldsymbol{\alpha}} \times \boldsymbol{\mathcal{X}}_{\boldsymbol{\beta}} \subset \mathbb{R}^{k_{\boldsymbol{\theta}}}$ and $k_{\boldsymbol{\theta}} := k_{\boldsymbol{\alpha}} + k_{\boldsymbol{\beta}}$. Elements of $\boldsymbol{\mathcal{X}}_{\boldsymbol{\alpha}}$ (resp. $\boldsymbol{\mathcal{X}}_{\boldsymbol{\beta}}$, $\boldsymbol{\mathcal{X}}_{\boldsymbol{\theta}}$) are usually denoted by $\boldsymbol{x}^{\boldsymbol{\alpha}}$ (resp. $\boldsymbol{x}^{\boldsymbol{\beta}}$, $\boldsymbol{x}^{\boldsymbol{\theta}}$).

Estimators for $\boldsymbol{\alpha}$ (resp. $\boldsymbol{\beta}$) are denoted by $\widehat{\boldsymbol{\alpha}}$ (resp. $\widehat{\boldsymbol{\beta}}$) and we let

$$\widehat{\boldsymbol{\theta}}(t) = (\widehat{\boldsymbol{\alpha}}(t)^\top, \widehat{\boldsymbol{\beta}}(t)^\top)^\top.$$

Often, we need derivatives with respect to the parameters. Suppose $g : \boldsymbol{\mathcal{X}}_{\boldsymbol{\theta}} \to \mathbb{R}$ is some function. For $\boldsymbol{x}^{\boldsymbol{\theta}} = (x_1^{\boldsymbol{\theta}}, \ldots, x_{k_{\boldsymbol{\theta}}}^{\boldsymbol{\theta}})^\top \in \boldsymbol{\mathcal{X}}_{\boldsymbol{\theta}}$, the operator ∇_j for $j \in \{1, \ldots, k_{\boldsymbol{\theta}}\}$ is defined by $\nabla_j g(\boldsymbol{x}^{\boldsymbol{\theta}}) = \frac{\partial}{\partial x_j^{\boldsymbol{\theta}}} g(x_1^{\boldsymbol{\theta}}, \ldots, x_{k_{\boldsymbol{\theta}}}^{\boldsymbol{\theta}})$. Furthermore, we define the row vector of operators $\nabla_{\boldsymbol{\theta}} := (\nabla_1, \ldots, \nabla_{k_{\boldsymbol{\theta}}})$, $\nabla_{\boldsymbol{\alpha}} := (\nabla_1, \ldots, \nabla_{k_{\boldsymbol{\alpha}}})$, and $\nabla_{\boldsymbol{\beta}} := (\nabla_{k_{\boldsymbol{\alpha}}+1}, \ldots, \nabla_{k_{\boldsymbol{\theta}}})$. For Hessian matrices, we also use the matrix of operators $\nabla_{\boldsymbol{\theta}}^\top \nabla_{\boldsymbol{\theta}}$.

The basis for our tests will be the stochastic process $T(\boldsymbol{c}, \widehat{\boldsymbol{\theta}}, \cdot)$, where for $\boldsymbol{\theta} = (\boldsymbol{\alpha}^\top, \boldsymbol{\beta}^\top)^\top \in \mathrm{bm}(\boldsymbol{\mathcal{X}}_{\boldsymbol{\theta}})$,

$$T(\boldsymbol{c}, \boldsymbol{\theta}, t) = n^{-\frac{1}{2}} \int_0^t \boldsymbol{c}(\boldsymbol{\theta}(s), s)^\top (\mathrm{d}\boldsymbol{N}(s) - \boldsymbol{f}(\boldsymbol{\alpha}(s), s)\,\mathrm{d}s)$$

and where $\boldsymbol{c}(\boldsymbol{\theta}(\cdot), \cdot)$ is an n-variate vector of predictable, locally bounded stochastic processes for each $\boldsymbol{\theta} \in \mathrm{bm}(\boldsymbol{\mathcal{X}}_{\boldsymbol{\theta}})$. We will use $T(\boldsymbol{c}, \boldsymbol{\theta}, t)$ throughout this chapter. Note that in contrast to Chapter 3,

$$n^{-\frac{1}{2}} \int_0^t \boldsymbol{c}(\boldsymbol{\theta}(s), s)^\top \boldsymbol{f}(\boldsymbol{\alpha}(s), s)\,\mathrm{d}s$$

need not be zero.

We assume that the estimator $\widehat{\boldsymbol{\theta}}$ converges to some $\boldsymbol{\theta}_0 = (\boldsymbol{\alpha}_0{}^\top, \boldsymbol{\beta}_0{}^\top)^\top \in \mathrm{bm}(\boldsymbol{\mathcal{X}}_{\boldsymbol{\theta}})$. More precisely, we require

(N1) (rate of convergence of $\widehat{\boldsymbol{\theta}}$) For some $0 < \nu \leq 1/2$,

$$\sup_{t \in [0,\tau]} \|\widehat{\boldsymbol{\theta}}(t) - \boldsymbol{\theta}_0(t)\| = O_{\mathrm{P}}(n^{-1/4-\nu/2}).$$

Implicitly, condition (N1) includes the assumption that the left hand side is measurable. Similar assumptions are understood to be included in all conditions that are concerned with sup and total variation. An example for a sufficient condition that $\sup_{t\in[0,\tau]} \|\widehat{\boldsymbol{\theta}}(t) - \boldsymbol{\theta}_0(t)\|$ is measurable is that $\widehat{\boldsymbol{\theta}}(\cdot) - \boldsymbol{\theta}_0(\cdot)$ is a càglàd or càdlàg function.

Condition (N1) is weaker than the requirement that $\widehat{\boldsymbol{\theta}}$ converges at the parametric rate $n^{1/2}$ in which case we could use $\nu = 1/2$. For example, if we use the procedure of Section 4.2, we can only guarantee a rate of $n^{1/3}$. In this case, we may choose $\nu = 1/6$.

We need another condition on the convergence of $\widehat{\boldsymbol{\theta}}$. In particular, when using $\widehat{\boldsymbol{\theta}}$ in integrands of integrals with respect to processes that converge to a Gaussian process, the following regularity condition is useful:

(N1') (total variation of $\widehat{\boldsymbol{\theta}} - \boldsymbol{\theta}_0$ vanishes)
The total variation of the components of $\widehat{\boldsymbol{\theta}} - \boldsymbol{\theta}_0$ converges stochastically to 0, i.e. for $j = 1, \dots, k_{\boldsymbol{\theta}}$,

$$\int_0^{\tau} |\,\mathrm{d}(\widehat{\theta}_j(s) - \theta_{0,j}(s))| \xrightarrow{\mathrm{P}} 0,$$

where $\theta_{0,j}$ denotes the jth component of $\boldsymbol{\theta}_0$.

Using the smoothing procedure of Section 4.2, (N1') can be ascertained.

If $\lambda_i(t) = f_i(\boldsymbol{\alpha}_0(t), t)$, then, under some regularity conditions, $T(\boldsymbol{c}, \boldsymbol{\theta}_0, \cdot)$ is a mean zero local martingale and, by a martingale central limit theorem, it converges to a mean zero Gaussian martingale. As mentioned in the introduction, to get $|T(\boldsymbol{c}, \widehat{\boldsymbol{\theta}}, t) - T(\boldsymbol{c}, \boldsymbol{\theta}_0, t)| \xrightarrow{\mathrm{P}} 0$, uniformly in $t \in [0, \tau]$, we use a Taylor expansion. Most terms vanish under mild regularity conditions. However, the following term does not vanish:

$$n^{-\frac{1}{2}} \int_0^t \boldsymbol{c}(\boldsymbol{\theta}_0(s), s)^{\top} (\nabla_{\boldsymbol{\alpha}} \boldsymbol{f})(\boldsymbol{\alpha}_0(s), s)(\widehat{\boldsymbol{\alpha}}(s) - \boldsymbol{\alpha}_0(s))\, \mathrm{d}s. \tag{5.3}$$

To let (5.3) vanish, we will place the following restrictions on $\boldsymbol{c}$:

(N2) (orthogonality condition) For $\mathrm{P} \otimes \lambda$-almost all $(\omega, s) \in \Omega \times [0, \tau]$,

$$\sum_{i=1}^{n} c_i(\omega, \boldsymbol{\theta}_0(s), s) \nabla_{\boldsymbol{\alpha}} f_i(\omega, \boldsymbol{\alpha}_0(s), s) = \boldsymbol{0}.$$

Since $\boldsymbol{\theta}_0$ is unknown, we cannot check (N2). Therefore, in applications, we impose the following stronger condition.

(N2') (strengthened orthogonality condition)
For $\mathrm{P} \otimes \lambda$-almost all $(\omega, s) \in \Omega \times [0, \tau]$,

$$\sum_{i=1}^{n} c_i(\omega, \boldsymbol{x}^{\boldsymbol{\theta}}, s) \nabla_{\boldsymbol{\alpha}} f_i(\omega, \boldsymbol{x}^{\boldsymbol{\alpha}}, s) = \boldsymbol{0}, \quad \text{for all } \boldsymbol{x}^{\boldsymbol{\theta}} = \begin{pmatrix} \boldsymbol{x}^{\boldsymbol{\alpha}} \\ \boldsymbol{x}^{\boldsymbol{\beta}} \end{pmatrix} \in \boldsymbol{\mathcal{X}}_{\boldsymbol{\theta}}.$$

It suffices to require some conditions on a compact, convex set $\mathcal{C} \subset \boldsymbol{\mathcal{X}}_{\boldsymbol{\theta}} \times [0, \tau]$ that contains $\boldsymbol{\theta}_0$ in its interior in the following sense:

$$\text{There exists } \epsilon > 0 \text{ such that } \{(\boldsymbol{x}^{\boldsymbol{\theta}}, t) \in \boldsymbol{\mathcal{X}}_{\boldsymbol{\theta}} \times [0, \tau] : \|\boldsymbol{x}^{\boldsymbol{\theta}} - \boldsymbol{\theta}_0(t)\| < \epsilon\} \subset \mathcal{C}. \tag{5.4}$$

For an illustration, see Figure 5.1. Without loss of generality, we will assume that $\mathcal{C}$ is the same for all conditions in which it appears. Define the section $\mathcal{C}_t$ of $\mathcal{C}$ at $t \in [0, \tau]$ by $\mathcal{C}_t := \{\boldsymbol{x}^{\boldsymbol{\theta}} \in \boldsymbol{\mathcal{X}}_{\boldsymbol{\theta}} : (\boldsymbol{x}^{\boldsymbol{\theta}}, t) \in \mathcal{C}\}$. Often, we have to deal with

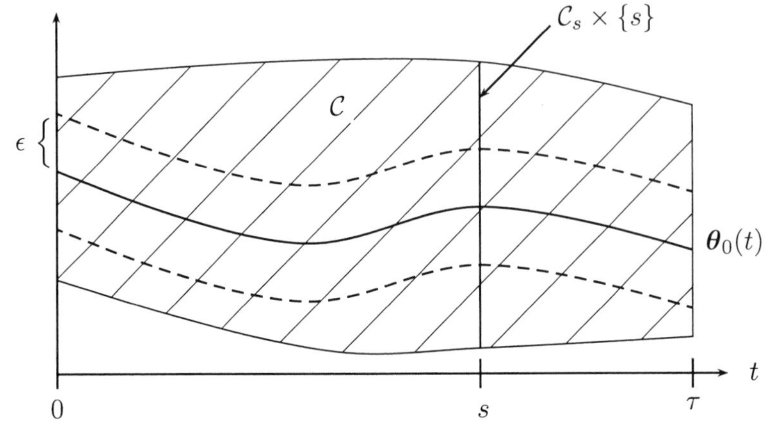

Figure 5.1: Sketch of (5.4)

functions $f : \mathcal{C} \to \mathbb{R}$. We say that f is equicontinuous at $\boldsymbol{\theta}_0$ if for each $\epsilon > 0$, there exists $\delta > 0$ such that for all $(\boldsymbol{x}^{\boldsymbol{\theta}}, t) \in \mathcal{C}$,

$$\|\boldsymbol{x}^{\boldsymbol{\theta}} - \boldsymbol{\theta}_0(t)\| < \epsilon \quad \text{implies} \quad |f(\boldsymbol{x}^{\boldsymbol{\theta}}, t) - f(\boldsymbol{\theta}_0(t), t)| \leq \delta.$$

We impose the following conditions on the weight process $\boldsymbol{c}$:

(N3) (differentiability of $\boldsymbol{c}$)
There are events A_n with $\mathrm{P}(A_n) \to 1$, such that on A_n, for all $i = 1, \dots, n, t \in [0, \tau]$, the mapping $c_i(\cdot, t) : \mathcal{C}_t \to \mathbb{R}$ is twice differentiable.

Outside of A_n, we set $\nabla_{\boldsymbol{\theta}} c_i(\boldsymbol{x}^{\boldsymbol{\theta}}, t) = \mathbf{0}$ and $\nabla_{\boldsymbol{\theta}}^\top \nabla_{\boldsymbol{\theta}} c_i(\boldsymbol{x}^{\boldsymbol{\theta}}, t) = \mathbf{0}$.

(N4) (predictability of $\boldsymbol{c}$ and $\nabla_{\boldsymbol{\theta}} \boldsymbol{c}$)
For all $\boldsymbol{\theta} \in \mathrm{bm}(\boldsymbol{\mathcal{X}}_{\boldsymbol{\theta}})$, such that $\{(\boldsymbol{\theta}(t), t) : t \in [0, \tau]\} \subset \mathcal{C}$, the process $\boldsymbol{c}(\boldsymbol{\theta}(\cdot), \cdot)$ is locally bounded and predictable. For $i = 1, \dots, n$, $l = 1, \dots, k_{\boldsymbol{\theta}}$, there exists a locally bounded and predictable process $g_{il}(\cdot)$ such that on the event A_n given in (N3), $g_{il}(\cdot) = \nabla_l c_i(\boldsymbol{\theta}_0(\cdot), \cdot)$.

(N5) (growth conditions) With ν from (N1) and A_n from (N3),

$$n^{-1/8-\nu/4} \sup_{\substack{i=1\dots n \\ (\boldsymbol{x}^{\boldsymbol{\theta}}, t) \in \mathcal{C}}} |c_i(\boldsymbol{x}^{\boldsymbol{\theta}}, t)| I(A_n) \xrightarrow{\mathrm{P}} 0,$$

$$n^{-1/8-\nu/4} \sup_{\substack{i=1\dots n \\ (\boldsymbol{x}^{\boldsymbol{\theta}}, t) \in \mathcal{C}}} \|\nabla_{\boldsymbol{\theta}} c_i(\boldsymbol{x}^{\boldsymbol{\theta}}, t)\| \xrightarrow{\mathrm{P}} 0,$$

$$n^{-\nu} \sup_{\substack{i=1\dots n \\ (\boldsymbol{x}^{\boldsymbol{\theta}}, t)\in\mathcal{C}}} \|\nabla_{\boldsymbol{\theta}}^\top \nabla_{\boldsymbol{\theta}} c_i(\boldsymbol{x}^{\boldsymbol{\theta}}, t)\| \xrightarrow{\mathrm{P}} 0.$$

Note that if $\nu = 1/6$ then all the required rates in (N5) are the same.

5.1 Asymptotics under the Null Hypothesis

In this section, we consider the asymptotic behavior of T under the null hypothesis, meaning that (5.2) holds true with parameter $\boldsymbol{\alpha} = \boldsymbol{\alpha}_0 \in \mathrm{bm}(\boldsymbol{\mathcal{X}}_{\boldsymbol{\alpha}})$. To derive the asymptotic distribution of T we need the following conditions.

(N6) (differentiability of $\boldsymbol{f}$)
For all $t \in [0, \tau]$, $i = 1, \dots, n$, the function $f_i(\cdot, t) : \mathcal{C}_t \to \mathbb{R}$ is twice continuously differentiable.
For all $(\boldsymbol{\alpha}, \boldsymbol{\beta}) \in \mathrm{bm}(\boldsymbol{\mathcal{X}}_{\boldsymbol{\theta}})$, such that $\{(\boldsymbol{\alpha}(t), \boldsymbol{\beta}(t), t) : t \in [0, \tau]\} \subset \mathcal{C}$, the stochastic processes $(\nabla_{\boldsymbol{\alpha}} f_i)(\boldsymbol{\alpha}(\cdot), \cdot)$ and $(\nabla_{\boldsymbol{\alpha}}^\top \nabla_{\boldsymbol{\alpha}} f_i)(\boldsymbol{\alpha}(\cdot), \cdot)$ are progressively measurable. The following stochastic processes are locally bounded: $\sup |f_i(\boldsymbol{x}^{\boldsymbol{\alpha}}, t)|$, $\sup \|(\nabla_{\boldsymbol{\alpha}} f_i)(\boldsymbol{x}^{\boldsymbol{\alpha}}, t)\|$, and $\sup \|(\nabla_{\boldsymbol{\alpha}}^\top \nabla_{\boldsymbol{\alpha}} f_i)(\boldsymbol{x}^{\boldsymbol{\alpha}}, t)\|$, where for each t the sup is over $(\boldsymbol{x}^{\boldsymbol{\alpha}}, \boldsymbol{x}^{\boldsymbol{\beta}}) \in \mathcal{C}_t$.

(N7) (asymptotic stability)
Let $\widetilde{\boldsymbol{c}} = \boldsymbol{c} I(A_n)$, where the event A_n is from (N3).
$\overline{\boldsymbol{f}}(\boldsymbol{\alpha}_0(\cdot), \cdot)$ and $\overline{\widetilde{\boldsymbol{c}} \boldsymbol{f} \widetilde{\boldsymbol{c}}}(\boldsymbol{\theta}_0(\cdot), \cdot)$ converge uip on $[0, \tau]$.
$\overline{(\nabla_i \boldsymbol{c}) \boldsymbol{f} (\nabla_j \boldsymbol{c})}(\boldsymbol{\theta}_0(\cdot), \cdot)$ converges uip on $[0, \tau]$ for $i, j = 1, \dots, k_{\boldsymbol{\theta}}$.
$n^{-\nu}\overline{(\nabla_i \boldsymbol{c})(\nabla_j \boldsymbol{f})}(\cdot, \cdot)$, $n^{-\nu}\overline{(\nabla_i \nabla_j \boldsymbol{c}) \boldsymbol{f}}(\cdot, \cdot)$, and $n^{-\nu}\overline{\widetilde{\boldsymbol{c}}(\nabla_i \nabla_j \boldsymbol{f})}(\cdot, \cdot)$ converge to zero in probability uniformly on $\mathcal{C}$ for $i, j = 1, \dots, k_{\boldsymbol{\theta}}$.

Theorem 5.1. *Suppose conditions (N1), (N1'), (N2)-(N7) hold true and that*

$$\lambda_i(\cdot) = f_i(\boldsymbol{\alpha}_0(\cdot), \cdot), i = 1, \dots, n.$$

Then in $D[0, \tau]$,

$$T(\boldsymbol{c}, \widehat{\boldsymbol{\theta}}, \cdot) \xrightarrow{d} X(\cdot),$$

where X is a mean zero Gaussian process with covariance $\mathrm{Cov}(X(s), X(t)) = \sigma^2(s \wedge t)$, *where*

$$\sigma^2(t) = \int_0^t \overrightarrow{\boldsymbol{c} \boldsymbol{f} \boldsymbol{c}}(\boldsymbol{\theta}_0(s), s) \, \mathrm{d}s.$$

Furthermore, uniformly in $t \in [0, \tau]$,

$$\hat{\sigma}^2(\boldsymbol{c}, t) := \frac{1}{n} \int_0^t \boldsymbol{c}(\widehat{\boldsymbol{\theta}}(s), s)^\top \mathrm{diag}(\mathrm{d}\boldsymbol{N}(s)) \boldsymbol{c}(\widehat{\boldsymbol{\theta}}(s), s) \xrightarrow{\mathrm{P}} \sigma^2(t).$$

Proof. We start by showing

$$Y^{(n)}(\cdot) := n^{-\frac{1}{2}} \int_0^{\cdot} \boldsymbol{c}(\boldsymbol{\theta}_0(s), s)^\top \mathrm{d}\boldsymbol{M}(s) \xrightarrow{\mathrm{d}} X(\cdot),$$

where, as defined in Section 2.2, $\boldsymbol{M}(t) = \boldsymbol{N}(t) - \int_0^t \boldsymbol{\lambda}(s)\,\mathrm{d}s$. By (N4), the process $\boldsymbol{c}(\boldsymbol{\theta}_0(\cdot), \cdot)$ is predictable and locally bounded, and hence, $Y^{(n)}$ is a locally square integrable martingale whose predictable covariation process

$$\left\langle Y^{(n)} \right\rangle (t) = \int_0^t \overline{\boldsymbol{c}\boldsymbol{f}\boldsymbol{c}}(\boldsymbol{\theta}_0(s), s)\,\mathrm{d}s$$

converges to $\sigma^2(t)$ by (N7). Since the jumps of $Y^{(n)}$ are asymptotically negligible by (N5), Rebolledo's central limit theorem (Theorem A.3) shows that $Y^{(n)} \xrightarrow{\mathrm{d}} X$ and

$$\frac{1}{n} \int_0^t \boldsymbol{c}(\boldsymbol{\theta}_0(s), s)^\top \operatorname{diag}(\mathrm{d}\boldsymbol{N}(s)) \boldsymbol{c}(\boldsymbol{\theta}_0(s), s) \xrightarrow{\mathrm{P}} \sigma^2(t)$$

uniformly in $t \in [0, \tau]$. Next, we show that uniformly in $t \in [0, \tau]$,

$$T(\boldsymbol{c}, \widehat{\boldsymbol{\theta}}, t) - Y^{(n)}(t) \xrightarrow{\mathrm{P}} 0.$$

For $a \in [0, 1]$, let $U(a, t) := T(\boldsymbol{c}, a\widehat{\boldsymbol{\theta}} + (1-a)\boldsymbol{\theta}_0, t)$, and hence, $U(1, t) = T(\boldsymbol{c}, \widehat{\boldsymbol{\theta}}, t)$ and $U(0, t) = Y^{(n)}(t)$. Let

$$\begin{aligned} B_n := & A_n \cap \{\widehat{\boldsymbol{\theta}}(t) \in \mathcal{C}_t \,\forall t \in [0, \tau]\} \\ & \cap \{ \sup_{\substack{i=1\dots n \\ (\boldsymbol{x}^{\boldsymbol{\theta}}, t) \in \mathcal{C}}} \max\{|c_i(\boldsymbol{x}^{\boldsymbol{\theta}}, t)|, \|\nabla_{\boldsymbol{\theta}} c_i(\boldsymbol{x}^{\boldsymbol{\theta}}, t)\|, \|\nabla_{\boldsymbol{\theta}}^\top \nabla_{\boldsymbol{\theta}} c_i(\boldsymbol{x}^{\boldsymbol{\theta}}, t)\|\} < \infty\}, \end{aligned} \tag{5.5}$$

where A_n is as given in (N3). By (5.4) and (N1),

$$\mathrm{P}((\widehat{\boldsymbol{\theta}}(t), t) \in \mathcal{C} \,\forall t \in [0, \tau]) \geq \mathrm{P}(\sup_{t \in [0,\tau]} \|\widehat{\boldsymbol{\theta}}(t) - \boldsymbol{\theta}_0(t)\| < \epsilon) \to 1.$$

Hence, (N3) and (N5) imply $\mathrm{P}(B_n) \to 1$. On B_n, a Taylor expansion of $U(a, t)$ around $a = 0$ for $a = 1$ yields

$$\begin{aligned} T(\boldsymbol{c}, \widehat{\boldsymbol{\theta}}, t) - Y^{(n)}(t) = & \\ = & \int_0^t (\widehat{\boldsymbol{\theta}}(s) - \boldsymbol{\theta}_0(s))^\top n^{-\frac{1}{2}} \sum_{i=1}^n \nabla_{\boldsymbol{\theta}} c_i(\boldsymbol{\theta}_0(s), s)^\top \mathrm{d}M_i(s) \\ & + n^{-\frac{1}{2}} \int_0^t \left[\sum_{i=1}^n c_i(\boldsymbol{\theta}_0(s), s) \nabla_{\boldsymbol{\alpha}} f_i(\boldsymbol{\alpha}_0(s), s) \right] (\widehat{\boldsymbol{\alpha}}(s) - \boldsymbol{\alpha}_0(s))\,\mathrm{d}s \\ & + \frac{1}{2} \left(\frac{\partial}{\partial a} \right)^2 U(\tilde{a}, t), \end{aligned} \tag{5.6}$$

for some $\tilde{a} \in [0, 1]$ which may depend on t. To justify the interchange of differentiation and integration, one can argue as follows: First note that we work with one fixed $\omega \in B_n$. Since integration with respect to $N_i(s)$ leads to a finite sum, it suffices to show

$$\frac{\partial}{\partial a}\int_0^t c_i(u(a,s),s) f_i(v(a,s),s)\,\mathrm{d}s = \int_0^t \frac{\partial}{\partial a}\Big(c_i(u(a,s),s) f_i(v(a,s),s)\Big)\,\mathrm{d}s, \quad (5.7)$$

where $u(a,s) = a\widehat{\boldsymbol{\theta}}(s) + (1-a)\boldsymbol{\theta}_0(s)$ and $v(a,s) = a\widehat{\boldsymbol{\alpha}}(s) + (1-a)\boldsymbol{\alpha}_0(s)$. The integrand on the right hand is equal to

$$\begin{aligned}(\nabla_{\boldsymbol{\theta}} c_i)(u(a,s),s) f_i(v(a,s),s)(\widehat{\boldsymbol{\theta}}(s) - \boldsymbol{\theta}_0(s))& \\ + c_i(u(a,s),s)(\nabla_{\boldsymbol{\alpha}} f_i)(v(a,s),s)(\widehat{\boldsymbol{\alpha}}(s) - \boldsymbol{\alpha}_0(s)).&\end{aligned}$$

For a fixed realization, $\boldsymbol{f}$ and its derivatives of first order are uniformly bounded by (N6). Since $\omega \in B_n$, the weights $\boldsymbol{c}$ and its derivatives of first order as well as the estimated parameters $\widehat{\boldsymbol{\theta}}$ are uniformly bounded. Hence, the integrand on the right hand side of (5.7) is uniformly bounded and thus we may use the differentiation lemma, see e.g. Bauer (1992, p. 102), to do the interchange of differentiation and integration needed to get (5.6). From now on, the order of differentiation and integration will be switched without further comment, as it can be justified similarly.

Next, we show that the terms on the right hand side of (5.6) vanish asymptotically. By (N4), on B_n,

$$n^{-\frac{1}{2}}\int_0^t \sum_{i=1}^n \nabla_{\boldsymbol{\theta}} c_i(\boldsymbol{\theta}_0(s),s)^\top \mathrm{d}M_i(s) = \left(n^{-\frac{1}{2}}\int_0^t \sum_{i=1}^n g_{il}(s)\,\mathrm{d}M_i(s)\right)_{l=1,\dots,k_{\boldsymbol{\theta}}}, \quad (5.8)$$

which is a locally square integrable martingale, whose predictable variation converges by (N7). Indeed,

$$\left\langle\left(n^{-\frac{1}{2}}\int_0^\cdot \sum_{i=1}^n g_{il}(s)\,\mathrm{d}M_i(s)\right)_{l=1,\dots,k_{\boldsymbol{\theta}}}\right\rangle(t) = \int_0^t \overline{\boldsymbol{gfg}}(\boldsymbol{\alpha}_0(s),s)\,\mathrm{d}s.$$

On A_n, the right hand side is equal to $\int_0^t \overline{(\nabla_{\boldsymbol{\theta}}\boldsymbol{c})\boldsymbol{f}(\nabla_{\boldsymbol{\theta}}\boldsymbol{c})}(\boldsymbol{\theta}_0(s),s)\,\mathrm{d}s$ which converges uniformly. Hence, by Lenglart's inequality (Lemma A.2), (5.8) is stochastically bounded, uniformly in t. By (N1) and (N1'), we may use Lemma 4.5 to get that the first term on the right hand side of (5.6) converges to zero uniformly in $t \in [0,\tau]$. The second term on the right hand side of (5.6) is identically 0 by (N2). We may write $\left(\frac{\partial}{\partial a}\right)^2 U(\tilde{a},t)$ as

$$\left(\frac{\partial}{\partial a}\right)^2 U(\tilde{a},t) = A(t) - B(t) + 2C(t) + D(t), \quad (5.9)$$

where

$$A(t)=n^{-\frac{1}{2}}\sum_{i=1}^{n}\int_0^t(\widehat{\boldsymbol{\theta}}(s)-\boldsymbol{\theta}_0(s))^\top(\nabla_{\boldsymbol{\theta}}^\top\nabla_{\boldsymbol{\theta}}c_i)(\widetilde{\boldsymbol{\theta}}(s),s)(\widehat{\boldsymbol{\theta}}(s)-\boldsymbol{\theta}_0(s))\,\mathrm{d}N_i(s),$$

$$B(t)=n^{-\frac{1}{2}}\sum_{i=1}^{n}\int_0^t(\widehat{\boldsymbol{\theta}}(s)-\boldsymbol{\theta}_0(s))^\top(\nabla_{\boldsymbol{\theta}}^\top\nabla_{\boldsymbol{\theta}}c_i)(\widetilde{\boldsymbol{\theta}}(s),s)(\widehat{\boldsymbol{\theta}}(s)-\boldsymbol{\theta}_0(s))f_i(\widetilde{\boldsymbol{\alpha}}(s),s)\,\mathrm{d}s,$$

$$C(t)=n^{-\frac{1}{2}}\sum_{i=1}^{n}\int_0^t(\widehat{\boldsymbol{\theta}}(s)-\boldsymbol{\theta}_0(s))^\top\Big(\nabla_{\boldsymbol{\theta}}c_i(\widetilde{\boldsymbol{\theta}}(s),s)\Big)^\top(\nabla_{\boldsymbol{\alpha}}f_i)(\widetilde{\boldsymbol{\alpha}}(s),s)(\widehat{\boldsymbol{\alpha}}(s)-\boldsymbol{\alpha}_0(s))\,\mathrm{d}s,$$

$$D(t)=n^{-\frac{1}{2}}\sum_{i=1}^{n}\int_0^t c_i(\widetilde{\boldsymbol{\theta}}(s),s)(\widehat{\boldsymbol{\alpha}}(s)-\boldsymbol{\alpha}_0(s))^\top\Big(\nabla_{\boldsymbol{\alpha}}^\top\nabla_{\boldsymbol{\alpha}}f_i\Big)(\widetilde{\boldsymbol{\alpha}}(s),s)(\widehat{\boldsymbol{\alpha}}(s)-\boldsymbol{\alpha}_0(s))\,\mathrm{d}s,$$

and $\widetilde{\boldsymbol{\theta}}(t):=\tilde{a}\widehat{\boldsymbol{\theta}}(t)+(1-\tilde{a})\boldsymbol{\theta}_0(t)$. We shall show that $A(t)$, $B(t)$, $C(t)$, and $D(t)$ converge to 0 uniformly in $t\in[0,\tau]$. We start with $A(t)$. Lenglart's inequality and (N7) imply $\overline{\boldsymbol{N}}(\tau)\xrightarrow{\mathrm{P}}\int_0^\tau\overrightarrow{\boldsymbol{f}}(\boldsymbol{\alpha}_0(s),s)\,\mathrm{d}s$. Hence, by (N1) and (N5),

$$\begin{aligned}|A(t)|\leq&n^{-\frac{1}{2}}\sum_{i=1}^{n}\int_0^\tau\left\|(\nabla_{\boldsymbol{\theta}}^\top\nabla_{\boldsymbol{\theta}}c_i)(\widetilde{\boldsymbol{\theta}}(s),s)\right\|\,\mathrm{d}N_i(s)(O_{\mathrm{P}}(n^{-\frac{1}{4}-\frac{\nu}{2}}))^2\\ \leq&n^{\frac{1}{2}}\sup_{\substack{i=1\dots n\\(\boldsymbol{x}^{\boldsymbol{\theta}},s)\in\mathcal{C}}}\|(\nabla_{\boldsymbol{\theta}}^\top\nabla_{\boldsymbol{\theta}}c_i)(\boldsymbol{x}^{\boldsymbol{\theta}},s)\|\overline{\boldsymbol{N}}(\tau)O_{\mathrm{P}}(n^{-\frac{1}{2}-\nu})\\ =&n^{-\nu}\sup_{\substack{i=1\dots n\\(\boldsymbol{x}^{\boldsymbol{\theta}},s)\in\mathcal{C}}}\|(\nabla_{\boldsymbol{\theta}}^\top\nabla_{\boldsymbol{\theta}}c_i)(\boldsymbol{x}^{\boldsymbol{\theta}},s)\|\overline{\boldsymbol{N}}(\tau)O_{\mathrm{P}}(1)\xrightarrow{\mathrm{P}}0\end{aligned}$$

uniformly in $t\in[0,\tau]$. Since

$$\begin{aligned}|B(t)|\leq&O_{\mathrm{P}}(n^{-\frac{1}{2}-\nu})n^{-\frac{1}{2}}\tau\sup_{\substack{i=1\dots n\\(\boldsymbol{x}^{\boldsymbol{\alpha}},\boldsymbol{x}^{\boldsymbol{\beta}},s)\in\mathcal{C}}}\left\|\sum_{i=1}^{n}\Big(\nabla_{\boldsymbol{\theta}}^\top\nabla_{\boldsymbol{\theta}}c_i\Big)(\boldsymbol{x}^{\boldsymbol{\alpha}},\boldsymbol{x}^{\boldsymbol{\beta}},s)f_i(\boldsymbol{x}^{\boldsymbol{\alpha}},s)\right\|\\ =&O_{\mathrm{P}}(n^{-\nu})\tau\sup_{\substack{i=1\dots n\\(\boldsymbol{x}^{\boldsymbol{\alpha}},\boldsymbol{x}^{\boldsymbol{\beta}},s)\in\mathcal{C}}}\left\|\frac{1}{n}\sum_{i=1}^{n}\Big(\nabla_{\boldsymbol{\theta}}^\top\nabla_{\boldsymbol{\theta}}c_i\Big)(\boldsymbol{x}^{\boldsymbol{\alpha}},\boldsymbol{x}^{\boldsymbol{\beta}},s)f_i(\boldsymbol{x}^{\boldsymbol{\alpha}},s)\right\|,\end{aligned}$$

(N7) implies that $B(t)$ converges to 0 uniformly in $t\in[0,\tau]$. The arguments for showing that $C(t)$ and $D(t)$ vanish asymptotically are similar.

It remains to show the convergence of $\hat{\sigma}^2(\boldsymbol{c},t)$. On B_n defined in (5.5), a Taylor expansion of $\hat{\sigma}^2(\boldsymbol{c},t)$ yields

$$\begin{aligned}&\hat{\sigma}^2(\boldsymbol{c},t)-\frac{1}{n}\sum_{i=1}^{n}\int_0^t c_i(\boldsymbol{\theta}_0(s),s)^2\,\mathrm{d}N_i(s)=\\ &\qquad=\frac{1}{n}\sum_{i=1}^{n}2\int_0^t c_i(\widetilde{\boldsymbol{\theta}}(s),s)(\nabla_{\boldsymbol{\theta}}c_i)(\widetilde{\boldsymbol{\theta}}(s),s)(\widehat{\boldsymbol{\theta}}(s)-\boldsymbol{\theta}_0(s))\,\mathrm{d}N_i(s)\end{aligned}$$

for some $\widetilde{\boldsymbol{\theta}}$ between $\widehat{\boldsymbol{\theta}}$ and $\boldsymbol{\theta}_0$. By (N1), (N5), and (N7),

$$\begin{aligned}
&\left|\frac{1}{n}\sum_{i=1}^{n} 2\int_0^t c_i(\widetilde{\boldsymbol{\theta}}(s),s)(\nabla_{\boldsymbol{\theta}} c_i)(\widetilde{\boldsymbol{\theta}}(s),s)(\widehat{\boldsymbol{\theta}}(s)-\boldsymbol{\theta}_0(s))\,\mathrm{d}N_i(s)\right| \\
&\quad\leq O_{\mathrm{P}}(n^{-\frac{1}{4}-\frac{\nu}{2}})\left(\sup_{\substack{i=1\ldots n\\(\boldsymbol{x}^{\boldsymbol{\theta}},s)\in\mathcal{C}}}|c_i(\boldsymbol{x}^{\boldsymbol{\theta}},s)|\right)\left(\sup_{\substack{i=1\ldots n\\(\boldsymbol{x}^{\boldsymbol{\theta}},s)\in\mathcal{C}}}\|\nabla_{\boldsymbol{\theta}} c_i(\boldsymbol{x}^{\boldsymbol{\theta}},s)\|\right)\overline{\boldsymbol{N}}(\tau) \\
&\quad\leq O_{\mathrm{P}}(1)\left(n^{-\frac{1}{8}-\frac{\nu}{4}}\sup_{\substack{i=1\ldots n\\(\boldsymbol{x}^{\boldsymbol{\theta}},s)\in\mathcal{C}}}|c_i(\boldsymbol{x}^{\boldsymbol{\theta}},s)|\right)\left(n^{-\frac{1}{8}-\frac{\nu}{4}}\sup_{\substack{i=1\ldots n\\(\boldsymbol{x}^{\boldsymbol{\theta}},s)\in\mathcal{C}}}\|\nabla_{\boldsymbol{\theta}} c_i(\boldsymbol{x}^{\boldsymbol{\theta}},s)\|\right)\overline{\boldsymbol{N}}(\tau) \\
&\quad\xrightarrow{\mathrm{P}} 0.
\end{aligned}$$

□

Remark 5.1. Recall that in the Aalen model the intensity has the following form:

$$\lambda_i(t) = f_i(\boldsymbol{\alpha}(t),t) = \sum_{j=1}^{k_{\boldsymbol{\alpha}}} Y_{ij}(t)\alpha_j(t).$$

Then $\nabla_j f_i(\boldsymbol{x}^{\boldsymbol{\alpha}}, t) = Y_{ij}(t)$ and hence, if (N2) holds true,

$$T(\boldsymbol{c},\boldsymbol{\theta},t) = n^{-\frac{1}{2}}\sum_{i=1}^{n}\int_0^t c_i(\boldsymbol{\theta}(s),s)\,\mathrm{d}N_i(s).$$

Furthermore, (N2) does not depend on $\boldsymbol{\alpha}$ and thus we may choose weights that do not depend on $\boldsymbol{\alpha}$, making the estimation of $\boldsymbol{\alpha}$ no longer necessary.

This holds more generally for one or more parameters that act affine linearly on the intensity. Consider the model

$$\lambda_i(t) = f_i(\boldsymbol{\alpha}(t),t) = \eta_i(\boldsymbol{\alpha}_{(2)}(t),t)\alpha_1(t) + \chi_i(\boldsymbol{\alpha}_{(3)}(t),t),$$

where $\boldsymbol{\alpha}(t) = (\alpha_1(t), \boldsymbol{\alpha}_{(2)}(t)^\top, \boldsymbol{\alpha}_{(3)}(t)^\top)^\top$. Then $\nabla_1 f_i(\boldsymbol{x}^{\boldsymbol{\alpha}}, t) = \eta_i(\boldsymbol{x}^{\boldsymbol{\alpha}}_{(2)}(t),t)$, where $\boldsymbol{x}^{\boldsymbol{\alpha}} = (x^{\boldsymbol{\alpha}}_1, {\boldsymbol{x}^{\boldsymbol{\alpha}}_{(2)}}^\top, {\boldsymbol{x}^{\boldsymbol{\alpha}}_{(3)}}^\top)^\top$ and the orthogonality conditions (N2) and (N2') do not depend on $x^{\boldsymbol{\alpha}}_1$. Furthermore, if (N2') is satisfied then

$$T(\boldsymbol{c},\boldsymbol{\alpha},\boldsymbol{\beta},t) = n^{-\frac{1}{2}}\sum_{i=1}^{n}\int_0^t c_i(\boldsymbol{\alpha}(s),\boldsymbol{\beta}(s),s)\left(\mathrm{d}N_i(s) - \chi_i(\boldsymbol{\alpha}_{(3)}(s),s)\,\mathrm{d}s\right).$$

Thus if one chooses weights that do not depend on α_1, no estimate of α_1 is needed.

Remark 5.2. In our applications, $\widehat{\boldsymbol{\theta}}(t)$ is usually not a predictable process. But if it is predictable, then we may drop the condition (N1') in Theorem 5.1. Indeed, we only need (N1') to show that the first term on the right hand side of (5.6) vanishes asymptotically. If $\widehat{\boldsymbol{\theta}}$ is predictable then we can argue as follows:

$$\left\langle \int_0^{\cdot} (\widehat{\boldsymbol{\theta}}(s) - \boldsymbol{\theta}_0(s))^\top n^{-\frac{1}{2}} \sum_{i=1}^n \nabla_{\boldsymbol{\theta}} c_i(\boldsymbol{\theta}_0(s), s)^\top \mathrm{d}M_i(s) \right\rangle (t) =$$

$$= \int_0^t (\widehat{\boldsymbol{\theta}}(s) - \boldsymbol{\theta}_0(s))^\top \overline{(\nabla_{\boldsymbol{\theta}} \boldsymbol{c}) \boldsymbol{f} (\nabla_{\boldsymbol{\theta}} \boldsymbol{c})}(\boldsymbol{\theta}_0(s), s)(\widehat{\boldsymbol{\theta}}(s) - \boldsymbol{\theta}_0(s))\,\mathrm{d}s$$

which converges uniformly to 0. Hence, by Lenglart's inequality (Lemma A.2),

$$\int_0^{\cdot} (\widehat{\boldsymbol{\theta}}(s) - \boldsymbol{\theta}_0(s))^\top n^{-\frac{1}{2}} \sum_{i=1}^n \nabla_{\boldsymbol{\theta}} c_i(\boldsymbol{\theta}_0(s), s)^\top \mathrm{d}M_i(s)$$

converges to 0 uniformly in probability.

5.2 Asymptotics under Fixed Alternatives

In this section, we consider the behavior of the test statistic if the null hypothesis (5.2) does not hold true. We consider the fixed alternative $\lambda_i = h_i$, where h_i is some predictable locally bounded process. For a result about the asymptotic behavior of T we need the following conditions.

(F1) (stability condition under fixed alternatives I)
$\overline{\boldsymbol{h}}$ and $\overline{\boldsymbol{chc}}(\boldsymbol{\theta}_0(\cdot), \cdot)$ converge uip on $[0, \tau]$.

(F2) (stability condition under fixed alternatives II)
$\overline{\boldsymbol{ch}}$ and $\overline{\boldsymbol{cf}}$ converge uip on $\mathcal{C}$. The mappings $\overrightarrow{\boldsymbol{ch}} : \mathcal{C} \to \mathbb{R}$ and $\overrightarrow{\boldsymbol{cf}} : \mathcal{C} \to \mathbb{R}$ are equicontinuous at $\boldsymbol{\theta}_0$.

Theorem 5.2. *Suppose conditions (N1), (N1'), (N3)-(N5), (F1), and (F2) hold true. If*

$$\lambda_i(t) = h_i(t), i = 1, \ldots, n,$$

then uniformly in $t \in [0, \tau]$,

$$n^{-\frac{1}{2}} T(\boldsymbol{c}, \widehat{\boldsymbol{\theta}}, t) \xrightarrow{\mathrm{P}} \int_0^t \left(\overrightarrow{\boldsymbol{ch}}(\boldsymbol{\theta}_0(s), s) - \overrightarrow{\boldsymbol{cf}}(\boldsymbol{\theta}_0(s), s) \right) \mathrm{d}s$$

and

$$\hat{\sigma}^2(\boldsymbol{c}, t) \xrightarrow{\mathrm{P}} \int_0^t \overrightarrow{\boldsymbol{chc}}(\boldsymbol{\theta}_0(s), s)\,\mathrm{d}s.$$

Proof. Adding and subtracting $\int_0^t \overline{\boldsymbol{ch}}(\widehat{\boldsymbol{\theta}}(s), s)\,\mathrm{d}s$ we get the decomposition

$$\begin{aligned} n^{-\frac{1}{2}} T(\boldsymbol{c}, \widehat{\boldsymbol{\theta}}, t) = &\int_0^t \left(\overline{\boldsymbol{ch}}(\widehat{\boldsymbol{\theta}}(s), s) - \overline{\boldsymbol{cf}}(\widehat{\boldsymbol{\theta}}(s), s)\right)\,\mathrm{d}s \\ &+ \frac{1}{n}\int_0^t \boldsymbol{c}(\widehat{\boldsymbol{\theta}}(s), s)^\top (\mathrm{d}\boldsymbol{N}(s) - \boldsymbol{h}(s)\,\mathrm{d}s)\,. \end{aligned} \tag{5.10}$$

By (F2), we can use Lemma A.5 to show that the first term on the right hand side converges uniformly to

$$\int_0^t \left(\overrightarrow{\boldsymbol{ch}}(\boldsymbol{\theta}_0(s), s) - \overrightarrow{\boldsymbol{cf}}(\boldsymbol{\theta}_0(s), s)\right)\,\mathrm{d}s.$$

Let $B_n := A_n \cap \{\widehat{\boldsymbol{\theta}}(t) \in \mathcal{C}_t \,\forall t \in [0, \tau]\}$, where A_n is from (N3). On B_n, by a Taylor expansion, the second term on the right hand side of (5.10) is equal to

$$\frac{1}{n}\int_0^t \boldsymbol{c}(\boldsymbol{\theta}_0(s), s)^\top \mathrm{d}\boldsymbol{M}(s) + \frac{1}{n}\int_0^t (\widehat{\boldsymbol{\theta}}(s) - \boldsymbol{\theta}_0(s))^\top (\nabla_{\boldsymbol{\theta}} \boldsymbol{c})(\widetilde{\boldsymbol{\theta}}(s), s)^\top \mathrm{d}\boldsymbol{M}(s), \tag{5.11}$$

where $\widetilde{\boldsymbol{\theta}}$ is between $\widehat{\boldsymbol{\theta}}$ and $\boldsymbol{\theta}_0$. The first term vanishes asymptotically by Lenglart's inequality. The absolute value of the second term can be bounded by

$$a_n \left(a_n \sup_{\substack{i=1\dots n \\ (\boldsymbol{x}^{\boldsymbol{\theta}}, s) \in \mathcal{C}}} \|\nabla_{\boldsymbol{\theta}} c_i(\boldsymbol{x}^{\boldsymbol{\theta}}, s)\| \right) \left(a_n^{-2} \sup_{s \in [0,\tau]} \|\widehat{\boldsymbol{\theta}}(s) - \boldsymbol{\theta}_0(s)\| \right) \left(\overline{\boldsymbol{N}}(\tau) + \int_0^\tau \overline{\boldsymbol{h}}(s)\,\mathrm{d}s \right),$$

where $a_n := n^{-\frac{1}{8} - \frac{\nu}{4}}$. Since by (F1) and Lenglart's inequality,

$$\overline{\boldsymbol{N}}(\tau) + \int_0^\tau \overline{\boldsymbol{h}}(s)\,\mathrm{d}s \xrightarrow{\mathrm{P}} 2\int_0^\tau \overrightarrow{\boldsymbol{h}}(s)\,\mathrm{d}s,$$

conditions (N5) and (N1) imply that the second term of (5.11) converges uniformly to 0.

As in the proof of Theorem 5.1, a consequence of Rebolledo's theorem (Theorem A.3), for which we need (N5) and (F2), applied to $n^{-1/2}\int_0^t \boldsymbol{c}(\boldsymbol{\theta}_0(s), s)^\top \mathrm{d}\boldsymbol{M}(s)$ is that uniformly in $t \in [0, \tau]$,

$$\frac{1}{n}\int_0^t \boldsymbol{c}(\boldsymbol{\theta}_0(s), s)^\top \operatorname{diag}(\mathrm{d}\boldsymbol{N}(s)) \boldsymbol{c}(\boldsymbol{\theta}_0(s), s) \xrightarrow{\mathrm{P}} \int_0^t \overrightarrow{\boldsymbol{chc}}(\boldsymbol{\theta}_0(s), s)\,\mathrm{d}s.$$

Similar to the arguments used in the proof of Theorem 5.1 it can be shown that

$$\hat{\sigma}^2(\boldsymbol{c}, t) - \frac{1}{n}\int_0^t \boldsymbol{c}(\boldsymbol{\theta}_0(s), s)^\top \operatorname{diag}(\mathrm{d}\boldsymbol{N}(s)) \boldsymbol{c}(\boldsymbol{\theta}_0(s), s) \xrightarrow{\mathrm{P}} 0,$$

uniformly in $t \in [0, \tau]$. Hence, $\hat{\sigma}^2(\boldsymbol{c}, t) \xrightarrow{\mathrm{P}} \int_0^t \overrightarrow{\boldsymbol{chc}}(\boldsymbol{\theta}_0(s), s)\,\mathrm{d}s$ uniformly in $t \in [0, \tau]$. □

5.3 Asymptotics under Local Alternatives

Consider the following sequence of local alternatives to the null hypothesis (5.2):

$$\lambda_i(t) = f_i(\boldsymbol{\alpha}_0(t), t) + n^{-\frac{1}{2}} g_i(t), \tag{5.12}$$

where g_i are locally bounded, predictable stochastic processes. We derive the asymptotic distribution of T if (5.12) holds true. For this, the following conditions are needed:

(L1) (stability conditions under local alternatives I)
$\overline{\boldsymbol{cg}}$ converges uip on $\mathcal{C}$. $\overrightarrow{\boldsymbol{cg}} : \mathcal{C} \to \mathbb{R}$ is equicontinuous at $\boldsymbol{\theta}_0$.

(L2) (stability conditions under local alternatives II)
$\overline{\boldsymbol{cgc}}(\boldsymbol{\theta}_0(\cdot), \cdot)$ converges uip on $[0, \tau]$.

Theorem 5.3. *Suppose conditions (N1), (N1'), (N2)-(N7), (L1), and (L2) hold true. If*

$$\lambda_i(\cdot) = f_i(\boldsymbol{\alpha}_0(\cdot), \cdot) + n^{-\frac{1}{2}} g_i(\cdot), i = 1, \dots n,$$

then in $D[0, \tau]$,

$$T(\boldsymbol{c}, \widehat{\boldsymbol{\theta}}, \cdot) \xrightarrow{d} X(\cdot),$$

where X is a Gaussian process with mean $\mathrm{E}[X(t)] = \mu(t) := \int_0^t \overrightarrow{\boldsymbol{cg}}(\boldsymbol{\theta}_0(s), s)\, \mathrm{d}s$ *and covariance* $\mathrm{Cov}(X(s), X(t)) = \sigma^2(s \wedge t)$, *where*

$$\sigma^2(t) = \int_0^t \overrightarrow{\boldsymbol{cfc}}(\boldsymbol{\theta}_0(s), s)\, \mathrm{d}s.$$

Furthermore, $\hat{\sigma}^2(\boldsymbol{c}, t) = \frac{1}{n} \int_0^t \boldsymbol{c}(\widehat{\boldsymbol{\theta}}(s), s)^\top \mathrm{diag}(\mathrm{d}\boldsymbol{N}(s)) \boldsymbol{c}(\widehat{\boldsymbol{\theta}}(s), s) \xrightarrow{\mathrm{P}} \sigma^2(t)$ *uniformly in $t \in [0, \tau]$.*

Proof. As in Theorem 5.1, we consider

$$Y^{(n)}(t) := n^{-\frac{1}{2}} \int_0^t \boldsymbol{c}(\boldsymbol{\theta}_0(s), s)^\top \mathrm{d}\boldsymbol{M}(s).$$

By (N7) and (L2),

$$\left\langle Y^{(n)} \right\rangle(t) = \int_0^t \overline{\boldsymbol{cfc}}(\boldsymbol{\theta}_0(s), s)\, \mathrm{d}s + \int_0^t n^{-\frac{1}{2}} \overline{\boldsymbol{cgc}}(\boldsymbol{\theta}_0(s), s)\, \mathrm{d}s \xrightarrow{\mathrm{P}} \sigma^2(t).$$

Hence, we can use Rebolledo's theorem (Theorem A.3) to show that $Y^{(n)}$ converges in distribution to a mean zero Gaussian process Z with $\mathrm{Cov}(Z(s), Z(t)) =$

$\sigma^2(s \wedge t)$. We may use the same Taylor expansion as in Theorem 5.1 to show that $Y^{(n)}$ and

$$T(\boldsymbol{c}, \widehat{\boldsymbol{\theta}}, t) - \int_0^t \overline{\boldsymbol{c}\boldsymbol{g}}(\widehat{\boldsymbol{\theta}}(s), s)\,\mathrm{d}s =$$
$$= n^{-\frac{1}{2}} \int_0^\tau \boldsymbol{c}(\widehat{\boldsymbol{\theta}}(s), s)^\top \left(\mathrm{d}\boldsymbol{N}(s) - \boldsymbol{f}(\boldsymbol{\alpha}_0(s), s)\,\mathrm{d}s - n^{-\frac{1}{2}}\boldsymbol{g}(s)\,\mathrm{d}s\right)$$

are asymptotically equivalent. By (L1), we can use Lemma A.5 to show that

$$\int_0^t \overline{\boldsymbol{c}\boldsymbol{g}}(\widehat{\boldsymbol{\theta}}(s), s)\,\mathrm{d}s \xrightarrow{\mathrm{P}} \mu(t)$$

uniformly in $t \in [0, \tau]$. Hence, by a Slutsky argument, $T(\boldsymbol{c}, \widehat{\boldsymbol{\theta}}, \cdot) \xrightarrow{\mathrm{d}} X(\cdot)$.

□

5.4 Least Squares Weights

In this section, we shall be concerned with a simple method to guarantee the orthogonality condition (N2). Actually, we ensure the stronger condition (N2'). The idea is the same as in Chapter 3: take some n-variate vector of stochastic processes $\boldsymbol{d}(\boldsymbol{x}^{\boldsymbol{\theta}}, t)$ depending on $\boldsymbol{x}^{\boldsymbol{\theta}} \in \boldsymbol{\mathcal{X}}_{\boldsymbol{\theta}}$, $t \in [0, \tau]$, and define $\boldsymbol{c}(\boldsymbol{x}^{\boldsymbol{\theta}}, t)$ as the projection in $\mathbb{R}^n$ of $\boldsymbol{d}(\boldsymbol{x}^{\boldsymbol{\theta}}, t)$ onto the space orthogonal to the columns of

$$\boldsymbol{B}(\boldsymbol{x}^{\boldsymbol{\theta}}, t) := (\nabla_{\boldsymbol{\alpha}} \boldsymbol{f})(\boldsymbol{x}^{\boldsymbol{\alpha}}, t),$$

where $\boldsymbol{x}^{\boldsymbol{\theta}} = (\boldsymbol{x}^{\boldsymbol{\alpha}\top}, \boldsymbol{x}^{\boldsymbol{\beta}\top})^\top$. Then (N2') is satisfied. In other words, if $\boldsymbol{d}(\boldsymbol{x}^{\boldsymbol{\theta}}, \cdot)$ is any n-dimensional vector of stochastic processes then

$$\boldsymbol{c}(\boldsymbol{x}^{\boldsymbol{\theta}}, \cdot) = \boldsymbol{Q}_{\mathbb{P}_n}^{\boldsymbol{B}(\boldsymbol{x}^{\boldsymbol{\theta}}, \cdot)} \boldsymbol{d}(\boldsymbol{x}^{\boldsymbol{\theta}}, \cdot) \tag{5.13}$$

satisfies (N2') because of properties of projections, see (3.3).

It remains to consider the other conditions needed for the asymptotic results under the null hypothesis (Theorem 5.1), under fixed alternatives (Theorem 5.2), and under local alternatives (Theorem 5.3). In the remainder of this section, we give a relatively simple i.i.d. setup, where the other conditions hold true.

(LS1) (several conditions for unweighted projections) There exists a compact, convex set $K \subset \boldsymbol{\mathcal{X}}_{\boldsymbol{\theta}}$ such that $\{\boldsymbol{\theta}_0(t) : t \in [0, \tau]\}$ is in its interior and such that the following holds true:

For all $t \in [0, \tau]$, $i = 1, \ldots, n$, the mapping $d_i(\cdot, t) : K \to \mathbb{R}$ is twice differentiable and the mapping $f_i(\cdot, t) : K \to \mathbb{R}$ is three times differentiable. Furthermore, d_i, $\nabla_j d_i$, $\nabla_j \nabla_\nu d_i$, f_i, $\nabla_j f_i$, $\nabla_j \nabla_\nu f_i$, and $\nabla_j \nabla_\nu \nabla_\mu f_i$,

$i = 1, \ldots, n$, $j, \nu, \mu = 1, \ldots, k_{\boldsymbol{\theta}}$, are càglàd, adapted, locally bounded stochastic processes with values in the space $C(K)$ of continuous mappings from K into $\mathbb{R}$ equipped with the supremum norm.

Remark 5.3. Suppose that $\boldsymbol{X}_i, i = 1, \ldots, n$, are ν-variate vectors of càglàd, adapted, locally bounded stochastic processes, $f : \mathbb{R}^\nu \times K \to \mathbb{R}$ is a function that is three times continuously differentiable with respect to its last components, and $d : \mathbb{R}^\nu \times K \to \mathbb{R}$ is twice continuously differentiable with respect to its last components. Let $f_i(\boldsymbol{x}^{\boldsymbol{\theta}}, t) := f(\boldsymbol{X}_i(t), \boldsymbol{x}^{\boldsymbol{\theta}})$ and $d_i(\boldsymbol{x}^{\boldsymbol{\theta}}, t) := d(\boldsymbol{X}_i(t), \boldsymbol{x}^{\boldsymbol{\theta}})$. Then (LS1) is satisfied by Lemma A.6 and Lemma A.7.

(LS2) (properties of $\overrightarrow{\boldsymbol{BB}}$) $\overrightarrow{\boldsymbol{BB}}(\boldsymbol{x}^{\boldsymbol{\alpha}}; t)$ is invertible for all $\boldsymbol{x}^{\boldsymbol{\alpha}} \in K, t \in [0, \tau]$, and continuous in $(\boldsymbol{x}^{\boldsymbol{\alpha}}; t) \in K \times [0, \tau]$.

The following ensures that with $\boldsymbol{c}$ given in (5.13), we can use the theorems of the previous section.

Theorem 5.4. *Suppose conditions (N1), (LS1), and (LS2) are satisfied,*

$$\boldsymbol{c}(\boldsymbol{x}^{\boldsymbol{\theta}}; s) = \boldsymbol{Q}_{\mathbb{P}_n}^{\boldsymbol{B}(\boldsymbol{x}^{\boldsymbol{\theta}}; s)} \boldsymbol{d}(\boldsymbol{x}^{\boldsymbol{\theta}}; s),$$

where $s \in [0, \tau]$ and $\boldsymbol{x}^{\boldsymbol{\theta}} \in \boldsymbol{\mathcal{X}}_{\boldsymbol{\theta}}$. Furthermore, suppose that (d_i, f_i) are i.i.d. and suppose that the following random elements have an l-th moment that is uniformly bounded in $(\boldsymbol{x}^{\boldsymbol{\alpha}}; \boldsymbol{x}^{\boldsymbol{\beta}}; t)$ on $K \times [0, \tau]$, where $l = \max(\nu^{-1}, (1/8 + \nu/4)^{-1})$:

$$d_1(\boldsymbol{x}^{\boldsymbol{\alpha}}; \boldsymbol{x}^{\boldsymbol{\beta}}; t), \nabla_{\boldsymbol{\theta}} d_1(\boldsymbol{x}^{\boldsymbol{\alpha}}; \boldsymbol{x}^{\boldsymbol{\beta}}; t), \nabla_{\boldsymbol{\theta}}^\top \nabla_{\boldsymbol{\theta}} d_1(\boldsymbol{x}^{\boldsymbol{\alpha}}; \boldsymbol{x}^{\boldsymbol{\beta}}; t),$$
$$f_1(\boldsymbol{x}^{\boldsymbol{\alpha}}; t), \nabla_j f_1(\boldsymbol{x}^{\boldsymbol{\alpha}}; t), \nabla_j \nabla_\nu f_1(\boldsymbol{x}^{\boldsymbol{\alpha}}; t), \nabla_j \nabla_\nu \nabla_\mu f_1(\boldsymbol{x}^{\boldsymbol{\alpha}}; t), \quad j, \nu, \mu = 1, \ldots, k_{\boldsymbol{\alpha}}.$$

Then the following conditions hold true: (N2') and (N3)-(N7).
Furthermore, we have the following:

(i) (fixed alternatives)
If in addition to the above, (d_i, f_i, h_i) are i.i.d., h_i are càglàd, locally bounded, adapted stochastic processes with uniformly bounded third moment and the mappings

$$\mathrm{E}[(\boldsymbol{Q}_{\mathrm{P}}^{\boldsymbol{B}_1(\cdot)} d_1(\cdot)) h_1(\cdot)] : K \times [0, \tau] \to \mathbb{R}$$

and

$$\mathrm{E}[(\boldsymbol{Q}_{\mathrm{P}}^{\boldsymbol{B}_1(\cdot)} d_1(\cdot)) f_1(\cdot)] : K \times [0, \tau] \to \mathbb{R}$$

are equicontinuous at $\boldsymbol{\theta}_0$ *then (F1) and (F2) hold true. In particular, if, in addition, (N1') holds true and* $\lambda_i(t) = h_i(t)$ *then uniformly in* $t \in [0, \tau]$,

$$n^{-\frac{1}{2}} T(\boldsymbol{c}, \widehat{\boldsymbol{\theta}}, t) \xrightarrow{\mathrm{P}} \int_0^t \mathrm{E}\left[\left\{\boldsymbol{Q}_{\mathrm{P}}^{\boldsymbol{B}_1(\boldsymbol{\alpha}_0(s),s)} d_1(\boldsymbol{\theta}_0(s), s)\right\} \{h_1(s) - f_1(\boldsymbol{\alpha}_0(s), s)\}\right] \mathrm{d}s$$

and

$$\hat{\sigma}^2(\boldsymbol{c}, t) \xrightarrow{\mathrm{P}} \int_0^t \mathrm{E}\left[\left(\boldsymbol{Q}_{\mathrm{P}}^{\boldsymbol{B}_1(\boldsymbol{\alpha}_0(s),s)} d_1(\boldsymbol{\theta}_0(s), s)\right)^2 h_1(s)\right] \mathrm{d}s.$$

(ii) *(local alternatives)*
If in addition to the above the following holds: (d_i, f_i, g_i) *are i.i.d.,* g_i *are càglàd, locally bounded, adapted stochastic processes with uniformly bounded third moment and the mapping*

$$\mathrm{E}[(\boldsymbol{Q}_{\mathrm{P}}^{\boldsymbol{B}_1(\cdot)} d_1(\cdot)) g_1(\cdot)] : K \times [0, \tau] \to \mathbb{R}$$

is equicontinuous at $\boldsymbol{\theta}_0$ *then (L1) and (L2) hold true.*

Note that since $0 < \nu \leq 1/2$, we have $l \geq 4$. We will not give a proof of Theorem 5.4, since it is a corollary of Theorem 5.6 which concerns weighted projections.

A reasonable choice against a fixed alternative given by $\boldsymbol{\lambda}(t) = \boldsymbol{h}(t)$ is $\boldsymbol{d}(\boldsymbol{x}^{\boldsymbol{\alpha}}; t) = \boldsymbol{h}(t) - \boldsymbol{f}(\boldsymbol{x}^{\boldsymbol{\alpha}}; t)$. This guarantees that under the fixed alternative, the limit of $n^{-1/2} T(\boldsymbol{c}, \widehat{\boldsymbol{\theta}}, \cdot)$ is nonnegative, leading to a consistent test. Usually, $\boldsymbol{h}$ has some free parameters. Suppose that the alternative intensity can be written as follows:

$$\lambda_i(t) = h_i(\boldsymbol{\beta}(t), t) \text{ for some } \boldsymbol{\beta} \in \mathrm{bm}(\boldsymbol{\mathcal{X}}_{\boldsymbol{\beta}}),$$

where $\boldsymbol{h}(\boldsymbol{\beta}(\cdot), \cdot)$ is some observable, predictable stochastic process for all $\boldsymbol{\beta} \in \mathrm{bm}(\boldsymbol{\mathcal{X}}_{\boldsymbol{\beta}})$. Under the assumptions of Theorem 5.4, if we use weights $\boldsymbol{c}$ given by

$$\boldsymbol{c}(\boldsymbol{x}^{\boldsymbol{\alpha}}; \boldsymbol{x}^{\boldsymbol{\beta}}; t) = \boldsymbol{Q}_{\mathbb{P}_n}^{\boldsymbol{B}(\boldsymbol{x}^{\boldsymbol{\alpha}}; t)} \left(\boldsymbol{h}(\boldsymbol{x}^{\boldsymbol{\beta}}; t) - \boldsymbol{f}(\boldsymbol{x}^{\boldsymbol{\alpha}}; t)\right) \tag{5.14}$$

and if the alternatives holds true, i.e. $\lambda_i(t) = h_i(\boldsymbol{\beta}_0(t), t)$ then

$$n^{-\frac{1}{2}} T(\boldsymbol{c}, \widehat{\boldsymbol{\theta}}, t) \xrightarrow{\mathrm{P}} \int_0^t \mathrm{E}\left[\left\{\boldsymbol{Q}_{\mathrm{P}}^{\boldsymbol{B}_1(\boldsymbol{\alpha}_0(s),s)} \left(h_1(\boldsymbol{\beta}_0(s), s) - f_1(\boldsymbol{\alpha}_0(s), s)\right)\right\}^2\right] \mathrm{d}s =: H(t).$$

Whenever $H(t) = 0$ for all $t \in [0, \tau]$, we cannot detect this particular alternative with the weights given by (5.14).

Hence, as $\boldsymbol{B} = \nabla_{\boldsymbol{\theta}} \boldsymbol{f}$, the following are equivalent if $\boldsymbol{c}$ is given by (5.14):

(i) $H(t) = 0$ for all $t \in [0, \tau]$.

(ii) For some $\boldsymbol{\gamma} : [0, \tau] \to \mathbb{R}^{k_\alpha}$, for almost all $t \in [0, \tau]$,

$$h_1(\boldsymbol{\beta}_0(t), t) = f_1(\boldsymbol{\alpha}_0(t), t) + \nabla_{\boldsymbol{\alpha}} f_1(\boldsymbol{\alpha}_0(t), t)\boldsymbol{\gamma}(t).$$

This characterizes the alternatives that cannot be detected for one particular value of $\boldsymbol{\alpha}_0$. Figure 5.2 gives a sketch. Heuristically, we cannot detect alternatives that are elements of the 'tangent' to the model space at $\boldsymbol{f}(\boldsymbol{\alpha}_0(\cdot), \cdot)$.

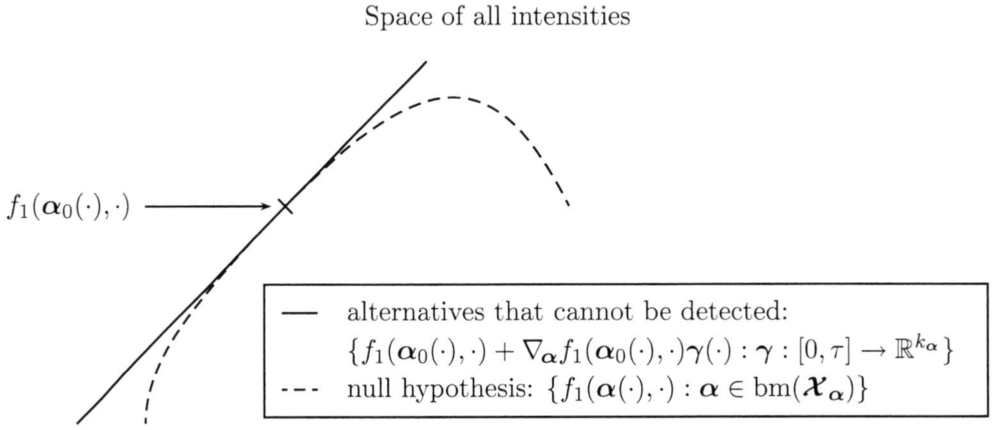

Figure 5.2: Sketch of alternatives that cannot be detected for one fixed $\boldsymbol{\alpha}_0$ by using (5.14).

The following heuristic argument suggests that if $\widehat{\boldsymbol{\alpha}}$ is a reasonable estimator and the alternative intensity $\boldsymbol{h}$ is the 'true' intensity, then $\boldsymbol{h}$ is never in this tangent. In other words, any alternative intensity $\boldsymbol{h}$ can be detected.

The 'true' intensity $\boldsymbol{\lambda}$ of $\boldsymbol{N}$ and the particular estimator $\widehat{\boldsymbol{\alpha}}$ we use determine $\boldsymbol{\alpha}_0$. So if $\boldsymbol{h}$ is the 'true' intensity then $\boldsymbol{h}$ determines $\boldsymbol{\alpha}_0$. So, to see if our test with the weights $\boldsymbol{c}$ given in (5.14) is consistent against a particular alternative given by $\boldsymbol{h}$, we need to start with $\boldsymbol{h}$, determine the appropriate $\boldsymbol{\alpha}_0$ and then check if $\boldsymbol{h}$ is an element of the 'tangent' to the model at $\boldsymbol{\alpha} = \boldsymbol{\alpha}_0$.

If the alternative intensity $\boldsymbol{h}$ is 'true', i.e. $\boldsymbol{\lambda} = \boldsymbol{h}$, then one expects $f_1(\boldsymbol{\alpha}_0(\cdot), \cdot)$ to be as close to h_1 as possible. If the space of intensities under the null hypothesis is 'smooth' then one expects that $h_1(\cdot) - f_1(\boldsymbol{\alpha}_0(\cdot), \cdot)$ is 'orthogonal' to $\nabla_{\boldsymbol{\alpha}} f_1(\boldsymbol{\alpha}_0(\cdot), \cdot)$ - implying that h_1 is not one of the alternatives that cannot be detected. In Section 10.1, we consider a purely parametric setup in which the above reasoning could be made rigorous.

5.5 Optimal Tests

5.5.1 Fixed Alternatives

We want to consider optimal tests against a fixed alternative given by $\lambda_i(t) = h_i(t)$ based on the test statistic

$$V(\boldsymbol{c}) := T(\boldsymbol{c}, \widehat{\boldsymbol{\alpha}}, \tau)/\sqrt{\hat{\sigma}^2(\boldsymbol{c}, \tau)},$$

where $\hat{\sigma}^2(\boldsymbol{c}, \tau)$ is as in Theorem 5.1. As in Subsection 3.6.1, the optimality criterion we use is the approximate Bahadur efficiency. If the null hypothesis (5.2) holds true then

$$V(\boldsymbol{c}) \xrightarrow{\mathrm{P}} N(0,1)$$

if $\sigma^2(\tau) \neq 0$, where $\sigma^2(\tau)$ is as in Theorem 5.1. If the fixed alternative $\lambda_i(t) = h_i(t)$ holds true then by Theorem 5.2,

$$V(\boldsymbol{c}) \xrightarrow{\mathrm{P}} \int_0^\tau \left(\overrightarrow{\boldsymbol{ch}}(\boldsymbol{\alpha}_0(s), s) - \overrightarrow{\boldsymbol{cf}}(\boldsymbol{\alpha}_0(s), s)\right) \mathrm{d}s \left(\int_0^\tau \overrightarrow{\boldsymbol{chc}}(\boldsymbol{\alpha}_0(s), s)\, \mathrm{d}s\right)^{-\frac{1}{2}} =: b(\boldsymbol{c}).$$

In our case, the approximate Bahadur slope is given by $b(\boldsymbol{c})^2$. In order to maximize $b(\boldsymbol{c})^2$, it suffices to maximize

$$Z(\boldsymbol{c}) := \int_0^\tau \left(\overline{\boldsymbol{ch}}(\boldsymbol{\alpha}_0(s), s) - \overline{\boldsymbol{cf}}(\boldsymbol{\alpha}_0(s), s)\right) \mathrm{d}s \left(\int_0^\tau \overline{\boldsymbol{chc}}(\boldsymbol{\alpha}_0(s), s)\, \mathrm{d}s\right)^{-\frac{1}{2}},$$

since $Z(\boldsymbol{c}) \xrightarrow{\mathrm{P}} b(\boldsymbol{c})$, $Z(\boldsymbol{c}) = -Z(-\boldsymbol{c})$, and $\boldsymbol{c}$ will be admissible iff $-\boldsymbol{c}$ is admissible. Instead of optimizing $\boldsymbol{c}$ in the class of weights satisfying all conditions for Theorem 5.1 and Theorem 5.2, we optimize only under the side condition (N2). It will turn out that the optimal weights use weighted orthogonal projections. In Section 5.6, we will give regularity conditions under which those weights satisfy the other conditions needed for the asymptotics of our test statistic.

Thus the optimization problem we are interested in solving is

$$\begin{cases} Z(\boldsymbol{c}) \to \max \\ \sum_{i=1}^n c_i(t)(\nabla_{\boldsymbol{\alpha}} f_i)(\boldsymbol{\alpha}_0(t), t) = \boldsymbol{0}, \quad \text{for } \lambda\text{-almost all } t \in [0, \tau], \end{cases} \tag{5.15}$$

where for simplicity we write $\boldsymbol{c}(t)$ instead of $\boldsymbol{c}(\boldsymbol{\alpha}_0(t), t)$. Note that (5.15) is an optimization problem for one fixed $\omega \in \Omega$. The key idea to solve this optimization problem is to use weighted projections. Recall the notation of Section 3.2.

Proposition 5.1. *Suppose that for all $i = 1, \ldots, n$, $t \in [0, \tau]$,*

$$h_i(t) = 0 \textit{ implies } (f_i(\boldsymbol{\alpha}_0(t), t) = 0 \textit{ and } \nabla_{\boldsymbol{\alpha}} f_i(\boldsymbol{\alpha}_0(t), t) = \boldsymbol{0}).$$

If $\frac{\boldsymbol{h}(\cdot)-\boldsymbol{f}(\boldsymbol{\alpha}_0(\cdot),\cdot)}{\boldsymbol{h}(\cdot)}$ *and* $\frac{\nabla_j \boldsymbol{f}(\boldsymbol{\alpha}_0(\cdot),\cdot)}{\boldsymbol{h}(\cdot)}$, $j = 1,\dots,k_{\boldsymbol{\alpha}}$, *are elements of* $L_2(\boldsymbol{h}\cdot(\mathbb{P}_n\otimes\lambda))$ *then for all* $\boldsymbol{c} \in L_2(\boldsymbol{h}\cdot(\mathbb{P}_n\otimes\lambda))$ *that are admissible for* (5.15)*, the following holds true:*

$$Z(\boldsymbol{c}^*) \geq Z(\boldsymbol{c}),$$

where the admissible $\boldsymbol{c}^* \in L_2(\boldsymbol{h}\cdot(\mathbb{P}_n\otimes\lambda))$ *is given by*

$$\boldsymbol{c}^*(t) = \boldsymbol{Q}_{\boldsymbol{h}(t)\cdot\mathbb{P}_n}^{\boldsymbol{B}(\boldsymbol{\alpha}_0(t),t)} \frac{\boldsymbol{h}(t) - \boldsymbol{f}(\boldsymbol{\alpha}_0(t),t)}{\boldsymbol{h}(t)}$$

and $\boldsymbol{B}(\boldsymbol{x}^{\boldsymbol{\alpha}}, t) = \left(\frac{\nabla_1 \boldsymbol{f}(\boldsymbol{x}^{\boldsymbol{\alpha}}, t)}{\boldsymbol{h}(t)},\dots,\frac{\nabla_{k_{\boldsymbol{\alpha}}} \boldsymbol{f}(\boldsymbol{x}^{\boldsymbol{\alpha}}, t)}{\boldsymbol{h}(t)}\right)$.

Proof. First note that $\boldsymbol{c}^* \in L_2(\boldsymbol{h}\cdot(\mathbb{P}_n\otimes\lambda))$ because

$$\begin{aligned}
\|\boldsymbol{c}^*\|^2_{\boldsymbol{h}\cdot(\mathbb{P}_n\otimes\lambda)} &= \int_0^\tau \left\| \boldsymbol{Q}_{\boldsymbol{h}(s)\cdot\mathbb{P}_n}^{\boldsymbol{B}(\boldsymbol{\alpha}_0(s),s)} \frac{\boldsymbol{h}(s) - \boldsymbol{f}(\boldsymbol{\alpha}_0(s),s)}{\boldsymbol{h}(s)} \right\|^2_{\boldsymbol{h}(s)\cdot\mathbb{P}_n} \mathrm{d}s \\
&\leq \int_0^\tau \left\| \frac{\boldsymbol{h}(s) - \boldsymbol{f}(\boldsymbol{\alpha}_0(s),s)}{\boldsymbol{h}(s)} \right\|^2_{\boldsymbol{h}(s)\cdot\mathbb{P}_n} \mathrm{d}s \\
&= \left\| \frac{\boldsymbol{h}(\cdot) - \boldsymbol{f}(\boldsymbol{\alpha}_0(\cdot),\cdot)}{\boldsymbol{h}(\cdot)} \right\|^2_{\boldsymbol{h}\cdot(\mathbb{P}_n\otimes\lambda)} < \infty.
\end{aligned}$$

The side condition can be rewritten as

$$< \boldsymbol{c}(t), \frac{\nabla_j \boldsymbol{f}(\boldsymbol{\alpha}_0(t),t)}{\boldsymbol{h}(t)} >_{\boldsymbol{h}(t)\cdot\mathbb{P}_n} = \boldsymbol{0}, \ \forall j = 1,\dots,k_{\boldsymbol{\alpha}}, \text{ and almost all } t \in [0,\tau].$$

Hence, $\boldsymbol{c}^*$ satisfies the side condition. Using properties of orthogonal projections (3.3), we get for all $\boldsymbol{c}$ that are admissible for (5.15), that for almost all s,

$$\begin{aligned}
n\left(\overline{\boldsymbol{ch}}(s) - \overline{\boldsymbol{cf}}(\boldsymbol{\alpha}_0(s),s)\right) &= < \boldsymbol{c}(s), \frac{\boldsymbol{h}(s) - \boldsymbol{f}(\boldsymbol{\alpha}_0(s),s)}{\boldsymbol{h}(s)} >_{\boldsymbol{h}(s)\cdot\mathbb{P}_n} \\
&= < \boldsymbol{Q}_{\boldsymbol{h}(s)\cdot\mathbb{P}_n}^{\boldsymbol{B}(\boldsymbol{\alpha}_0(s),s)} \boldsymbol{c}(s), \frac{\boldsymbol{h}(s) - \boldsymbol{f}(\boldsymbol{\alpha}_0(s),s)}{\boldsymbol{h}(s)} >_{\boldsymbol{h}(s)\cdot\mathbb{P}_n} \\
&= < \boldsymbol{c}(s), \boldsymbol{c}^*(s) >_{\boldsymbol{h}(s)\cdot\mathbb{P}_n}.
\end{aligned}$$

Hence, by the the Cauchy-Schwarz inequality,

$$\begin{aligned}
n\int_0^\tau \left(\overline{\boldsymbol{ch}}(s) - \overline{\boldsymbol{cf}}(\boldsymbol{\alpha}_0(s),s)\right) \mathrm{d}s &\leq \int_0^\tau \|\boldsymbol{c}(s)\|_{\boldsymbol{h}(s)\cdot\mathbb{P}_n} \|\boldsymbol{c}^*(s)\|_{\boldsymbol{h}(s)\cdot\mathbb{P}_n} \mathrm{d}s \\
&\leq \left(\int_0^\tau \|\boldsymbol{c}(s)\|^2_{\boldsymbol{h}(s)\cdot\mathbb{P}_n} \mathrm{d}s\right)^{\frac{1}{2}} \left(\int_0^\tau \|\boldsymbol{c}^*(s)\|^2_{\boldsymbol{h}(s)\cdot\mathbb{P}_n} \mathrm{d}s\right)^{\frac{1}{2}} \\
&= \left(n\int_0^\tau \overline{\boldsymbol{chc}}(s)\,\mathrm{d}s\right)^{\frac{1}{2}} \|\boldsymbol{c}^*\|_{\boldsymbol{h}\cdot(\mathbb{P}_n\otimes\lambda)},
\end{aligned}$$

with equality if $\boldsymbol{c} = \boldsymbol{c}^*$. By the assumptions, $\int_0^\tau \overline{\boldsymbol{chc}}(s)\,\mathrm{d}s = 0$ implies $\int_0^\tau \left(\overline{\boldsymbol{ch}}(s) - \overline{\boldsymbol{cf}}(s)\right)\mathrm{d}s = 0$. Thus, since we use the convention $0/0 = 0$,

$$Z(\boldsymbol{c}) \leq n^{-\frac{1}{2}} \|\boldsymbol{c}^*\|_{\boldsymbol{h}\cdot(\mathbb{P}_n \otimes \lambda)},$$

with equality if $\boldsymbol{c} = \boldsymbol{c}^*$. □

5.5.2 Local Alternatives

Optimal weights against local alternatives of type (5.12) in the sense of Pitman efficiency can be derived similarly. In fact, Theorem 5.3 can be used to derive optimal tests against local alternatives of type (5.12) within the class of one-sided tests based on

$$V(\boldsymbol{c}) := T(\boldsymbol{c}, \widehat{\boldsymbol{\alpha}}, \tau)/\sqrt{\hat{\sigma}^2(\boldsymbol{c}, \tau)},$$

where $\hat{\sigma}^2(\boldsymbol{c}, \tau)$ is as in Theorem 5.1. By a Slutsky argument,

$$V(\boldsymbol{c}) \xrightarrow{\mathrm{d}} N(\mu(\tau)/\sqrt{\sigma^2(\tau)}, 1)$$

if $\sigma^2(\tau) \neq 0$, where $\sigma^2(\tau)$ is given in Theorem 5.3. As in Subsection 3.6.2, we want to maximize $\mu(\tau)/\sqrt{\sigma^2(\tau)}$. Since

$$\int_0^\tau \overline{\boldsymbol{cg}}(\boldsymbol{\alpha}_0(s), s)\,\mathrm{d}s \xrightarrow{\mathrm{P}} \mu(\tau) \quad \text{and} \quad \int_0^\tau \overline{\boldsymbol{cfc}}(\boldsymbol{\alpha}_0(s), s)\,\mathrm{d}s \xrightarrow{\mathrm{P}} \sigma^2(\tau),$$

it suffices to maximize

$$Z_l(\boldsymbol{c}) := \left(\int_0^\tau \overline{\boldsymbol{cfc}}(\boldsymbol{\alpha}_0(s), s)\,\mathrm{d}s\right)^{-\frac{1}{2}} \int_0^\tau \overline{\boldsymbol{cg}}(\boldsymbol{\alpha}_0(s), s)\,\mathrm{d}s.$$

As for fixed alternatives, we shall only consider the side condition (N2). The optimization problem we want to solve is

$$\begin{cases} Z_l(\boldsymbol{c}) \to \max \\ \sum_{i=1}^n c_i(t)(\nabla_{\boldsymbol{\alpha}} f_i)(\boldsymbol{\alpha}_0(t), t) = \mathbf{0}, \quad \text{for } \lambda\text{-almost all } t \in [0, \tau]. \end{cases} \tag{5.16}$$

To ease notation, we use the abbreviation $\boldsymbol{f}_0(\cdot) := \boldsymbol{f}(\boldsymbol{\alpha}_0(\cdot), \cdot)$.

Theorem 5.5. *Suppose that for all $i = 1, \ldots, n$, $t \in [0, \tau]$,*

$$f_i(\boldsymbol{\alpha}_0(t), t) = 0 \textit{ implies } (g_i(t) = 0 \textit{ and } \nabla_{\boldsymbol{\alpha}} f_i(\boldsymbol{\alpha}_0(t), t) = \mathbf{0}).$$

Suppose that $\frac{\boldsymbol{g}(\cdot)}{\boldsymbol{f}_0(\cdot)}, \frac{\nabla_1 \boldsymbol{f}(\boldsymbol{\alpha}_0(\cdot), \cdot)}{\boldsymbol{f}_0(\cdot)}, \ldots, \frac{\nabla_{k_{\boldsymbol{\alpha}}} \boldsymbol{f}(\boldsymbol{\alpha}_0(\cdot), \cdot)}{\boldsymbol{f}_0(\cdot)}$ are elements of $L_2(\boldsymbol{f}_0 \cdot (\mathbb{P}_n \otimes \lambda))$. A solution of (5.16) in the class of all $\boldsymbol{c} \in L_2(\boldsymbol{f}_0 \cdot (\mathbb{P}_n \otimes \lambda))$ is given by

$$\tilde{\boldsymbol{c}}(s) := \boldsymbol{Q}_{\boldsymbol{f}_0(s)\cdot \mathbb{P}_n}^{\boldsymbol{B}(\boldsymbol{\alpha}_0(s), s)} \frac{\boldsymbol{g}(s)}{\boldsymbol{f}_0(s)},$$

where $\boldsymbol{B}(\boldsymbol{x}^{\boldsymbol{\alpha}}, s) = \left(\frac{\nabla_1 \boldsymbol{f}(\boldsymbol{x}^{\boldsymbol{\alpha}}, s)}{\boldsymbol{f}(\boldsymbol{x}^{\boldsymbol{\alpha}}, s)}, \ldots, \frac{\nabla_{k_{\boldsymbol{\alpha}}} \boldsymbol{f}(\boldsymbol{x}^{\boldsymbol{\alpha}}, s)}{\boldsymbol{f}(\boldsymbol{x}^{\boldsymbol{\alpha}}, s)}\right)$.

Proof. First note that

$$\begin{aligned}
\|\tilde{\boldsymbol{c}}\|^2_{\boldsymbol{f}_0\cdot(\mathbb{P}_n\otimes\lambda)} &= \int_0^\tau \left\| \boldsymbol{Q}^{\boldsymbol{B}(\boldsymbol{\alpha}_0(s),s)}_{\boldsymbol{f}_0(s)\cdot\mathbb{P}_n} \frac{\boldsymbol{g}(s)}{\boldsymbol{f}_0(s)} \right\|^2_{\boldsymbol{f}_0(s)\cdot\mathbb{P}_n} \mathrm{d}s \\
&\le \int_0^\tau \left\| \frac{\boldsymbol{g}(s)}{\boldsymbol{f}_0(s)} \right\|^2_{\boldsymbol{f}_0(s)\cdot\mathbb{P}_n} \mathrm{d}s = \left\| \frac{\boldsymbol{g}(\cdot)}{\boldsymbol{f}_0(\cdot)} \right\|^2_{\boldsymbol{f}_0\cdot(\mathbb{P}_n\otimes\lambda)} < \infty.
\end{aligned}$$

The side condition is satisfied by $\tilde{\boldsymbol{c}}$, since the side condition can be rewritten as

$$<\boldsymbol{c}(s), \frac{\nabla_j \boldsymbol{f}(\boldsymbol{\alpha}_0(s), s)}{\boldsymbol{f}_0(s)}>_{\boldsymbol{f}_0(s)\cdot\mathbb{P}_n} = 0,\ \forall j = 1, \ldots, k_{\boldsymbol{\alpha}},\ \text{almost all}\ s \in [0, \tau].$$

Using the properties (3.3) of projection matrices and twice the Cauchy-Schwarz inequality, we get for any $\boldsymbol{c}$ which is admissible for (5.16) that the following holds true:

$$\begin{aligned}
n\int_0^\tau \overline{\boldsymbol{c}\boldsymbol{g}}(s)\,\mathrm{d}s &= \int_0^\tau <\boldsymbol{c}(s), \frac{\boldsymbol{g}(s)}{\boldsymbol{f}_0(s)}>_{\boldsymbol{f}_0(s)\cdot\mathbb{P}_n} \mathrm{d}s \\
&= \int_0^\tau <\boldsymbol{Q}^{\boldsymbol{B}(\boldsymbol{\alpha}_0(s),s)}_{\boldsymbol{f}_0(s)\cdot\mathbb{P}_n}\boldsymbol{c}(s), \frac{\boldsymbol{g}(s)}{\boldsymbol{f}_0(s)}>_{\boldsymbol{f}_0(s)\cdot\mathbb{P}_n} \mathrm{d}s \\
&= \int_0^\tau <\boldsymbol{c}(s), \tilde{\boldsymbol{c}}(s)>_{\boldsymbol{f}_0(s)\cdot\mathbb{P}_n} \mathrm{d}s \\
&\le \int_0^\tau \|\boldsymbol{c}(s)\|_{\boldsymbol{f}_0(s)\cdot\mathbb{P}_n} \|\tilde{\boldsymbol{c}}(s)\|_{\boldsymbol{f}_0(s)\cdot\mathbb{P}_n} \mathrm{d}s \\
&\le \left(\int_0^\tau \|\boldsymbol{c}(s)\|^2_{\boldsymbol{f}_0(s)\cdot\mathbb{P}_n} \mathrm{d}s\right)^{\frac{1}{2}} \left(\int_0^\tau \|\tilde{\boldsymbol{c}}(s)\|^2_{\boldsymbol{f}_0(s)\cdot\mathbb{P}_n} \mathrm{d}s\right)^{\frac{1}{2}} \\
&= \left(n\int_0^\tau \overline{\boldsymbol{c}\boldsymbol{f}_0\boldsymbol{c}}(s)\,\mathrm{d}s\right)^{\frac{1}{2}} \|\tilde{\boldsymbol{c}}\|_{\boldsymbol{f}_0\cdot(\mathbb{P}_n\otimes\lambda)},
\end{aligned}$$

with equality if $\boldsymbol{c} = \tilde{\boldsymbol{c}}$. □

5.6 Optimal Weights are Admissible

In the previous sections, we have seen that optimal weights against fixed or local alternatives are given by certain weighted orthogonal projections. The goal of this section is to show that weights defined in this way actually satisfy the conditions needed for the asymptotic results under the null hypothesis (Theorem 5.1), under fixed alternatives (Theorem 5.2), and under local alternatives (Theorem 5.3).

First, we sketch the basic idea to satisfy the orthogonality condition (N2'). If $\boldsymbol{d}$ is an n-dimensional vector of stochastic processes and $\boldsymbol{w}$ is an n-dimensional vector of nonnegative stochastic processes, both possibly depending on $\boldsymbol{x}^{\boldsymbol{\theta}} \in \boldsymbol{\mathcal{X}}_{\boldsymbol{\theta}}$, then the weights

$$\boldsymbol{c}(\boldsymbol{x}^{\boldsymbol{\theta}}, \cdot) = \boldsymbol{Q}^{\boldsymbol{B}(\boldsymbol{x}^{\boldsymbol{\theta}}, \cdot)}_{\boldsymbol{w}(\boldsymbol{x}^{\boldsymbol{\theta}}, \cdot)\cdot\mathbb{P}_n} \boldsymbol{d}(\boldsymbol{x}^{\boldsymbol{\theta}}, \cdot), \tag{5.17}$$

satisfy (N2'), where for $(\boldsymbol{x}^{\boldsymbol{\alpha}\top}, \boldsymbol{x}^{\boldsymbol{\beta}\top})^\top \in \boldsymbol{\mathcal{X}}_{\boldsymbol{\theta}}$,

$$\boldsymbol{B}(\boldsymbol{x}^{\boldsymbol{\alpha}}, \boldsymbol{x}^{\boldsymbol{\beta}}, t) = \left(\frac{\nabla_1 \boldsymbol{f}(\boldsymbol{x}^{\boldsymbol{\alpha}}, t)}{\boldsymbol{w}(\boldsymbol{x}^{\boldsymbol{\alpha}}, \boldsymbol{x}^{\boldsymbol{\beta}}, t)}, \dots, \frac{\nabla_{k_{\boldsymbol{\alpha}}} \boldsymbol{f}(\boldsymbol{x}^{\boldsymbol{\alpha}}, t)}{\boldsymbol{w}(\boldsymbol{x}^{\boldsymbol{\alpha}}, \boldsymbol{x}^{\boldsymbol{\beta}}, t)} \right). \tag{5.18}$$

Indeed, for $j = 1, \dots, k_{\boldsymbol{\alpha}}$,

$$\sum_{i=1}^n c_i(\boldsymbol{x}^{\boldsymbol{\theta}}, t) \nabla_j f_i(\boldsymbol{x}^{\boldsymbol{\alpha}}, t) = < \boldsymbol{c}(\boldsymbol{x}^{\boldsymbol{\theta}}, t), \frac{\nabla_j \boldsymbol{f}(\boldsymbol{x}^{\boldsymbol{\alpha}}, t)}{\boldsymbol{w}(\boldsymbol{x}^{\boldsymbol{\theta}}, t)} >_{\boldsymbol{w}(\boldsymbol{x}^{\boldsymbol{\theta}}, t)\cdot \mathbb{P}_n} = 0.$$

Next, we give a setup in which the other conditions needed for the asymptotic results are satisfied. We define $\boldsymbol{c}$ in a slightly more general way to be able to reuse the result in the next chapter. We focus on an i.i.d. setup. Suppose that $\boldsymbol{d}$ is an n-dimensional vector of stochastic processes, $\boldsymbol{w}$ is an n-dimensional vector of nonnegative stochastic processes, and $\boldsymbol{B}$ is an $n \times k$-matrix of stochastic processes. $\boldsymbol{B}$, $\boldsymbol{w}$, and $\boldsymbol{d}$ may all depend on $\boldsymbol{x}^{\boldsymbol{\theta}} \in \boldsymbol{\mathcal{X}}_{\boldsymbol{\theta}}$. The matrix $\boldsymbol{B}$ need not be defined by (5.18). We consider $\boldsymbol{c}$ defined by

$$\boldsymbol{c}(\boldsymbol{x}^{\boldsymbol{\theta}}, \cdot) = \boldsymbol{Q}^{\boldsymbol{B}(\boldsymbol{x}^{\boldsymbol{\theta}}, \cdot)}_{\boldsymbol{w}(\boldsymbol{x}^{\boldsymbol{\theta}}, \cdot)\cdot\mathbb{P}_n} \boldsymbol{d}(\boldsymbol{x}^{\boldsymbol{\theta}}, \cdot). \tag{5.19}$$

If $w_i(\boldsymbol{x}^{\boldsymbol{\theta}}, t) = 0$ we set $c_i(\boldsymbol{x}^{\boldsymbol{\theta}}, t) = 0$. The last requirement is needed since (5.19) only defines $\boldsymbol{c}(\boldsymbol{x}^{\boldsymbol{\theta}}, t)$ up to a $\boldsymbol{w}(\boldsymbol{x}^{\boldsymbol{\theta}}, t) \cdot \mathbb{P}_n$-null set. To show that $\boldsymbol{c}$ satisfies certain properties of stochastic processes like e.g. predictability we need this more precise definition.

We use the following conditions:

(LSW1) (several conditions for weighted projections) There exists a compact, convex set $K \subset \boldsymbol{\mathcal{X}}_{\boldsymbol{\theta}}$ such that $\{\boldsymbol{\theta}_0(t) : t \in [0, \tau]\}$ is in its interior and such that the following holds true:

For all $t \in [0, \tau]$, $i = 1, \dots, n$, $j = 1, \dots, k$, the mappings $d_i(\cdot, t) : K \to \mathbb{R}$, $B_{ij}(\cdot, t) : K \to \mathbb{R}$, and $w_i(\cdot, t) : K \to \mathbb{R}$ are twice differentiable. Furthermore, d_i, $\nabla_j d_i$, $\nabla_j \nabla_\nu d_i$, w_i, $\nabla_j w_i$, $\nabla_j \nabla_\nu w_i$, w_i^{-1}, $B_{i\mu}$, $\nabla_j B_{i\mu}$, and $\nabla_j \nabla_\nu B_{i\mu}$, $i = 1, \dots, n$, $j, \nu = 1, \dots, k_{\boldsymbol{\theta}}$, $\mu = 1, \dots, k$, are càglàd, adapted, locally bounded stochastic processes with values in the space $C(K)$ of all continuous mappings from K into $\mathbb{R}$ equipped with the supremum norm.

(LSW2) (properties of $\overrightarrow{\boldsymbol{BwB}}$) $\overrightarrow{\boldsymbol{BwB}}(\boldsymbol{x}^{\boldsymbol{\theta}}, t)$ is invertible for all $\boldsymbol{x}^{\boldsymbol{\theta}} \in K$, $t \in [0, \tau]$, and continuous in $(\boldsymbol{x}^{\boldsymbol{\theta}}, t) \in K \times [0, \tau]$.

Theorem 5.6. *Suppose conditions (N1), (LSW1), and (LSW2) are satisfied,*

$$\boldsymbol{c}(\boldsymbol{x}^{\boldsymbol{\theta}}, \cdot) = \boldsymbol{Q}^{\boldsymbol{B}(\boldsymbol{x}^{\boldsymbol{\theta}}, \cdot)}_{\boldsymbol{w}(\boldsymbol{x}^{\boldsymbol{\theta}}, \cdot)\cdot\mathbb{P}_n} \boldsymbol{d}(\boldsymbol{x}^{\boldsymbol{\theta}}, \cdot),$$

where $\boldsymbol{x}^{\boldsymbol{\theta}} \in \boldsymbol{\mathcal{X}}_{\boldsymbol{\theta}}$. Furthermore, suppose that $(d_i, \boldsymbol{B}_i, w_i, f_i)$ are i.i.d. and suppose that the following random elements have an l-th moment that is uniformly bounded in $(\boldsymbol{x}^{\boldsymbol{\theta}}, t) = (\boldsymbol{x}^{\boldsymbol{\alpha}}, \boldsymbol{x}^{\beta}, t)$ on $K \times [0, \tau]$, where $l = \max(\nu^{-1}, (1/8 + \nu/4)^{-1})$:

$$\begin{aligned}
&d_1(\boldsymbol{x}^{\boldsymbol{\theta}}, t), \nabla_{\boldsymbol{\theta}} d_1(\boldsymbol{x}^{\boldsymbol{\theta}}, t), \nabla_{\boldsymbol{\theta}}^{\top} \nabla_{\boldsymbol{\theta}} d_1(\boldsymbol{x}^{\boldsymbol{\theta}}, t),\\
&w_1(\boldsymbol{x}^{\boldsymbol{\theta}}, t), \nabla_{\boldsymbol{\theta}} w_1(\boldsymbol{x}^{\boldsymbol{\theta}}, t), \nabla_{\boldsymbol{\theta}}^{\top} \nabla_{\boldsymbol{\theta}} w_1(\boldsymbol{x}^{\boldsymbol{\theta}}, t),\\
&\boldsymbol{B}_1(\boldsymbol{x}^{\boldsymbol{\theta}}, t), \nabla_j \boldsymbol{B}_1(\boldsymbol{x}^{\boldsymbol{\theta}}, t), \nabla_j \nabla_{\nu} \boldsymbol{B}_1(\boldsymbol{x}^{\boldsymbol{\theta}}, t), j, \nu = 1, \ldots, k_{\boldsymbol{\theta}},\\
&f_1(\boldsymbol{x}^{\boldsymbol{\alpha}}, t), \nabla_{\boldsymbol{\alpha}} f_1(\boldsymbol{x}^{\boldsymbol{\alpha}}, t), \nabla_{\boldsymbol{\alpha}}^{\top} \nabla_{\boldsymbol{\alpha}} f_1(\boldsymbol{x}^{\boldsymbol{\alpha}}, t).
\end{aligned}$$

Then the following conditions hold true: (N3)-(N7). Furthermore, the asymptotic variance $\sigma^2(t)$ from Theorem 5.1 is given by

$$\sigma^2(t) = \int_0^t \left\| \boldsymbol{Q}_{w_1(\boldsymbol{\theta}_0(s),s)\cdot \mathrm{P}}^{\boldsymbol{B}_1(\boldsymbol{\theta}_0(s),s)} d_1(\boldsymbol{\theta}_0(s), s) \right\|_{f_1(\boldsymbol{\alpha}_0(s),s)\cdot \mathrm{P}}^2 \mathrm{d}s.$$

Moreover, we have the following:

(i) (fixed alternatives)
If in addition to the above, $(d_i, \boldsymbol{B}_i, w_i, h_i)$ are i.i.d., h_i are càglàd, locally bounded, adapted stochastic processes with uniformly bounded third moment and the mappings

$$\mathrm{E}[(\boldsymbol{Q}_{w_1(\cdot)\cdot \mathrm{P}}^{\boldsymbol{B}_1(\cdot)} d_1(\cdot)) h_1(\cdot)] : K \times [0, \tau] \to \mathbb{R}$$

and

$$\mathrm{E}[(\boldsymbol{Q}_{w_1(\cdot)\cdot \mathrm{P}}^{\boldsymbol{B}_1(\cdot)} d_1(\cdot)) f_1(\cdot)] : K \times [0, \tau] \to \mathbb{R}$$

are equicontinuous at $\boldsymbol{\theta}_0$ then (F1) and (F2) hold true. In particular, if, in addition, (N1') holds true and $\lambda_i(t) = h_i(t)$ then uniformly in $t \in [0, \tau]$,

$$n^{-\frac{1}{2}} T(\boldsymbol{c}, \widehat{\boldsymbol{\theta}}, t) \xrightarrow{\mathrm{P}} \int_0^t \mathrm{E}\left[\left\{\boldsymbol{Q}_{w_1(\boldsymbol{\theta}_0(s),s)\cdot \mathrm{P}}^{\boldsymbol{B}_1(\boldsymbol{\theta}_0(s),s)} d_1(\boldsymbol{\theta}_0(s), s)\right\} \{h_1(s) - f_1(\boldsymbol{\alpha}_0(s), s)\}\right] \mathrm{d}s$$

and

$$\hat{\sigma}^2(\boldsymbol{c}, t) \xrightarrow{\mathrm{P}} \int_0^t \mathrm{E}\left[\left(\boldsymbol{Q}_{w_1(\boldsymbol{\theta}_0(s),s)\cdot \mathrm{P}}^{\boldsymbol{B}_1(\boldsymbol{\theta}_0(s),s)} d_1(\boldsymbol{\theta}_0(s), s)\right)^2 h_1(s)\right] \mathrm{d}s.$$

(ii) (local alternatives)
If in addition to the above the following holds: $(d_i, \boldsymbol{B}_i, w_i, f_i, g_i)$ are i.i.d., g_i are càglàd, locally bounded, adapted stochastic processes with uniformly bounded third moment and the mapping

$$\mathrm{E}[(\boldsymbol{Q}_{\mathrm{P}}^{\boldsymbol{B}_1(\cdot)} d_1(\cdot)) g_1(\cdot)] : K \times [0, \tau] \to \mathbb{R}$$

is equicontinuous at $\boldsymbol{\theta}_0$ then (L1) and (L2) hold true.

In the above, we have $l \geq 4$ since $0 < \nu \leq 1/2$. If the covariates are bounded then the l-th moment condition is trivially satisfied.

Proof. Let $\mathcal{C} = K \times [0, \tau]$ and let the event A_n be defined by

$$A_n := \{(\widehat{\boldsymbol{\theta}}(t), t) \in \mathcal{C}\, \forall\, t \in [0,\tau]\} \cap \{\det(\overline{\boldsymbol{BwB}}(\boldsymbol{x}^{\boldsymbol{\theta}}, t)) > L\, \forall\, (\boldsymbol{x}^{\boldsymbol{\theta}}, t) \in \mathcal{C}\} \quad (5.20)$$

with $L := 1/2 \inf_{(\boldsymbol{x}^{\boldsymbol{\theta}}, t) \in \mathcal{C}} \det(\overrightarrow{\boldsymbol{BwB}}(\boldsymbol{x}^{\boldsymbol{\theta}}, t))$. Note that $L > 0$ since by (LSW2), $\overrightarrow{\boldsymbol{BwB}}$ is invertible on the compact set $\mathcal{C}$. By (N1), by (LSW2), and by the integrability of $\boldsymbol{B}_1^\top w_1 \boldsymbol{B}_1$, (recall that $l \geq 4$), we have $\mathrm{P}(A_n) \to 1$.

Now we show (N3). On A_n, the matrix $\overline{\boldsymbol{BwB}}(\boldsymbol{x}^{\boldsymbol{\theta}}, t)$ is invertible for $(\boldsymbol{x}^{\boldsymbol{\theta}}, t) \in \mathcal{C}$, and hence, dropping the dependence on $(\boldsymbol{x}^{\boldsymbol{\theta}}, t) \in \mathcal{C}$,

$$\boldsymbol{c} = \boldsymbol{Q}^{\boldsymbol{B}}_{\boldsymbol{w}\cdot\mathbb{P}_n} \boldsymbol{d} = \boldsymbol{d} - \boldsymbol{B}(\overline{\boldsymbol{BwB}})^{-1}\overline{\boldsymbol{Bwd}}. \quad (5.21)$$

By (LS1), the components of $\boldsymbol{B}(\cdot, t)$, $\boldsymbol{d}(\cdot, t)$, and $\boldsymbol{w}(\cdot, t)$ are twice differentiable and thus $\boldsymbol{Q}^{\boldsymbol{B}(\cdot,t)}_{\boldsymbol{w}(\cdot,t)\cdot\mathbb{P}_n} \boldsymbol{d}(\cdot, t) : \mathcal{C}_t \to \mathbb{R}^n$ is twice differentiable. This shows (N3). Furthermore, on A_n, the first derivative of $\boldsymbol{c}$ can be written as (dropping the dependence on $(\boldsymbol{x}^{\boldsymbol{\theta}}, t) \in \mathcal{C}$):

$$\begin{aligned}
\nabla_j \boldsymbol{Q}^{\boldsymbol{B}}_{\boldsymbol{w}\cdot\mathbb{P}_n} \boldsymbol{d} =& \nabla_j \left(\boldsymbol{d} - \boldsymbol{B}\left(\overline{\boldsymbol{BwB}}\right)^{-1} \overline{\boldsymbol{Bwd}}\right) \\
=& \nabla_j \boldsymbol{d} - (\nabla_j \boldsymbol{B})\left(\overline{\boldsymbol{BwB}}\right)^{-1} \overline{\boldsymbol{Bwd}} \\
&+ \boldsymbol{B}\left(\overline{\boldsymbol{BwB}}\right)^{-1} \left[\left(\nabla_j \overline{\boldsymbol{BwB}}\right)\left(\overline{\boldsymbol{BwB}}\right)^{-1} \overline{\boldsymbol{Bwd}} - \nabla_j \overline{\boldsymbol{Bwd}}\right]
\end{aligned} \quad (5.22)$$

and the second derivative $\nabla_\nu \nabla_j \boldsymbol{Q}^{\boldsymbol{B}}_{\mathbb{P}_n} \boldsymbol{d}$ of $\boldsymbol{c}$ equals

$$\begin{aligned}
&\nabla_\nu \nabla_j \boldsymbol{d} - (\nabla_\nu \nabla_j \boldsymbol{B})\left(\overline{\boldsymbol{BwB}}\right)^{-1} \overline{\boldsymbol{Bwd}} + (\nabla_j \boldsymbol{B})\nabla_\nu \left(\left(\overline{\boldsymbol{BwB}}\right)^{-1} \overline{\boldsymbol{Bwd}}\right) \\
&+ (\nabla_\nu \boldsymbol{B})\left(\overline{\boldsymbol{BwB}}\right)^{-1} \left[\left(\nabla_j \overline{\boldsymbol{BwB}}\right)\left(\overline{\boldsymbol{BwB}}\right)^{-1} \overline{\boldsymbol{Bwd}} - \nabla_j \overline{\boldsymbol{Bwd}}\right] \\
&+ \boldsymbol{B}\nabla_\nu \left(\left(\overline{\boldsymbol{BwB}}\right)^{-1} \left[\left(\nabla_j \overline{\boldsymbol{BwB}}\right)\left(\overline{\boldsymbol{BwB}}\right)^{-1} \overline{\boldsymbol{Bwd}} - \nabla_j \overline{\boldsymbol{Bwd}}\right]\right).
\end{aligned} \quad (5.23)$$

Next, we prove that (N4) holds. Let $\boldsymbol{\theta} \in \mathrm{bm}(K)$. By Lemma A.6, $\boldsymbol{d}(\boldsymbol{\theta}(\cdot), \cdot)$, $\boldsymbol{w}(\boldsymbol{\theta}(\cdot), \cdot)$, and $\boldsymbol{B}(\boldsymbol{\theta}(\cdot), \cdot)$ are predictable. Furthermore, since d_i, w_i, and B_{ij} are locally bounded as stochastic process into the space $C(K)$ of continuous functions equipped with the supremum norm, $d_i(\boldsymbol{\theta}(\cdot), \cdot)$, $w_i(\boldsymbol{\theta}(\cdot), \cdot)$, and $B_{ij}(\boldsymbol{\theta}(\cdot), \cdot)$ are locally bounded as real-valued stochastic processes. Furthermore, we assumed in (LSW1) that w_i^{-1} is locally bounded. Thus by Lemma 3.7, the process $\boldsymbol{c}(\boldsymbol{\theta}(\cdot), \cdot)$ is locally bounded and predictable. Let

$$g_{ij}(t) := I\{\det(\overline{\boldsymbol{BwB}}(\boldsymbol{\theta}_0(t), t)) > L\} \nabla_j \left(\boldsymbol{Q}^{\boldsymbol{B}(\boldsymbol{\theta}_0(t),t)}_{\boldsymbol{w}(\boldsymbol{\theta}_0(t),t)\cdot\mathbb{P}_n} \boldsymbol{d}(\boldsymbol{\theta}_0(t), t)\right)_i.$$

Clearly, on A_n, $g_{ij}(t) = (\nabla_j c_i)(\boldsymbol{\alpha}_0(t), \boldsymbol{\beta}_0(t), t)$. Using Cramer's rule and the local boundedness of $\boldsymbol{B}$, $\boldsymbol{w}$, and $\boldsymbol{d}$, one can show that g_{ij} is locally bounded.

Since the determinant is continuous and hence measurable, g_{ij} is predictable as a sum/product/quotient of predictable processes.

Next, we show that the growth conditions given in (N5) hold. On A_n,

$$|c_i| \leq |d_i| - \sqrt{k} \max_{j=1,\dots,k} |B_{ij}| \| (\overline{\boldsymbol{BwB}})^{-1} \| \| \overline{\boldsymbol{Bwd}} \|.$$

By the integrability condition, and Lemma A.4,

$$\sup_{\substack{i=1,\dots,n \\ (\boldsymbol{x}^{\boldsymbol{\theta}},t)\in\mathcal{C}}} |d_i(\boldsymbol{x}^{\boldsymbol{\theta}}, t)| = o_{\mathrm{P}}(n^{1/8+\nu/4}) \quad \text{and} \quad \sup_{\substack{i=1,\dots,n \\ (\boldsymbol{x}^{\boldsymbol{\theta}},t)\in\mathcal{C}}} |B_{ij}(\boldsymbol{x}^{\boldsymbol{\theta}}, t)| = o_{\mathrm{P}}(n^{1/8+\nu/4}).$$

By the strong law of large numbers (Theorem A.5), $\overline{\boldsymbol{BwB}}$ and $\overline{\boldsymbol{Bwd}}$ converge in probability to bounded limits, uniformly on $\mathcal{C}$. By Lemma 3.3 this also holds for $(\overline{\boldsymbol{BwB}})^{-1}$. Hence,

$$\sup_{\substack{i=1\dots n \\ (\boldsymbol{x}^{\boldsymbol{\theta}},t)\in\mathcal{C}}} |c_i(\boldsymbol{x}^{\boldsymbol{\theta}}, t) I(A_n)| = o_{\mathrm{P}}(n^{1/8+\nu/4}).$$

Using the formulas for $\nabla_j \boldsymbol{Q}^{\boldsymbol{B}}_{\boldsymbol{w}\cdot\mathbb{P}_n} \boldsymbol{d}$ and $\nabla_\nu \nabla_j \boldsymbol{Q}^{\boldsymbol{B}}_{\boldsymbol{w}\cdot\mathbb{P}_n} \boldsymbol{d}$ on A_n, the remaining conditions of (N5) can be shown similarly.

Differentiability of $f_i(\cdot, t)$ is a part of (LS1). The rest of (N6) is a direct consequence of the remark after Lemma A.6.

To see (N7), one uses the expressions for $\boldsymbol{c}$, $\nabla_j \boldsymbol{c}$, and $\nabla_j \nabla_\nu \boldsymbol{c}$ on A_n and the two laws of large numbers: Theorem A.4 and Theorem A.5.

Next, we show (i). The convergence of $\overline{\boldsymbol{h}}$ is a consequence of the law of large numbers given in Theorem A.4. The convergence of $\overline{\boldsymbol{chc}}$ follows by (5.22) and the same law of large numbers. This shows (F1). To see (F2), one uses (5.22) and Theorem A.5. By the same theorem and (LSW2), $\overrightarrow{\boldsymbol{ch}}(\boldsymbol{x}^{\boldsymbol{\theta}}, s) = \mathrm{E}[(\boldsymbol{Q}^{\boldsymbol{B}_1(\boldsymbol{x}^{\boldsymbol{\theta}},s)}_{w_1(\boldsymbol{x}^{\boldsymbol{\theta}},s)\cdot \mathrm{P}} \boldsymbol{d}(\boldsymbol{x}^{\boldsymbol{\theta}}, s)]$ and similarly for $\overrightarrow{\boldsymbol{cf}}$. Hence, the equicontinuity conditions of (F1) are part of the conditions in (i). The limits of $n^{-1/2} T(\boldsymbol{c}, \widehat{\boldsymbol{\theta}}, t)$ and $\hat{\sigma}^2(\boldsymbol{c}, t)$ are consequences of Theorem A.4 and Theorem 5.2.

(ii) can be shown similarly to (i).

□

Chapter 6

Checking a Semiparametric Model

In Chapter 3 and Chapter 5, we only considered models with time-dependent parameters. However, many models in survival analysis require that certain parameters are constant over time. The prime example is the classical Cox model (2.1), where the regression parameter $\boldsymbol{\beta}$ does not depend on time. We can ignore that some parameters do not depend on time and use the techniques of Chapter 5. In fact, we shall do this to construct a check of the Cox model in Chapter 7. But of course, ignoring that parameters are constant over time does not use all information about the model.

The goal of this chapter is to extend the techniques of Chapter 5 to make use of the information that some parameters are constant over time. Basically, we want to check the model (1.4) mentioned in the introduction, which, with some more precision than in the introduction, can be written as follows:

$$\lambda_i(t) = f(\boldsymbol{X}_i(t), \boldsymbol{\alpha}^v(t), \boldsymbol{\alpha}^c) \text{ for some } \boldsymbol{\alpha}^v \in \text{bm}(\boldsymbol{\mathcal{X}}_{\boldsymbol{\alpha}^v}), \boldsymbol{\alpha}^c \in \boldsymbol{\mathcal{X}}_{\boldsymbol{\alpha}^c}, \tag{6.1}$$

where $\boldsymbol{\mathcal{X}}_{\boldsymbol{\alpha}^v} \subset \mathbb{R}^{k_{\alpha^v}}$ and $\boldsymbol{\mathcal{X}}_{\boldsymbol{\alpha}^c} \subset \mathbb{R}^{k_{\alpha^c}}$ are convex sets, f is a known continuous function and the observable covariates $\boldsymbol{X}_i$ are vectors of locally bounded predictable stochastic processes.

Similar to Chapter 5, we shall not be working with (6.1) but with the following, slightly more general model:

$$\lambda_i(t) = f_i(\boldsymbol{\alpha}^v(t), \boldsymbol{\alpha}^c, t) \text{ for some } \boldsymbol{\alpha}^v \in \text{bm}(\boldsymbol{\mathcal{X}}_{\boldsymbol{\alpha}^v}), \boldsymbol{\alpha}^c \in \boldsymbol{\mathcal{X}}_{\boldsymbol{\alpha}^c}, \tag{6.2}$$

where $\boldsymbol{\mathcal{X}}_{\boldsymbol{\alpha}^v} \subset \mathbb{R}^{k_{\alpha^v}}$, $\boldsymbol{\mathcal{X}}_{\boldsymbol{\alpha}^c} \subset \mathbb{R}^{k_{\alpha^c}}$ are convex sets, and for each $\boldsymbol{\alpha}^v \in \text{bm}(\boldsymbol{\mathcal{X}}_{\boldsymbol{\alpha}^v})$ and $\boldsymbol{\alpha}^c \in \boldsymbol{\mathcal{X}}_{\boldsymbol{\alpha}^c}$, the observable stochastic processes $f_i(\boldsymbol{\alpha}^v(\cdot), \boldsymbol{\alpha}^c, \cdot)$ are predictable and locally bounded.

To see that every model of type (6.1) can be written as (6.2), one only has to let $f_i(\boldsymbol{\alpha}^v(t), \boldsymbol{\alpha}^c, t) := f(\boldsymbol{X}_i(t), \boldsymbol{\alpha}^v(t), \boldsymbol{\alpha}^c)$ and note that under the assumptions of

(6.1), Lemma A.6 and Lemma A.7 imply that for all $\boldsymbol{\alpha}^v \in \mathrm{bm}(\boldsymbol{\mathcal{X}}_{\boldsymbol{\alpha}^v})$, $\boldsymbol{\alpha}^c \in \boldsymbol{\mathcal{X}}_{\boldsymbol{\alpha}^c}$, the stochastic process $f(\boldsymbol{X}_i(t), \boldsymbol{\alpha}^v(t), \boldsymbol{\alpha}^c)$ is locally bounded and predictable.

6.1 Orthogonality Conditions

To use that $\boldsymbol{\alpha}^c$ does not depend on time, we work with our test statistic T at time $t = \tau$. Thus we use

$$T(\boldsymbol{c}, \boldsymbol{\theta}) = n^{-\frac{1}{2}} \int_0^\tau \boldsymbol{c}(\boldsymbol{\theta}(t), t)^\top (\mathrm{d}\boldsymbol{N}(t) - \boldsymbol{f}(\boldsymbol{\alpha}(t), t)\,\mathrm{d}t)\,.$$

Note that since in this chapter, we always consider $T(\boldsymbol{c}, \boldsymbol{\theta}, t)$ at time $t = \tau$, we remove time from the list of arguments of T and write $T(\boldsymbol{c}, \boldsymbol{\theta})$.

The parameter $\boldsymbol{\theta}$ is composed of several components, some from the null hypothesis (denoted by $\boldsymbol{\alpha} = (\boldsymbol{\alpha}^v, \boldsymbol{\alpha}^c)$), some which are used to direct the test (denoted by $\boldsymbol{\gamma}$) and some (this is different from Chapter 5) that are used to satisfy the orthogonality conditions (denoted by $\boldsymbol{\eta}$). Before describing the details of these components we first give an overview how $\boldsymbol{\theta}$ is composed:

$$\boldsymbol{\theta} = (\underbrace{\overbrace{\boldsymbol{\alpha}^v, \boldsymbol{\alpha}^c}^{=\boldsymbol{\alpha}}, \overbrace{\boldsymbol{\gamma}}^{=\boldsymbol{\beta}}}_{=\boldsymbol{\psi}}, \boldsymbol{\eta}) \in \boldsymbol{\Theta} := \underbrace{\mathrm{bm}(\boldsymbol{\mathcal{X}}_{\boldsymbol{\alpha}^v}) \times \boldsymbol{\mathcal{X}}_{\boldsymbol{\alpha}^c} \times \mathrm{bm}(\boldsymbol{\mathcal{X}}_{\boldsymbol{\gamma}})}_{=:\boldsymbol{\Psi}} \times \boldsymbol{\mathcal{X}}_{\boldsymbol{\eta}},$$

$$\boldsymbol{\theta}(t) = (\underbrace{\overbrace{\boldsymbol{\alpha}^v(t)^\top, \boldsymbol{\alpha}^{c\top}}^{=\boldsymbol{\alpha}(t)^\top}, \overbrace{\boldsymbol{\gamma}(t)^\top}^{=\boldsymbol{\beta}(t)^\top}}_{=\boldsymbol{\psi}(t)^\top}, \boldsymbol{\eta}^\top)^\top \in \boldsymbol{\mathcal{X}}_{\boldsymbol{\theta}} = \underbrace{\overbrace{\boldsymbol{\mathcal{X}}_{\boldsymbol{\alpha}^v} \times \boldsymbol{\mathcal{X}}_{\boldsymbol{\alpha}^c}}^{=:\boldsymbol{\mathcal{X}}_{\boldsymbol{\alpha}}} \times \overbrace{\boldsymbol{\mathcal{X}}_{\boldsymbol{\gamma}}}^{=:\boldsymbol{\mathcal{X}}_{\boldsymbol{\beta}}}}_{=:\boldsymbol{\mathcal{X}}_{\boldsymbol{\psi}}} \times \boldsymbol{\mathcal{X}}_{\boldsymbol{\eta}} \subset \mathbb{R}^{k_{\boldsymbol{\theta}}}.$$

We require $\boldsymbol{\mathcal{X}}_{\boldsymbol{\theta}}$ to be convex. Similarly to the previous chapter, $k_{\boldsymbol{\theta}}$ (resp. $k_{\boldsymbol{\psi}}, \ldots$) denotes the dimension and $\boldsymbol{x}^{\boldsymbol{\theta}}$ (resp. $\boldsymbol{x}^{\boldsymbol{\psi}}, \ldots$) denotes an element of $\boldsymbol{\mathcal{X}}_{\boldsymbol{\theta}}$ (resp. $\boldsymbol{\mathcal{X}}_{\boldsymbol{\psi}}, \ldots$). As in Chapter 5, we will use the operator $\nabla_j = \frac{\partial}{\partial x_j^{\boldsymbol{\theta}}}$ for $j = 1, \ldots, k_{\boldsymbol{\theta}}$, and the row vectors of operators $\nabla_{\boldsymbol{\theta}}$, $\nabla_{\boldsymbol{\alpha}}$, and $\nabla_{\boldsymbol{\beta}}$. The following row vectors of operators are defined analogously: $\nabla_{\boldsymbol{\psi}}$, $\nabla_{\boldsymbol{\alpha}^v}$, $\nabla_{\boldsymbol{\alpha}^c}$, and $\nabla_{\boldsymbol{\eta}}$.

Using the test statistic only at time $t = \tau$ enables us to work with the relaxed orthogonality condition (1.8). We will use it in the following modified version: We assume that we have a stochastic process

$$\widehat{\boldsymbol{\eta}} : \Omega \times \boldsymbol{\Psi} \to \boldsymbol{\mathcal{X}}_{\boldsymbol{\eta}},$$

indexed by $\boldsymbol{\Psi} := \mathrm{bm}(\boldsymbol{\mathcal{X}}_{\boldsymbol{\alpha}^v}) \times \boldsymbol{\mathcal{X}}_{\boldsymbol{\alpha}^c} \times \mathrm{bm}(\boldsymbol{\mathcal{X}}_{\boldsymbol{\gamma}})$ such that the following condition is satisfied:

(SP1a) (orthogonality condition) For all $\boldsymbol{\psi} = (\boldsymbol{\alpha}^v, \boldsymbol{\alpha}^c, \boldsymbol{\gamma}) \in \boldsymbol{\Psi}$,

$$\int_0^\tau \sum_{i=1}^n c_i(\boldsymbol{\psi}(s), \widehat{\boldsymbol{\eta}}(\boldsymbol{\psi}), s) \nabla_{\boldsymbol{\alpha}^c} f_i(\boldsymbol{\alpha}^v(s), \boldsymbol{\alpha}^c, s)\, \mathrm{d}s = \mathbf{0}.$$

By plugging in $\widehat{\boldsymbol{\eta}}(\boldsymbol{\psi})$ into $\boldsymbol{c}$ we can satisfy the orthogonality condition (SP1a) as replacement for (1.8) and still require that for $\boldsymbol{\theta} \in \boldsymbol{\Theta}$ the stochastic process $\boldsymbol{c}(\boldsymbol{\theta}(\cdot), \cdot)$ is predictable.

The test statistic we use is $T(\boldsymbol{c}, \widehat{\boldsymbol{\theta}})$, where $\widehat{\boldsymbol{\theta}} := (\widehat{\boldsymbol{\psi}}, \widehat{\boldsymbol{\eta}}(\widehat{\boldsymbol{\psi}}))$, and where $\widehat{\boldsymbol{\psi}}$ is some estimator for $\boldsymbol{\psi}$.

To tie in with the notation of Chapter 5 and to use results of that chapter, all parameters besides those from the null hypothesis on which $\widehat{\boldsymbol{c}}$ depends are denoted by $\boldsymbol{\beta}$. The first components of $\boldsymbol{\beta}$, denoted by $\boldsymbol{\gamma} \in \mathrm{bm}(\boldsymbol{\mathcal{X}}_{\boldsymbol{\gamma}})$, where $\boldsymbol{\mathcal{X}}_{\boldsymbol{\gamma}} \subset \mathbb{R}^{k_{\boldsymbol{\gamma}}}$, will be used to direct the test. We will interpret $\widehat{\boldsymbol{\eta}}$ as an estimator of the last components of $\boldsymbol{\beta}$. These last components of $\boldsymbol{\beta}$ will be denoted by $\boldsymbol{\eta}$ and thus we shall write $\boldsymbol{\beta} = (\boldsymbol{\gamma}, \boldsymbol{\eta})$.

Using this notation, we may state the other orthogonality condition (1.6) as follows:

(SP1b) (orthogonality condition) For $\mathrm{P} \otimes \lambda$ almost all (ω, t),

$$\sum_{i=1}^n c_i(\omega, \boldsymbol{x}^{\boldsymbol{\theta}}, t) \nabla_{\boldsymbol{\alpha}^v} f_i(\omega, \boldsymbol{x}^{\boldsymbol{\alpha}^v}, \boldsymbol{x}^{\boldsymbol{\alpha}^c}, t) = \mathbf{0}, \quad \forall \boldsymbol{x}^{\boldsymbol{\theta}} = \begin{pmatrix} \boldsymbol{x}^{\boldsymbol{\alpha}^v} \\ \boldsymbol{x}^{\boldsymbol{\alpha}^c} \\ \boldsymbol{x}^{\boldsymbol{\beta}} \end{pmatrix} \in \boldsymbol{\mathcal{X}}_{\boldsymbol{\theta}}.$$

6.2 Asymptotics under the Null Hypothesis

In order to show asymptotic normality of $T(\boldsymbol{c}, \widehat{\boldsymbol{\theta}})$ under the null hypothesis, we reuse parts of the proof of Theorem 5.1. Therefore we need to consider the conditions (N1)-(N7) which were assumed in Theorem 5.1. Conditions (N3)-(N7) of Theorem 5.1 will be part of the assumptions of the next theorem. Condition (N2) will be replaced by the conditions (SP1a) and (SP1b) which we already introduced. Instead of (N1), we use the following conditions:

(SP2a) (convergence of $\widehat{\boldsymbol{\psi}}$)
There exists $\boldsymbol{\psi}_0 = (\boldsymbol{\alpha}_0, \boldsymbol{\gamma}_0) \in \boldsymbol{\Psi}$ such that for some $0 < \nu \leq 1/2$,

$$\sup_{t \in [0,\tau]} \|\widehat{\boldsymbol{\psi}}(t) - \boldsymbol{\psi}_0(t)\| = O_{\mathrm{P}}(n^{-1/4-\nu/2})$$

and such that the total variation of components of $\widehat{\boldsymbol{\psi}} - \boldsymbol{\psi}_0$ satisfies

$$\int_0^\tau |\,\mathrm{d}(\widehat{\psi}_j(t) - \psi_{0,j}(t))| \xrightarrow{\mathrm{P}} 0, \quad j = 1, \ldots, k_{\boldsymbol{\psi}},$$

where $\psi_{0,j}$ denotes the jth component of $\boldsymbol{\psi}_0$.

Condition (SP2a) can be ensured by using the smoothing procedure of Section 4.2.

(SP2b) (convergence of $\widehat{\boldsymbol{\eta}}(\widehat{\boldsymbol{\psi}})$ and $\widehat{\boldsymbol{\eta}}(\boldsymbol{\psi}_0)$)
There exists a deterministic vector $\boldsymbol{\eta}_0(\boldsymbol{\psi}_0) \in \mathbb{R}^{k_\eta}$ such that $\widehat{\boldsymbol{\eta}}(\widehat{\boldsymbol{\psi}}) - \boldsymbol{\eta}_0(\boldsymbol{\psi}_0) = O_{\mathrm{P}}(n^{-1/4-\nu/2})$ and $\widehat{\boldsymbol{\eta}}(\boldsymbol{\psi}_0) - \boldsymbol{\eta}_0(\boldsymbol{\psi}_0) = O_{\mathrm{P}}(n^{-1/4-\nu/2})$.

So under (SP2a) and (SP2b), the combined parameter estimator $\widehat{\boldsymbol{\theta}} = (\widehat{\boldsymbol{\psi}}, \widehat{\boldsymbol{\eta}}(\widehat{\boldsymbol{\psi}}))$ converges to $\boldsymbol{\theta}_0 := (\boldsymbol{\psi}_0, \boldsymbol{\eta}_0(\boldsymbol{\psi}_0))$.

If the null hypothesis (6.2) holds true the following theorem gives asymptotic normality of $T(\boldsymbol{c}, \widehat{\boldsymbol{\theta}})$.

Theorem 6.1. *Suppose conditions (SP1a), (SP1b), (SP2a), (SP2b), (N3) - (N7) hold true and that*

$$\lambda_i(\cdot) = f_i(\boldsymbol{\alpha}_0^v(\cdot), \boldsymbol{\alpha}_0^c, \cdot), i = 1, \dots n,$$

where $\boldsymbol{\alpha}_0(t) = (\boldsymbol{\alpha}_0^v(t)^\top, \boldsymbol{\alpha}_0^{c\top})^\top$. *Then*

$$T(\boldsymbol{c}, \widehat{\boldsymbol{\psi}}, \widehat{\boldsymbol{\eta}}(\widehat{\boldsymbol{\psi}})) \xrightarrow{d} N(0, \sigma^2(\boldsymbol{c})),$$

where

$$\sigma^2(\boldsymbol{c}) = \int_0^\tau \overrightarrow{\boldsymbol{cfc}}(\boldsymbol{\theta}_0(s), s)\, \mathrm{d}s.$$

Furthermore,

$$\hat{\sigma}^2(\boldsymbol{c}) := \frac{1}{n} \int_0^\tau \boldsymbol{c}(\widehat{\boldsymbol{\theta}}(s), s)^\top \operatorname{diag}(\mathrm{d}\boldsymbol{N}(s)) \boldsymbol{c}(\widehat{\boldsymbol{\theta}}(s), s) \xrightarrow{\mathrm{P}} \sigma^2(\boldsymbol{c}).$$

Proof. First, we show that (N1) and (N1') hold true. By (SP2a), (SP2b), we have $\sup_{t\in[0,\tau]} \|\widehat{\boldsymbol{\theta}}(t) - \boldsymbol{\theta}_0(t)\| = O_{\mathrm{P}}(n^{-1/4-\nu/2})$, since by definition, $\widehat{\boldsymbol{\theta}} = (\widehat{\boldsymbol{\psi}}, \widehat{\boldsymbol{\eta}}(\widehat{\boldsymbol{\psi}}))$ and $\boldsymbol{\theta}_0 = (\boldsymbol{\psi}_0, \boldsymbol{\eta}_0(\boldsymbol{\psi}_0))$. Since $\widehat{\boldsymbol{\eta}}(\widehat{\boldsymbol{\psi}}) - \boldsymbol{\eta}_0(\boldsymbol{\theta}_0)$ does not depend on time, the condition on the total variance of $\widehat{\boldsymbol{\theta}} - \boldsymbol{\theta}_0$ reduces to the condition of the total variance of $\widehat{\boldsymbol{\psi}} - \boldsymbol{\psi}_0$ given in (SP2b).

The remainder of the proof can be done along the lines of the proof of Theorem 5.1. The only difference occurs when we show that the second term on the right hand side of (5.6) vanishes. This term is

$$n^{-\frac{1}{2}} \int_0^t \left[\sum_{i=1}^n c_i(\boldsymbol{\theta}_0(s), s) \nabla_{\boldsymbol{\alpha}} f_i(\boldsymbol{\alpha}_0(s), s) \right] (\widehat{\boldsymbol{\alpha}}(s) - \boldsymbol{\alpha}_0(s))\, \mathrm{d}s. \tag{6.3}$$

We only need to consider (6.3) for $t = \tau$. We may rewrite (6.3) for $t = \tau$ as the sum of

$$n^{-\frac{1}{2}} \int_0^\tau \left[\sum_{i=1}^n c_i(\boldsymbol{\theta}_0(s), s) \nabla_{\boldsymbol{\alpha}^v} f_i(\boldsymbol{\alpha}_0(s), s) \right] (\widehat{\boldsymbol{\alpha}^v}(s) - \boldsymbol{\alpha}_0^v(s))\, \mathrm{d}s \tag{6.4}$$

and

$$n^{-\frac{1}{2}} \int_0^\tau \left[\sum_{i=1}^n c_i(\boldsymbol{\theta}_0(s), s) \nabla_{\boldsymbol{\alpha}^c} f_i(\boldsymbol{\alpha}_0(s), s) \right] \mathrm{d}s (\widehat{\boldsymbol{\alpha}^c} - \boldsymbol{\alpha}_0^c). \tag{6.5}$$

By (SP1b), the term (6.4) is zero. Because of (SP1a), we may rewrite the first part of (6.5) as

$$n^{-\frac{1}{2}} \int_0^\tau \left[\sum_{i=1}^n \left(c_i(\boldsymbol{\theta}_0(s), s) - c_i(\boldsymbol{\psi}_0(s), \widehat{\boldsymbol{\eta}}(\boldsymbol{\psi}_0), s) \right) \nabla_{\boldsymbol{\alpha}^c} f_i(\boldsymbol{\alpha}_0(s), s) \right] \mathrm{d}s,$$

which by a Taylor expansion is equal to

$$n^{-\frac{1}{2}} (\boldsymbol{\eta}_0(\boldsymbol{\psi}_0) - \widehat{\boldsymbol{\eta}}(\boldsymbol{\psi}_0))^\top \int_0^\tau \left[\sum_{i=1}^n (\nabla_{\boldsymbol{\eta}} c_i)(\boldsymbol{\psi}_0(s), \widetilde{\boldsymbol{\eta}}, s)^\top \nabla_{\boldsymbol{\alpha}^c} f_i(\boldsymbol{\alpha}_0(s), s) \right] \mathrm{d}s$$

for some $\widetilde{\boldsymbol{\eta}}$ between $\boldsymbol{\eta}_0(\boldsymbol{\psi}_0)$ and $\widehat{\boldsymbol{\eta}}(\boldsymbol{\psi}_0)$. Hence, as $\|\widehat{\boldsymbol{\alpha}^c} - \boldsymbol{\alpha}_0^c\| \|\boldsymbol{\eta}_0(\boldsymbol{\psi}_0) - \widehat{\boldsymbol{\eta}}(\boldsymbol{\psi}_0)\| = O_{\mathrm{P}}(n^{-1/2-\nu})$ and as $n^{-\nu}\overline{(\nabla_{\boldsymbol{\eta}} \boldsymbol{c})(\nabla_{\boldsymbol{\alpha}^c} \boldsymbol{f})}$ converges uniformly on $\mathcal{C}$ to zero by (N7), the term (6.5) converges to 0 in probability.

□

6.3 Least Squares Projections

Similar to Section 5.4, we first describe a simple way of ensuring the orthogonality conditions (SP1a) and (SP1b). Again, we take some n-variate vector of stochastic processes $\boldsymbol{d}(\boldsymbol{x}^{\boldsymbol{\theta}}, t)$, depending on $\boldsymbol{x}^{\boldsymbol{\theta}} \in \boldsymbol{\mathcal{X}}_{\boldsymbol{\theta}}$, and define $\boldsymbol{c}$ by an orthogonal projection of $\boldsymbol{d}$. Since condition (SP1a) involves all of $[0, \tau]$, this projection should involve all time-points.

Recall the space $L_2(\mathbb{P}_n \otimes \lambda\!\!\!\lambda)$ from Section 3.2. The conditions (SP1a) and (SP1b) are equivalent to the statement that

$$< \boldsymbol{c}(\boldsymbol{\theta}(\cdot), \cdot), \boldsymbol{u} >_{\mathbb{P}_n \otimes \lambda\!\!\!\lambda} = 0 \quad \forall\, \boldsymbol{u} \in U(\boldsymbol{\alpha}), \boldsymbol{\theta} = (\boldsymbol{\alpha}, \boldsymbol{\beta}) \in \boldsymbol{\Theta},$$

where $U(\boldsymbol{\alpha}) := U_v(\boldsymbol{\alpha}) + U_c(\boldsymbol{\alpha})$, $U_c(\boldsymbol{\alpha}) = \{\text{columns of } (\nabla_{\boldsymbol{\alpha}^c} \boldsymbol{f})(\boldsymbol{\alpha}(\cdot), \cdot)\}$, and

$$U_v(\boldsymbol{\alpha}) = \left\{ \widetilde{\boldsymbol{f}}(\cdot) g(\cdot) : g \in \mathrm{bm}([0, \tau]) \text{ and } \widetilde{\boldsymbol{f}}(\cdot) \text{ column of } (\nabla_{\boldsymbol{\alpha}^v} \boldsymbol{f})(\boldsymbol{\alpha}(\cdot), \cdot) \right\}.$$

Recall from Section 3.2, that $\boldsymbol{Q}_{\mathbb{P}_n \otimes \lambda\!\!\!\lambda}^{U(\boldsymbol{\alpha})}$ is the orthogonal projection onto $U(\boldsymbol{\alpha})^\perp$ with respect to $< \cdot, \cdot >_{\mathbb{P}_n \otimes \lambda\!\!\!\lambda}$. As weights we want to use

$$\boldsymbol{c}(\boldsymbol{\theta}(\cdot), \cdot) = \boldsymbol{Q}_{\mathbb{P}_n \otimes \lambda\!\!\!\lambda}^{U(\boldsymbol{\alpha})} \boldsymbol{d}(\boldsymbol{\theta}(\cdot), \cdot). \tag{6.6}$$

Equation (6.6) only defines $\boldsymbol{c}(\boldsymbol{\theta}(\cdot), \cdot)$ as an equivalence class of functions that are equal almost everywhere with respect to $\mathbb{P}_n \otimes \lambda\!\!\!\lambda$. However, in our test statistic T, we will integrate $\boldsymbol{c}(\boldsymbol{\theta}(\cdot),\cdot)$ with respect to the counting process $\boldsymbol{N}(t)$. Consider the set B of jumps of $\boldsymbol{N}$, i.e. $B = \{(i,t) \in \{1,\dots,n\} \times [0,\tau] : N_i(t) \neq N_i(t-)\}$. Clearly, $(\mathbb{P}_n \otimes \lambda)(B) = 0$. Hence, (6.6) does not define $\boldsymbol{c}(\boldsymbol{\theta}(\cdot),\cdot)$ on A. Thus we need to pick a particular element of this equivalence class. For the one that we pick, we need to be able to verify the conditions of Theorem 6.1. We shall do so for weighted projections in Section 6.5.

For now, we contend ourselves by giving an explicit formula for $\boldsymbol{Q}^{U(\boldsymbol{\alpha})}_{\mathbb{P}_n\otimes\lambda}$. The orthogonal projection onto $U_v(\boldsymbol{\alpha})^\perp$ can be reduced to ordinary orthogonal projections in $\mathbb{R}^n$ as follows: For $t \in [0,\tau]$, let $\boldsymbol{A}(t) := (\nabla_{\boldsymbol{\alpha}^v} \boldsymbol{f})(\boldsymbol{\alpha}(t),t)$ and let $\boldsymbol{Q}^{\boldsymbol{A}(t)}_{\mathbb{P}_n}$ be the orthogonal projection matrix onto $\{\text{columns of } \boldsymbol{A}(t)\}^\perp$ in $\mathbb{R}^n$. If $\boldsymbol{A}(t)$ has full column rank then $\boldsymbol{Q}^{\boldsymbol{A}(t)}_{\mathbb{P}_n} = \boldsymbol{I} - \boldsymbol{A}(t)(\boldsymbol{A}(t)^\top \boldsymbol{A}(t))^{-1}\boldsymbol{A}(t)^\top$. Let $\boldsymbol{Q}^{U_v(\boldsymbol{\alpha})}_{\mathbb{P}_n\otimes\lambda} : L_2(\mathbb{P}_n\otimes\lambda) \to L_2(\mathbb{P}_n\otimes\lambda)$ be given by

$$\boldsymbol{Q}^{U_v(\boldsymbol{\alpha})}_{\mathbb{P}_n\otimes\lambda}(\boldsymbol{x})(t) = \boldsymbol{Q}^{\boldsymbol{A}(t)}_{\mathbb{P}_n}\boldsymbol{x}(t).$$

By Lemma B.2, $\boldsymbol{Q}^{U_v(\boldsymbol{\alpha})}_{\mathbb{P}_n\otimes\lambda}$ is the projection onto $U_v(\boldsymbol{\alpha})^\perp$. Define the set

$$\tilde{U}_c(\boldsymbol{\alpha}) = \boldsymbol{Q}^{U_v(\boldsymbol{\alpha})}_{\mathbb{P}_n\otimes\lambda} U_c(\boldsymbol{\alpha}) = \{\boldsymbol{Q}^{U_v(\boldsymbol{\alpha})}_{\mathbb{P}_n\otimes\lambda}\boldsymbol{y} : \boldsymbol{y} \text{ column of } (\nabla_{\boldsymbol{\alpha}^c}\boldsymbol{f})(\boldsymbol{\alpha}(\cdot),\cdot)\},$$

and let $\boldsymbol{Q}^{\tilde{U}_c(\boldsymbol{\alpha})}_{\mathbb{P}_n\otimes\lambda}$ be the orthogonal projection onto $\tilde{U}_c(\boldsymbol{\alpha})^\perp$. By Lemma B.3, $\boldsymbol{Q}^{U(\boldsymbol{\alpha})}_{\mathbb{P}_n\otimes\lambda} = \boldsymbol{Q}^{U_v(\boldsymbol{\alpha})}_{\mathbb{P}_n\otimes\lambda}\boldsymbol{Q}^{\tilde{U}_c(\boldsymbol{\alpha})}_{\mathbb{P}_n\otimes\lambda} = \boldsymbol{Q}^{\tilde{U}_c(\boldsymbol{\alpha})}_{\mathbb{P}_n\otimes\lambda}\boldsymbol{Q}^{U_v(\boldsymbol{\alpha})}_{\mathbb{P}_n\otimes\lambda}$. Using that $\tilde{U}_c(\boldsymbol{\alpha})$ is a finite set, we immediately get the following result:

Proposition 6.1. *Let* $\boldsymbol{B} := \int_0^\tau \tilde{\boldsymbol{f}}(s)^\top \tilde{\boldsymbol{f}}(s)\,\mathrm{d}s$, $\tilde{\boldsymbol{f}}(t) := \boldsymbol{Q}^{\boldsymbol{A}(t)}_{\mathbb{P}_n}(\nabla_{\boldsymbol{\alpha}^c}\boldsymbol{f})(\boldsymbol{\alpha}(t),t)$, *and* $\boldsymbol{A}(t) = (\nabla_{\boldsymbol{\alpha}^v}\boldsymbol{f})(\boldsymbol{\alpha}(t),t)$. *If* $\boldsymbol{B}$ *is invertible then for* $\boldsymbol{a} \in L_2(\mathbb{P}_n\otimes\lambda)$,

$$\boldsymbol{Q}^{U(\boldsymbol{\alpha})}_{\mathbb{P}_n\otimes\lambda}(\boldsymbol{a})(t) = \boldsymbol{Q}^{\boldsymbol{A}(t)}_{\mathbb{P}_n}\boldsymbol{a}(t) - \tilde{\boldsymbol{f}}(t)\boldsymbol{B}^{-1}\int_0^\tau \tilde{\boldsymbol{f}}(s)^\top \boldsymbol{Q}^{\boldsymbol{A}(s)}_{\mathbb{P}_n}\boldsymbol{a}(s)\,\mathrm{d}s. \tag{6.7}$$

6.4 Optimal Weights under Alternatives

For brevity, we restrict ourselves to fixed alternatives given by $\lambda_i(t) = h_i(t)$. The modification to local alternatives can be done in a similar way to Subsection 5.5.2. We want to consider optimal tests based on

$$V(\boldsymbol{c}) := T(\boldsymbol{c},\hat{\boldsymbol{\theta}})/\sqrt{\hat{\sigma}^2(\boldsymbol{c})},$$

where $\hat{\sigma}^2(\boldsymbol{c})$ is as in Theorem 6.1. As in Subsection 3.6.1 and in Subsection 5.5.1, the optimality criterion we use is the approximate Bahadur efficiency, for which we need convergence results under the null hypothesis as well as under the fixed alternative. Under the null hypothesis we use Theorem 6.1. Since the orthogonality conditions are not part of Theorem 5.2, we may reuse this theorem for convergence under the alternative. One can argue as in Subsection 5.5.1, to see that it suffices to maximize

$$Z(\boldsymbol{c}) := \int_0^\tau \left(\overline{\boldsymbol{ch}}(\boldsymbol{\alpha}_0(s), s) - \overline{\boldsymbol{cf}}(\boldsymbol{\alpha}_0(s), s)\right) \mathrm{d}s \left(\int_0^\tau \overline{\boldsymbol{chc}}(\boldsymbol{\alpha}_0(s), s)\,\mathrm{d}s\right)^{-\frac{1}{2}}.$$

Instead of optimizing $\boldsymbol{c}$ in the class of all weights satisfying all conditions for Theorem 6.1 and Theorem 5.2, we optimize only under the orthogonality condition (SP1a) and (SP1b). Actually, we only require (SP1a) and (SP1b) for the limit $\boldsymbol{\alpha}_0$ of the parameter estimator and we are looking for weights $\boldsymbol{c}$ that do not depend on $\boldsymbol{\alpha}$. At the end of this section, we extend the optimal weights to weights depending on $\boldsymbol{\alpha}$ that satisfy (SP1a) and (SP1b). Thus, the optimization problem we are interested in solving is

$$\begin{cases} Z(\boldsymbol{c}) \to \max \\ \sum_{i=1}^n c_i(t)\nabla_{\boldsymbol{\alpha}^v} f_i(\boldsymbol{\alpha}_0(t), t) = \boldsymbol{0}, \quad \text{for almost all } t \in [0,\tau]. \\ \int_0^\tau \sum_{i=1}^n c_i(s)\nabla_{\boldsymbol{\alpha}^c} f_i(\boldsymbol{\alpha}_0(s), s)\,\mathrm{d}s = \boldsymbol{0}. \end{cases} \tag{6.8}$$

It will turn out that the solution of (6.8) is given by a certain orthogonal projection onto the orthogonal space to $V(\boldsymbol{\alpha}_0)$ where $V(\boldsymbol{\alpha})$ can be defined as follows: $V(\boldsymbol{\alpha}) := V_v(\boldsymbol{\alpha}) + V_c(\boldsymbol{\alpha})$, $V_c(\boldsymbol{\alpha}) = \left\{\frac{\widetilde{\boldsymbol{f}}(\cdot)}{\boldsymbol{h}(\cdot)} : \widetilde{\boldsymbol{f}}(\cdot) \text{ column of } (\nabla_{\boldsymbol{\alpha}^c}\boldsymbol{f})(\boldsymbol{\alpha}(\cdot), \cdot)\right\}$, and

$$V_v(\boldsymbol{\alpha}) = \left\{\frac{\widetilde{\boldsymbol{f}}(\cdot)}{\boldsymbol{h}(\cdot)} g(\cdot) : \widetilde{\boldsymbol{f}}(\cdot) \text{ column of } (\nabla_{\boldsymbol{\alpha}^v}\boldsymbol{f})(\boldsymbol{\alpha}(\cdot), \cdot) \text{ and } g \in \mathrm{bm}([0,\tau])\right\}.$$

Proposition 6.2. *Suppose that for all $i = 1, \ldots, n$, $t \in [0, \tau]$,*

$$h_i(t) = 0 \text{ implies } (f_i(\boldsymbol{\alpha}_0(t), t) = 0 \text{ and } \nabla_{\boldsymbol{\alpha}} f_i(\boldsymbol{\alpha}_0(t), t) = \boldsymbol{0}).$$

If $\frac{\boldsymbol{h}(\cdot) - \boldsymbol{f}(\boldsymbol{\alpha}_0(\cdot), \cdot)}{\boldsymbol{h}(\cdot)}$ and $\frac{(\nabla_j \boldsymbol{f})(\boldsymbol{\alpha}_0(\cdot), \cdot)}{\boldsymbol{h}(\cdot)}$, $j = 1, \ldots, k_{\boldsymbol{\alpha}}$, are elements of $L_2(\boldsymbol{h} \cdot (\mathbb{P}_n \otimes \lambda\!\!\lambda))$ then for all $\boldsymbol{c} \in L_2(\boldsymbol{h} \cdot (\mathbb{P}_n \otimes \lambda\!\!\lambda))$ that are admissible for (6.8), the following holds:

$$Z(\boldsymbol{c}^*) \geq Z(\boldsymbol{c}),$$

where $\boldsymbol{c}^* \in L_2(\boldsymbol{h} \cdot (\mathbb{P}_n \otimes \lambda\!\!\lambda))$ is given by

$$\boldsymbol{c}^* = \boldsymbol{Q}_{\boldsymbol{h} \cdot (\mathbb{P}_n \otimes \lambda\!\!\lambda)}^{V(\boldsymbol{\alpha}_0)} \frac{\boldsymbol{h}(\cdot) - \boldsymbol{f}(\boldsymbol{\alpha}_0(\cdot), \cdot)}{\boldsymbol{h}(\cdot)}.$$

Proof. We can rewrite the first side condition as follows:

$$< \boldsymbol{c}(t), \frac{\boldsymbol{v}(t)}{\boldsymbol{h}(t)} >_{\boldsymbol{h}(t) \cdot \mathbb{P}_n} = \boldsymbol{0}, \ \forall \text{ columns } \boldsymbol{v}(t) \text{ of } (\nabla_{\boldsymbol{\alpha}^v}\boldsymbol{f})(\boldsymbol{\alpha}_0(t), t) \text{ and almost all } t.$$

We can rewrite this further as

$$<\boldsymbol{c}, \boldsymbol{v}>_{\boldsymbol{h}\cdot(\mathbb{P}_n\otimes\lambda)}= 0 \quad \forall \boldsymbol{v} \in V_v(\boldsymbol{\alpha}_0).$$

We can rewrite the second side condition as follows:

$$<\boldsymbol{c}, \boldsymbol{v}>_{\boldsymbol{h}\cdot(\mathbb{P}_n\otimes\lambda)}= 0 \quad \forall \boldsymbol{v} \in V_c(\boldsymbol{\alpha}_0).$$

Thus, the side conditions can be written in a combined way as follows:

$$<\boldsymbol{c}, \boldsymbol{v}>_{\boldsymbol{h}\cdot(\mathbb{P}_n\otimes\lambda)}= 0, \ \forall \boldsymbol{v} \in V(\boldsymbol{\alpha}_0).$$

Using properties of orthogonal projections and the Cauchy-Schwarz inequality, we get for all $\boldsymbol{c}$ that are admissible for (6.8) that

$$\begin{aligned} n\int_0^\mathcal{T} \left(\overline{\boldsymbol{ch}}(s) - \overline{\boldsymbol{cf}}(\boldsymbol{\alpha}_0(s), s)\right) \mathrm{d}s &=< \boldsymbol{c}, \frac{\boldsymbol{h}(\cdot) - \boldsymbol{f}(\boldsymbol{\alpha}_0, \cdot)}{\boldsymbol{h}(\cdot)} >_{\boldsymbol{h}\cdot(\mathbb{P}_n\otimes\lambda)} \\ &=< Q^{V(\boldsymbol{\alpha}_0)}_{\boldsymbol{h}\cdot(\mathbb{P}_n\otimes\lambda)}\boldsymbol{c}, \frac{\boldsymbol{h}(\cdot) - \boldsymbol{f}(\boldsymbol{\alpha}_0, \cdot)}{\boldsymbol{h}(\cdot)} >_{\boldsymbol{h}\cdot(\mathbb{P}_n\otimes\lambda)} \\ &=< \boldsymbol{c}, \boldsymbol{c}^* >_{\boldsymbol{h}\cdot(\mathbb{P}_n\otimes\lambda)} \\ &\leq \|\boldsymbol{c}\|_{\boldsymbol{h}\cdot(\mathbb{P}_n\otimes\lambda)} \|\boldsymbol{c}^*\|_{\boldsymbol{h}\cdot(\mathbb{P}_n\otimes\lambda)} \\ &= \left(n\int_0^\mathcal{T} \overline{\boldsymbol{chc}}(s)\,\mathrm{d}s\right)^{\frac{1}{2}} \|\boldsymbol{c}^*\|_{\boldsymbol{h}\cdot(\mathbb{P}_n\otimes\lambda)} \end{aligned}$$

with equality if $\boldsymbol{c} = \boldsymbol{c}^*$. By the assumptions, $\int_0^\mathcal{T} \overline{\boldsymbol{chc}}(s)\,\mathrm{d}s = 0$ implies that $\int_0^\mathcal{T} \left(\overline{\boldsymbol{ch}}(s) - \overline{\boldsymbol{cf}}(\boldsymbol{\alpha}_0(s), s)\right) \mathrm{d}s = 0$ and hence, $Z(\boldsymbol{c}^*) \leq n^{-1/2} \|\boldsymbol{c}^*\|_{\boldsymbol{h}\cdot(\mathbb{P}_n\otimes\lambda)}$, with equality if $\boldsymbol{c} = \boldsymbol{c}^*$. □

If we define $\boldsymbol{c}(\boldsymbol{\alpha}) = Q^{V(\boldsymbol{\alpha})}_{\boldsymbol{h}\cdot(\mathbb{P}_n\otimes\lambda)} \frac{\boldsymbol{h}(\cdot) - \boldsymbol{f}(\boldsymbol{\alpha}(\cdot), \cdot)}{\boldsymbol{h}(\cdot)}$ then $\boldsymbol{c}(\boldsymbol{\alpha}_0) = \boldsymbol{c}^*$ and $\boldsymbol{c}$ satisfies the conditions (SP1a) and (SP1b). Of course, the definition of $\boldsymbol{c}$ is only up to $\boldsymbol{h}\cdot(\mathbb{P}_n\otimes\lambda)$-null sets, which is not precise enough for the other conditions we need for the convergence results. In the next section, we shall give more details, especially concerning $\widehat{\boldsymbol{\eta}}$ and show that under suitable conditions an element of the equivalence class $\boldsymbol{c}(\boldsymbol{\alpha})$ satisfies the remaining conditions we need for convergence under the null hypothesis as well as under fixed alternatives.

6.5 Weighted Projections are Admissible

In the previous section, we have seen that optimal weights are given by certain weighted orthogonal projections. The goal of this section is to show that one can choose a stochastic process that is an element of the equivalence class defined by weighted orthogonal projections, that can be shown to satisfy the other conditions needed for the asymptotic results.

Suppose $\boldsymbol{d}$ and $\boldsymbol{w}$ are n-variate vectors of stochastic processes and $\boldsymbol{B}^c$ is an $n \times k^c$-matrix of stochastic processes and $\boldsymbol{B}^v$ is an $n \times k^v$-matrix of stochastic processes. All of these stochastic processes may depend on $\boldsymbol{x}^{\boldsymbol{\psi}} \in \boldsymbol{\mathcal{X}}_{\boldsymbol{\psi}}$. We want to pick an element of the equivalence class defined by

$$\boldsymbol{Q}^{V(\boldsymbol{\psi})}_{\boldsymbol{w}(\boldsymbol{\psi}(\cdot),\cdot)\cdot(\mathbb{P}_n\otimes\lambda\!\!\lambda)}\boldsymbol{d}(\boldsymbol{\psi}(\cdot),\cdot), \tag{6.9}$$

where $V(\boldsymbol{\psi}) := V_v(\boldsymbol{\psi}) + V_c(\boldsymbol{\psi})$, $V_c(\boldsymbol{\psi}) = \{\text{columns of } \boldsymbol{B}^c(\boldsymbol{\psi}(\cdot),\cdot)\}$, and

$$V_v(\boldsymbol{\psi}) = \{\boldsymbol{y}(\cdot)g(\cdot) : \boldsymbol{y} \text{ column of } \boldsymbol{B}^v(\boldsymbol{\psi}(\cdot),\cdot) \text{ and } g \in \mathrm{bm}([0,\tau])\}.$$

Remark 6.1. Consider the optimal weights of the previous section. As the alternative is completely known, for these weights we have $\boldsymbol{\psi} = \boldsymbol{\alpha}$. Furthermore, the optimal weights are given by (6.9) with

$$\boldsymbol{w}(\boldsymbol{x}^{\boldsymbol{\alpha}},t) = \boldsymbol{h}(t), \qquad \boldsymbol{d}(\boldsymbol{x}^{\boldsymbol{\alpha}},t) = \frac{\boldsymbol{h}(t) - \boldsymbol{f}(\boldsymbol{x}^{\boldsymbol{\alpha}},t)}{\boldsymbol{h}(t)},$$

$$\boldsymbol{B}^c(\boldsymbol{x}^{\boldsymbol{\alpha}},t) = \frac{\nabla_{\boldsymbol{\alpha}^c}\boldsymbol{f}(\boldsymbol{x}^{\boldsymbol{\alpha}},t)}{\boldsymbol{h}(t)}, \text{ and} \qquad \boldsymbol{B}^v(\boldsymbol{x}^{\boldsymbol{\alpha}},t) = \frac{\nabla_{\boldsymbol{\alpha}^v}\boldsymbol{f}(\boldsymbol{x}^{\boldsymbol{\alpha}},t)}{\boldsymbol{h}(t)}.$$

The element of the equivalence class (6.9) we pick is $\boldsymbol{c}(\boldsymbol{\psi}(\cdot),\widehat{\boldsymbol{\eta}}(\boldsymbol{\psi}),\cdot)$, where

$$\boldsymbol{c}(\boldsymbol{x}^{\boldsymbol{\psi}},\boldsymbol{\eta},t) = \boldsymbol{Q}^{\boldsymbol{B}^v(\boldsymbol{x}^{\boldsymbol{\psi}},t)}_{\boldsymbol{w}(\boldsymbol{x}^{\boldsymbol{\psi}},t)\cdot\mathbb{P}_n}\left(\boldsymbol{d}(\boldsymbol{x}^{\boldsymbol{\psi}},t) - \boldsymbol{B}^c(\boldsymbol{x}^{\boldsymbol{\psi}},t)\boldsymbol{\eta}\right), \tag{6.10}$$

$$\widehat{\boldsymbol{\eta}}(\boldsymbol{\psi}) = \left(\int_0^\tau \overline{\tilde{\boldsymbol{B}}^c\boldsymbol{w}\tilde{\boldsymbol{B}}^c}(\boldsymbol{\psi}(s),s)\,\mathrm{d}s\right)^{-1}\int_0^\tau \overline{\tilde{\boldsymbol{B}}^c\boldsymbol{w}\boldsymbol{d}}(\boldsymbol{\psi}(s),s)\,\mathrm{d}s, \tag{6.11}$$

and $\tilde{\boldsymbol{B}}^c(\boldsymbol{x}^{\boldsymbol{\psi}},t) = \boldsymbol{Q}^{\boldsymbol{B}^v(\boldsymbol{x}^{\boldsymbol{\psi}},t)}_{\boldsymbol{w}(\boldsymbol{x}^{\boldsymbol{\psi}},t)\cdot\mathbb{P}_n}\boldsymbol{B}^c(\boldsymbol{x}^{\boldsymbol{\psi}},t)$. As before, we set

$$c_i(\boldsymbol{x}^{\boldsymbol{\psi}},\boldsymbol{\eta},t) = 0 \text{ if } w_i(\boldsymbol{x}^{\boldsymbol{\psi}},t) = 0.$$

By Lemma B.2 and Lemma B.3, $\boldsymbol{c}(\boldsymbol{\psi}(\cdot),\widehat{\boldsymbol{\eta}}(\boldsymbol{\psi}),\cdot)$ is an element of the equivalence class defined by (6.9).

As already mentioned, we want to consider the conditions needed for the asymptotic results, i.e. Theorem 6.1 for convergence under the null hypothesis and Theorem 5.2 for convergence under fixed alternatives.

Whether (SP1a) and (SP1b) are satisfied depends on the combination of $\boldsymbol{B}^c$, $\boldsymbol{B}^v$, and $\boldsymbol{w}$. In general, if $\mathrm{diag}(\boldsymbol{w})\boldsymbol{B}^c = \nabla_{\boldsymbol{\alpha}^c}\boldsymbol{f}$ and $\mathrm{diag}(\boldsymbol{w})\boldsymbol{B}^v = \nabla_{\boldsymbol{\alpha}^v}\boldsymbol{f}$ then (SP1a) and (SP1b) hold true.

Condition (SP2a) is concerned with the convergence of $\widehat{\boldsymbol{\psi}}$. This depends on the particular estimators used in $\widehat{\boldsymbol{\psi}}$. In the following, we assume that (SP2a) holds true.

Mainly, we need to show that condition (SP2b), which concerns the convergence of $\widehat{\boldsymbol{\eta}}$, holds true. After that, we may use Theorem 5.6 with $\boldsymbol{d}$ in that theorem replaced by $\boldsymbol{d}(\boldsymbol{x}^{\boldsymbol{\psi}}, t) - \boldsymbol{B}^c(\boldsymbol{x}^{\boldsymbol{\psi}}, t)\boldsymbol{\eta}$ to get (N3)-(N7), (F1), (F2), (L1), and (L2). For this to work, the following conditions suffice. Similarly to (SP1) and (SP2), they are not minimal conditions but represent reasonable assumptions.

(SPW1) (conditions for weighted projections) There exists a compact, convex set $K \subset \boldsymbol{\mathcal{X}}_{\boldsymbol{\psi}}$ such that $\{\boldsymbol{\psi}_0(t) : t \in [0,\tau]\}$ is in its interior and such that the following holds true:

For all $t \in [0,\tau]$, $i = 1,\ldots,n$, the mappings $d_i(\cdot,t)$, $B^c_{ij}(\cdot,t), j = 1,\ldots,k^c$, $B^v_{i\nu}(\cdot,t), \nu = 1,\ldots,k^v$, and $w_i(\cdot,t)$ from K into $\mathbb{R}$ are twice differentiable. Furthermore, d_i, $\nabla_j d_i$, $\nabla_j \nabla_\nu d_i$, w_i, $\nabla_j w_i$, $\nabla_j \nabla_\nu w_i$, w_i^{-1}, $B^c_{i\mu}$, $\nabla_j B^c_{i\mu}$ $\nabla_j \nabla_\nu B^c_{i\mu}$, $\mu = 1,\ldots,k^c$, and $B^v_{i\mu}$, $\nabla_j B^v_{i\mu}$, $\nabla_j \nabla_\nu B^v_{i\mu}$, $\mu = 1,\ldots,k^v$, $i = 1,\ldots,n$, $j,\nu = 1,\ldots,k_{\boldsymbol{\theta}}$, are càglàd, adapted, locally bounded stochastic processes with values in the space $C(K)$ of all continuous mappings from K into $\mathbb{R}$ equipped with the supremum norm.

(SPW2) (properties of certain matrices) $\overrightarrow{\boldsymbol{B}^v\boldsymbol{w}\boldsymbol{B}^v}(\boldsymbol{x}^{\boldsymbol{\psi}}, t)$ is invertible for all $\boldsymbol{x}^{\boldsymbol{\psi}} \in K$, $t \in [0,\tau]$, and continuous in $(\boldsymbol{x}^{\boldsymbol{\psi}}, t) \in K \times [0,\tau]$. The matrix $\int_0^\tau \mathrm{E}[\tilde{\boldsymbol{B}}^c_1(\boldsymbol{\psi}(s), s)^\top w_1(\boldsymbol{\psi}(s), s)\tilde{\boldsymbol{B}}^c_1(\boldsymbol{\psi}(s), s)]\,\mathrm{d}s$ is invertible for $\boldsymbol{\psi} \in \mathrm{bm}(K)$, where $\tilde{\boldsymbol{B}}^c_1(\boldsymbol{x}^{\boldsymbol{\psi}}, t) = \boldsymbol{Q}^{\boldsymbol{B}^v_1(\boldsymbol{x}^{\boldsymbol{\psi}}, t)}_{w_1(\boldsymbol{x}^{\boldsymbol{\psi}}, t)\cdot \mathrm{P}}\boldsymbol{B}^c_1(\boldsymbol{x}^{\boldsymbol{\psi}}, t)$.

Theorem 6.2. *Suppose conditions (SP2a), (SPW1), and (SPW2) are satisfied,* $\boldsymbol{c}$ *is defined by* (6.10), *and* $\widehat{\boldsymbol{\eta}}$ *is defined by* (6.11). *Furthermore, suppose that* $(d_i, \boldsymbol{B}^v_i, \boldsymbol{B}^c_i, w_i, f_i)$ *are i.i.d. and suppose that the following random elements have an* l*-th moment that is uniformly bounded in* $(\boldsymbol{x}^{\boldsymbol{\theta}}, t) = (\boldsymbol{x}^{\boldsymbol{\alpha}}, \boldsymbol{x}^{\boldsymbol{\beta}}, t)$ *on* $K \times [0,\tau]$, *where* $l = \max(\nu^{-1}, (1/8 + \nu/4)^{-1}, 6)$*:*

$$\begin{aligned}
&d_1(\boldsymbol{x}^{\boldsymbol{\theta}}, t), \nabla_{\boldsymbol{\theta}} d_1(\boldsymbol{x}^{\boldsymbol{\theta}}, t), \nabla_{\boldsymbol{\theta}}^\top \nabla_{\boldsymbol{\theta}} d_1(\boldsymbol{x}^{\boldsymbol{\theta}}, t),\\
&w_1(\boldsymbol{x}^{\boldsymbol{\theta}}, t), \nabla_{\boldsymbol{\theta}} w_1(\boldsymbol{x}^{\boldsymbol{\theta}}, t), \nabla_{\boldsymbol{\theta}}^\top \nabla_{\boldsymbol{\theta}} w_1(\boldsymbol{x}^{\boldsymbol{\theta}}, t),\\
&\boldsymbol{B}^c_1(\boldsymbol{x}^{\boldsymbol{\theta}}, t), \nabla_j \boldsymbol{B}^c_1(\boldsymbol{x}^{\boldsymbol{\theta}}, t), \nabla_j \nabla_\nu \boldsymbol{B}^c_1(\boldsymbol{x}^{\boldsymbol{\theta}}, t), j,\nu = 1,\ldots,k_{\boldsymbol{\theta}},\\
&\boldsymbol{B}^v_1(\boldsymbol{x}^{\boldsymbol{\theta}}, t), \nabla_j \boldsymbol{B}^v_1(\boldsymbol{x}^{\boldsymbol{\theta}}, t), \nabla_j \nabla_\nu \boldsymbol{B}^v_1(\boldsymbol{x}^{\boldsymbol{\theta}}, t), j,\nu = 1,\ldots,k_{\boldsymbol{\theta}},\\
&f_1(\boldsymbol{x}^{\boldsymbol{\alpha}}, t), \nabla_{\boldsymbol{\alpha}} f_1(\boldsymbol{x}^{\boldsymbol{\alpha}}, t), \nabla_{\boldsymbol{\alpha}}^\top \nabla_{\boldsymbol{\alpha}} f_1(\boldsymbol{x}^{\boldsymbol{\alpha}}, t).
\end{aligned}$$

Then the following conditions hold true: (SP2b), (N1), and (N3)-(N7).
Furthermore, we have the following result concerning fixed alternatives:
If in addition to the above, $(d_i, \boldsymbol{B}^v_i, \boldsymbol{B}^c_i, w_i, f_i, h_i)$ *are i.i.d.,* h_i *are càglàd, locally bounded, adapted stochastic processes with uniformly bounded third moment and the mappings*

$$\mathrm{E}[(\boldsymbol{Q}^{\boldsymbol{B}^v_1(\cdot)}_{w_1(\cdot)\cdot \mathrm{P}} d_1(\cdot))h_1(\cdot)] : K \times [0,\tau] \to \mathbb{R}$$

and

$$\mathrm{E}[(\boldsymbol{Q}_{w_1(\cdot)\cdot\mathrm{P}}^{\boldsymbol{B}_1^v(\cdot)} d_1(\cdot)) f_1(\cdot)] : K \times [0,\tau] \to \mathbb{R}$$

are equicontinuous at $\boldsymbol{\psi}_0$ *then (F1) and (F2) hold true. In particular, if, in addition,* $\lambda_i(t) = h_i(t)$ *then*

$$n^{-\frac{1}{2}} T(\boldsymbol{c}, \widehat{\boldsymbol{\psi}}, \widehat{\boldsymbol{\eta}}(\widehat{\boldsymbol{\psi}})) \xrightarrow{\mathrm{P}} \int_0^\tau \mathrm{E}\left[\widetilde{d}_1(\boldsymbol{\psi}_0(s), s)\left\{h_1(s) - f_1(\boldsymbol{\alpha}_0(s), s)\right\}\right] \mathrm{d}s \tag{6.12}$$

and

$$\hat{\sigma}^2(\boldsymbol{c}) \xrightarrow{\mathrm{P}} \int_0^\tau \mathrm{E}\left[\left(\widetilde{d}_1(\boldsymbol{\psi}_0(s), s)\right)^2 h_1(s)\right] \mathrm{d}s \tag{6.13}$$

uniformly in $t \in [0,\tau]$, *where* $\widetilde{d}_1 = \boldsymbol{Q}^U_{w_1(\boldsymbol{\psi}_0(\cdot),\cdot)\cdot(\mathrm{P}\otimes\lambda)} d_1$, $U = U_v + U_c$, $U_c = \{$*elements of* $\boldsymbol{B}_1^c(\boldsymbol{\psi}_0(\cdot),\cdot)\}$, *and*

$$U_v = \{y(\cdot)g(\cdot) : y(\cdot) \text{ element of } \boldsymbol{B}_1^v(\boldsymbol{\psi}_0(\cdot),\cdot) \text{ and } g \in \mathrm{bm}([0,\tau])\}.$$

Proof. We start by proving that (SP2b) holds. Let

$$\boldsymbol{\eta}_0(\boldsymbol{\psi}) = \left(\int_0^\tau \boldsymbol{A}(\boldsymbol{\psi}(s), s)\,\mathrm{d}s\right)^{-1}\left(\int_0^\tau \boldsymbol{D}(\boldsymbol{\psi}(s), s)\,\mathrm{d}s\right),$$

where

$$\begin{aligned}\boldsymbol{A}(\boldsymbol{x}^{\boldsymbol{\psi}}, s) &= \mathrm{E}[\boldsymbol{U}(\boldsymbol{x}^{\boldsymbol{\psi}}, s)^\top w_1(\boldsymbol{x}^{\boldsymbol{\psi}}, s)\boldsymbol{U}(\boldsymbol{x}^{\boldsymbol{\psi}}, s)],\\ \boldsymbol{D}(\boldsymbol{x}^{\boldsymbol{\psi}}, s) &= \mathrm{E}[\boldsymbol{U}(\boldsymbol{x}^{\boldsymbol{\psi}}, s)^\top w_1(\boldsymbol{x}^{\boldsymbol{\psi}}, s) d_1(\boldsymbol{x}^{\boldsymbol{\psi}}, s)],\end{aligned}$$

and $\boldsymbol{U}(\boldsymbol{x}^{\boldsymbol{\psi}}, t) = \boldsymbol{Q}_{w_1(\boldsymbol{x}^{\boldsymbol{\psi}},t)\cdot\mathrm{P}}^{\boldsymbol{B}_1^v(\boldsymbol{x}^{\boldsymbol{\psi}},t)} \boldsymbol{B}_1^c(\boldsymbol{x}^{\boldsymbol{\psi}}, t)$. Consider the decomposition

$$\widehat{\boldsymbol{\eta}}(\widehat{\boldsymbol{\psi}}) - \boldsymbol{\eta}_0(\boldsymbol{\psi}_0) = (\widehat{\boldsymbol{\eta}}(\widehat{\boldsymbol{\psi}}) - \widehat{\boldsymbol{\eta}}(\boldsymbol{\psi}_0)) + (\widehat{\boldsymbol{\eta}}(\boldsymbol{\psi}_0) - \boldsymbol{\eta}_0(\boldsymbol{\psi}_0)).$$

First, we show $\widehat{\boldsymbol{\eta}}(\widehat{\boldsymbol{\psi}}) - \widehat{\boldsymbol{\eta}}(\boldsymbol{\psi}_0) = O_{\mathrm{P}}(n^{-1/4-\nu/2})$. By a Taylor expansion,

$$\begin{aligned}&\int_0^\tau \overline{\tilde{\boldsymbol{B}}^c \boldsymbol{w} \tilde{\boldsymbol{B}}^c}(\widehat{\boldsymbol{\psi}}(s), s)\,\mathrm{d}s - \int_0^\tau \overline{\tilde{\boldsymbol{B}}^c \boldsymbol{w} \tilde{\boldsymbol{B}}^c}(\boldsymbol{\psi}_0(s), s)\,\mathrm{d}s =\\ &\quad= \int_0^\tau \nabla_{\boldsymbol{\psi}} \overline{\tilde{\boldsymbol{B}}^c \boldsymbol{w} \tilde{\boldsymbol{B}}^c}(\widetilde{\boldsymbol{\psi}}(s), s)\left(\widehat{\boldsymbol{\psi}}(s) - \boldsymbol{\psi}_0(s)\right)\mathrm{d}s\\ &\quad\le \left(\sup_{\substack{\boldsymbol{x}^{\boldsymbol{\psi}} \in K\\ s\in[0,\tau]}} \|\nabla_{\boldsymbol{\psi}} \overline{\tilde{\boldsymbol{B}}^c \boldsymbol{w} \tilde{\boldsymbol{B}}^c}(\boldsymbol{x}^{\boldsymbol{\psi}}, s)\|\right)\left(\sup_{s\in[0,\tau]} \left\|\widehat{\boldsymbol{\psi}}(s) - \boldsymbol{\psi}_0(s)\right\|\right)\end{aligned}$$

for some $\widetilde{\boldsymbol{\psi}}$ between $\widehat{\boldsymbol{\psi}}$ and $\boldsymbol{\psi}_0$. By Theorem A.5, (SPW2), and Lemma 3.3 on page 44, the first term is stochastically bounded. Hence, by (SP2a),

$$\int_0^\tau \overline{\tilde{\boldsymbol{B}}^c \boldsymbol{w} \tilde{\boldsymbol{B}}^c}(\widehat{\boldsymbol{\psi}}(s), s)\,\mathrm{d}s - \int_0^\tau \overline{\tilde{\boldsymbol{B}}^c \boldsymbol{w} \tilde{\boldsymbol{B}}^c}(\boldsymbol{\psi}_0(s), s)\,\mathrm{d}s = O_{\mathrm{P}}(n^{-1/4-\nu/2}).$$

Hence, using Lemma 3.3 and Lemma 3.5,

$$\left(\int_0^\tau \overline{\tilde{\boldsymbol{B}}^c \boldsymbol{w} \tilde{\boldsymbol{B}}^c}(\widehat{\boldsymbol{\psi}}(s), s)\,\mathrm{d}s\right)^{-1} - \left(\int_0^\tau \overline{\tilde{\boldsymbol{B}}^c \boldsymbol{w} \tilde{\boldsymbol{B}}^c}(\boldsymbol{\psi}_0(s), s)\,\mathrm{d}s\right)^{-1} = O_{\mathrm{P}}(n^{-1/4-\nu/2}).$$

Similarly, one can show that

$$\int_0^\tau \overline{\tilde{\boldsymbol{B}}^c \boldsymbol{w} \boldsymbol{d}}(\widehat{\boldsymbol{\psi}}(s), s)\,\mathrm{d}s - \int_0^\tau \overline{\tilde{\boldsymbol{B}}^c \boldsymbol{w} \boldsymbol{d}}(\boldsymbol{\psi}_0(s), s)\,\mathrm{d}s = O_{\mathrm{P}}(n^{-1/4-\nu/2}).$$

Next, we show $\widehat{\boldsymbol{\eta}}(\boldsymbol{\psi}_0) - \boldsymbol{\eta}_0(\boldsymbol{\psi}_0) = O_{\mathrm{P}}(n^{-1/2})$. We only show

$$\int_0^\tau \overline{\tilde{\boldsymbol{B}}^c \boldsymbol{w} \boldsymbol{d}}(\boldsymbol{\psi}_0(s), s)\,\mathrm{d}s - \int_0^\tau \boldsymbol{D}(\boldsymbol{\psi}_0(s), s)\,\mathrm{d}s = O_{\mathrm{P}}(n^{-\frac{1}{2}}). \tag{6.14}$$

The proof of

$$\int_0^\tau \overline{\tilde{\boldsymbol{B}}^c \boldsymbol{w} \tilde{\boldsymbol{B}}^c}(\boldsymbol{\psi}_0(s), s)\,\mathrm{d}s - \int_0^\tau \boldsymbol{A}(\boldsymbol{\psi}_0(s), s)\,\mathrm{d}s = O_{\mathrm{P}}(n^{-\frac{1}{2}})$$

is similar. The left hand side of (6.14) can be written as

$$\int_0^\tau (\overline{\boldsymbol{a}} - \mathrm{E}\,\boldsymbol{a}_1^\top)\,\mathrm{d}s - \int_0^\tau \left(\overline{\boldsymbol{b}}(\overline{\boldsymbol{F}})^{-1}\overline{\boldsymbol{g}} - \mathrm{E}[\boldsymbol{b}_1](\mathrm{E}\,\boldsymbol{F}_1)^{-1}\,\mathrm{E}\,\boldsymbol{g}_1^\top\right)\,\mathrm{d}s, \tag{6.15}$$

where

$$\boldsymbol{a}_i := \boldsymbol{B}_i^c w_i d_i, \qquad \boldsymbol{b}_i := \boldsymbol{B}_i^{c\top} w_i \boldsymbol{B}_i^v, \qquad \boldsymbol{F}_i := \boldsymbol{B}_i^{v\top} w_i \boldsymbol{B}_i^v, \qquad \boldsymbol{g}_i := \boldsymbol{B}_i^v w_i d_i,$$

$\overline{\boldsymbol{F}} := \frac{1}{n}\sum_{i=1}^n \boldsymbol{F}_i$, and $\overline{\boldsymbol{b}} := \frac{1}{n}\sum_{i=1}^n \boldsymbol{b}_i$. Since we assume $l \geq 6$, using the multivariate central limit theorem and dropping the dependence on $\boldsymbol{\psi}_0(s)$ and s,

$$n^{\frac{1}{2}} \int_0^\tau \left(\overline{\boldsymbol{a}} - \mathrm{E}\,\boldsymbol{a}_1^\top\right)\,\mathrm{d}s = n^{-\frac{1}{2}} \sum_{i=1}^n \int_0^\tau \left(\boldsymbol{a}_i - \mathrm{E}\,\boldsymbol{a}_1^\top\right)\,\mathrm{d}s = O_{\mathrm{P}}(1).$$

The integrand of the second term of (6.15) can be rewritten as follows:

$$\begin{aligned} &\overline{\boldsymbol{b}}\left(\overline{\boldsymbol{F}}^{-1} - (\mathrm{E}\,\boldsymbol{F}_1)^{-1}\right)\overline{\boldsymbol{g}} + \left(\overline{\boldsymbol{b}} - \mathrm{E}\,\boldsymbol{b}_1\right)(\mathrm{E}\,\boldsymbol{F}_1)^{-1}\overline{\boldsymbol{g}} + \\ &+ \mathrm{E}\,\boldsymbol{b}_1(\mathrm{E}\,\boldsymbol{F}_1)^{-1}\left(\overline{\boldsymbol{g}} - \mathrm{E}\,\boldsymbol{g}_1^\top\right). \end{aligned} \tag{6.16}$$

Hence, the absolute value of the second term of (6.15) is less than

$$\begin{aligned} &\int_0^\tau \left\|\overline{\boldsymbol{F}}^{-1} - (\mathrm{E}\,\boldsymbol{F}_1)^{-1}\right\|\,\mathrm{d}s \sup_{s\in[0,\tau]} \|\overline{\boldsymbol{g}}\|\|\overline{\boldsymbol{b}}\| \\ &+ \int_0^\tau \left\|\overline{\boldsymbol{b}} - \mathrm{E}\,\boldsymbol{b}_1\right\|\,\mathrm{d}s \sup_{s\in[0,\tau]} \|(\mathrm{E}\,\boldsymbol{F}_1)^{-1}\|\|\overline{\boldsymbol{g}}\| + \\ &+ \int_0^\tau \left\|\overline{\boldsymbol{g}} - \mathrm{E}\,\boldsymbol{g}_1^\top\right\|\,\mathrm{d}s \sup_{s\in[0,\tau]} \|\mathrm{E}\,\boldsymbol{b}_1\|\|(\mathrm{E}\,\boldsymbol{F}_1)^{-1}\|. \end{aligned}$$

The suprema are stochastically bounded by the assumptions, a law of large numbers (Theorem A.4), and Lemma 3.3. We may interpret the components $\boldsymbol{b}_i$ as elements of $L_2(\lambda\!\!\lambda)$. Using a central limit theorem for random variables in the Hilbert space $L_2(\lambda\!\!\lambda)$ (Theorem A.6), we get that the components of

$$n^{-\frac{1}{2}} \sum_{i=1}^{n} (\boldsymbol{b}_i - \mathrm{E}\, \boldsymbol{b}_1) \tag{6.17}$$

converge weakly to a random element of $L_2(\lambda\!\!\lambda)$. Hence, the components of (6.17) are stochastically bounded. By these arguments and similar reasoning for $\boldsymbol{F}_i$ and $\boldsymbol{g}_i$,

$$\left(\int_0^{\tau} \|\overline{\boldsymbol{b}} - \mathrm{E}\, \boldsymbol{b}_1\|^2 \,\mathrm{d}s\right)^{\frac{1}{2}} = O_{\mathrm{P}}(n^{-\frac{1}{2}}), \left(\int_0^{\tau} \|\overline{\boldsymbol{F}} - \mathrm{E}\, \boldsymbol{F}_1\|^2 \,\mathrm{d}s\right)^{\frac{1}{2}} = O_{\mathrm{P}}(n^{-\frac{1}{2}}),$$

$$\text{and} \quad \left(\int_0^{\tau} \|\overline{\boldsymbol{g}} - \mathrm{E}\, \boldsymbol{g}_1^{\top}\|^2 \,\mathrm{d}s\right)^{\frac{1}{2}} = O_{\mathrm{P}}(n^{-\frac{1}{2}}).$$

Using Lemma 3.5 we get

$$\left(\int_0^{\tau} \|\overline{\boldsymbol{F}}^{-1} - (\mathrm{E}\, \boldsymbol{F}_1)^{-1}\|^2 \,\mathrm{d}s\right)^{\frac{1}{2}} = O_{\mathrm{P}}(n^{-\frac{1}{2}}).$$

Since $\int_0^{\tau} \|\overline{\boldsymbol{b}} - \mathrm{E}\, \boldsymbol{b}_1\| \,\mathrm{d}s \leq \left(\int_0^{\tau} \|\overline{\boldsymbol{b}} - \mathrm{E}\, \boldsymbol{b}_1\|^2 \,\mathrm{d}s\right)^{\frac{1}{2}} \sqrt{\tau}$, the second term of (6.15) is of order $n^{-1/2}$. This finishes the proof that (SP2b) holds.

As already mentioned before the theorem, we may use Theorem 5.6 with $\boldsymbol{d}(\boldsymbol{x}^{\Psi}, \boldsymbol{\eta}, t) = \boldsymbol{d}(\boldsymbol{x}^{\Psi}, t) - \boldsymbol{B}^c(\boldsymbol{x}^{\Psi}, t)\boldsymbol{\eta}$ replacing $\boldsymbol{d}$ of that lemma to get (N3)-(N7), (F1), (F2), (L1), and (L2). It remains to show that the limits in (6.12) and (6.13) are indeed the terms on the right hand side. This follows from the limits given in Theorem 5.6 and by considering Lemma B.3. □

Chapter 7

Competing Models that are not Separated from the Null Hypothesis - The Problem of a Vanishing Variance

The tests we use are always based on the test statistic T. In Theorem 3.1, Theorem 5.1, and Theorem 6.1 we discussed the asymptotic distribution of the test statistic under the null hypothesis. To apply these results we need to standardize T as can be seen e.g. in Section 3.4. For this standardization, we need that the asymptotic variance of T is not zero. The goal of this chapter is to characterize the case in which the asymptotic variance vanishes and to suggest several approaches how to deal with this problem.

We focus on a particular test for a Cox-type model as null hypothesis, using the optimal weights against fixed alternatives given in Subsection 5.5.1. First, in Section 7.1, we introduce the setup we will be dealing with in more detail. After that, in Section 7.2, we give equivalent conditions for the asymptotic variance to vanish. In Section 7.3, it is sketched what type of asymptotic distribution of $n^{1/2}T$ to expect if the asymptotic variance of T is zero. In Section 7.4, we describe a test to detect whether the asymptotic variance is 0. We suggest to combine this test with our goodness-of-fit test to get a sequential test that is asymptotically conservative, irrespective of whether the asymptotic variance is zero or not. Another approach to deal with the problem of a vanishing variance is to use bootstrap procedures. This approach is mentioned briefly in Section 7.5.

7.1 A Nonparametric Check for the Cox Model

We want to check the Cox-type model (2.5), i.e. the model given by the intensity

$$\lambda_i(t) = \alpha^v(t)\rho_i(\boldsymbol{\alpha}^c, t), \tag{7.1}$$

where the baseline α^v and the regression parameter $\boldsymbol{\alpha}^c \in \boldsymbol{\mathcal{X}}_{\boldsymbol{\alpha}^c} \subset \mathbb{R}^{k_{\alpha^c}}$ are unknown, and where the nonnegative stochastic processes ρ_i are observable.

Interpreting $\boldsymbol{\alpha}^c$ as functions that are constant over time, we use the techniques of Chapter 5 with $\boldsymbol{\alpha}(t) = (\alpha^v(t), \boldsymbol{\alpha}^{c\top})^\top$ and

$$f_i(\boldsymbol{x}^{\boldsymbol{\alpha}}; t) = x^{\alpha^v}\rho_i(\boldsymbol{x}^{\boldsymbol{\alpha}^c}; t),$$

for $\boldsymbol{x}^{\boldsymbol{\alpha}} = (x^{\alpha^v}, \boldsymbol{x}^{\boldsymbol{\alpha}^c\top})^\top$. This approach does not use all information about the null hypothesis, since it ignores that $\boldsymbol{\alpha}^c$ is constant over time. Hence, if (7.1) holds true with a time-dependent parameter $\boldsymbol{\alpha}^c$, we will not be able to reject. But the approach has the advantage of a simpler form of the test and the advantage that no estimate of the baseline $\alpha^v(t)$ is needed, as we shall see shortly.

The derivative of $\boldsymbol{f}$ with respect to the parameters is given by

$$\frac{\partial}{\partial \boldsymbol{x}^{\boldsymbol{\alpha}}} f_i(\boldsymbol{x}^{\boldsymbol{\alpha}}; t) = \left(\rho_i(\boldsymbol{x}^{\boldsymbol{\alpha}^c}; t), x^{\alpha^v}\frac{\partial}{\partial \boldsymbol{x}^{\boldsymbol{\alpha}^c}}\rho_i(\boldsymbol{x}^{\boldsymbol{\alpha}^c}; t)\right).$$

If the weights $\boldsymbol{c}$ satisfy (N2') then, considering the first component of the derivative of f_i, the test statistic simplifies to

$$n^{-\frac{1}{2}}\int_0^t \boldsymbol{c}(s)^\top \mathrm{d}\boldsymbol{N}(s).$$

We focus on tests directed to be optimal against another Cox-type model given by

$$\lambda_i(t) = \beta^v(t)h_i(\boldsymbol{\beta}^c, t), \tag{7.2}$$

where the baseline β^v and the regression parameter $\boldsymbol{\beta}^c \in \boldsymbol{\mathcal{X}}_{\boldsymbol{\beta}^c} \subset \mathbb{R}^{k_{\beta^c}}$ are unknown and the nonnegative stochastic process h_i is observable.

The results of Subsection 5.5.1 suggest using the weights

$$\boldsymbol{c}(t) = \boldsymbol{Q}^{\widetilde{\boldsymbol{B}}(t)}_{(\beta^v(t)\boldsymbol{h}(\boldsymbol{\beta}^c,t))\cdot\mathbb{P}_n}\frac{\beta^v(t)\boldsymbol{h}(\boldsymbol{\beta}^c,t) - \alpha^v(t)\boldsymbol{\rho}(\boldsymbol{\alpha}^c,t)}{\beta^v(t)\boldsymbol{h}(\boldsymbol{\beta}^c,t)} = \boldsymbol{Q}^{\widetilde{\boldsymbol{B}}(t)}_{(\beta^v(t)\boldsymbol{h}(\boldsymbol{\beta}^c,t))\cdot\mathbb{P}_n}\mathbf{1},$$

where

$$\widetilde{\boldsymbol{B}}(t) = \left(\frac{\boldsymbol{\rho}(\boldsymbol{\alpha}^c,t)}{\beta^v(t)\boldsymbol{h}(\boldsymbol{\beta}^c,t)}, \frac{\frac{\partial}{\partial\boldsymbol{\alpha}^c}\boldsymbol{\rho}(\boldsymbol{\alpha}^c,t)\alpha^v(t)}{\beta^v(t)\boldsymbol{h}(\boldsymbol{\beta}^c,t)}\right).$$

This simplifies to

$$\boldsymbol{c}(\boldsymbol{\alpha}^c, \boldsymbol{\beta}^c, t) = \boldsymbol{Q}_{\boldsymbol{h}(\boldsymbol{\beta}^c,t)\cdot\mathbb{P}_n}^{\boldsymbol{B}(\boldsymbol{\alpha}^c,\boldsymbol{\beta}^c,t)}\mathbf{1} \tag{7.3}$$

and

$$\boldsymbol{B}(\boldsymbol{\alpha}^c, \boldsymbol{\beta}^c, t) = \left(\frac{\boldsymbol{\rho}(\boldsymbol{\alpha}^c, t)}{\boldsymbol{h}(\boldsymbol{\beta}^c, t)}, \frac{\frac{\partial}{\partial \boldsymbol{\alpha}^c}\boldsymbol{\rho}(\boldsymbol{\alpha}^c, t)}{\boldsymbol{h}(\boldsymbol{\beta}^c, t)} \right). \tag{7.4}$$

Note that $\boldsymbol{c}$ does not depend on $\alpha^v(t)$ and $\beta^v(t)$. If $\overline{\boldsymbol{BhB}}(\boldsymbol{\alpha}^c, \boldsymbol{\beta}^c, t)$ is invertible then

$$\boldsymbol{c}(\boldsymbol{\alpha}^c, \boldsymbol{\beta}^c, t) = \mathbf{1} - \boldsymbol{B}(\boldsymbol{\alpha}^c, \boldsymbol{\beta}^c, t) \left(\overline{\boldsymbol{BhB}}(\boldsymbol{\alpha}^c, \boldsymbol{\beta}^c, t)\right)^{-1} \overline{\boldsymbol{Bh}}(\boldsymbol{\alpha}^c, \boldsymbol{\beta}^c, t).$$

Similarly to the previous chapters, we adopt the following notation in this chapter: Since we do not need to estimate α^v and β^v, the combined parameter $\boldsymbol{\theta}$ denotes the remaining parameters, i.e. $\boldsymbol{\theta} = (\boldsymbol{\alpha}^{c\top}, \boldsymbol{\beta}^{c\top})^\top$. An estimator of $\boldsymbol{\theta}$ is denoted by $\widehat{\boldsymbol{\theta}} = (\widehat{\boldsymbol{\alpha}^c}^\top, \widehat{\boldsymbol{\beta}^c}^\top)^\top$. For $j = 1, \ldots, k_{\boldsymbol{\theta}} := k_{\boldsymbol{\alpha}^c} + k_{\boldsymbol{\beta}^c}$, ∇_j denotes $\frac{\partial}{\partial \theta_j}$, i.e. for $j = 1, \ldots, k_{\boldsymbol{\alpha}^c}$, ∇_j denotes $\frac{\partial}{\partial \alpha^c_j}$ and for $j = 1, \ldots, k_{\boldsymbol{\beta}^c}$, $\nabla_{j+k_{\boldsymbol{\alpha}^c}}$ denotes $\frac{\partial}{\partial \beta^c_j}$. Furthermore, $\nabla_{\boldsymbol{\theta}} := (\nabla_1, \ldots, \nabla_{k_{\boldsymbol{\theta}}})$, $\nabla_{\boldsymbol{\alpha}^c} := (\nabla_1, \ldots, \nabla_{k_{\boldsymbol{\alpha}^c}})$, and $\nabla_{\boldsymbol{\beta}^c} := (\nabla_{k_{\boldsymbol{\alpha}^c}+1}, \ldots, \nabla_{k_{\boldsymbol{\theta}}})$.

In the remainder of this chapter, we assume that the null hypothesis holds true, i.e. we have

$$\lambda_i(t) = \alpha_0^v(t)\rho_i(\boldsymbol{\alpha}_0^c, t), \tag{7.5}$$

for some α_0^v, $\boldsymbol{\alpha}_0^c$. As always, the limit of $\widehat{\boldsymbol{\beta}^c}$ is denoted by $\boldsymbol{\beta}_0^c$.

7.2 When does the Variance Vanish Asymptotically?

We can use the results about the convergence of T to a normal distribution or to a Gaussian process only if the asymptotic variance is positive under the null hypothesis. We consider the test of the previous section using the weight $\boldsymbol{c}$ given in (7.3).

Assume that we are in the i.i.d. setup of Theorem 5.6 with $d_1(t) = 1$, $\boldsymbol{B}_1(\boldsymbol{\theta}, t) = \left(\frac{\rho_1(\boldsymbol{\alpha}^c,t)}{h_1(\boldsymbol{\beta}^c,t)}, \frac{\nabla_{\boldsymbol{\alpha}^c}\rho_1(\boldsymbol{\alpha}^c,t)}{h_1(\boldsymbol{\beta}^c,t)}\right)$, and $w_1(\boldsymbol{\theta}, t) = h_1(\boldsymbol{\beta}^c, t)$. Furthermore, assume that (7.5) holds true, that that $\alpha_0^v(t) > 0 \,\forall\, t$, that ($h_1(\boldsymbol{\beta}_0^c, t) = 0$ iff $\rho_1(\boldsymbol{\alpha}_0^c, t) = 0$), and that $h_1(\boldsymbol{\beta}_0^c, t) = 0$ implies $\nabla_{\boldsymbol{\alpha}^c}\rho_1(\boldsymbol{\alpha}_0^c, t) = \mathbf{0}$. By Theorem 5.6, the asymptotic variance of $T(\boldsymbol{c}, \widehat{\boldsymbol{\theta}}, \tau)$ under the null hypothesis is given by

$$\sigma^2(\tau) = \int_0^\tau \left\| \boldsymbol{Q}_{h_1(\boldsymbol{\beta}_0^c,s)\cdot \mathrm{P}}^{\boldsymbol{B}_1(\boldsymbol{\theta}_0,s)} 1 \right\|^2_{(\alpha_0^v(s)\rho_1(\boldsymbol{\alpha}_0^c,s))\cdot \mathrm{P}} \mathrm{d}s.$$

By the assumptions, we can rewrite $\sigma^2(\tau)$ as follows:

$$\sigma^2(\tau) = \int_{\Omega\times[0,\tau]} \left(\boldsymbol{Q}^{\boldsymbol{B}_1(\boldsymbol{\theta}_0,s)}_{h_1(\boldsymbol{\beta}_0^c,s)\cdot \mathrm{P}} 1 \right)^2 \alpha_0^v(s)\rho_1(\boldsymbol{\alpha}_0^c,s)\,\mathrm{d}\mu(\omega,s),$$

where $\mu = I\{h_1(\boldsymbol{\beta}_0^c,\cdot)\neq 0\}\cdot(\mathrm{P}\otimes\lambda\!\!\!\lambda)$. As $\rho_1 \geq 0$, the asymptotic variance $\sigma^2(\tau)$ equals 0 iff for μ-almost all $(\omega,t)\in\Omega\times[0,\tau]$,

$$(\boldsymbol{Q}^{\boldsymbol{B}_1(\boldsymbol{\theta}_0,t)}_{h_1(\boldsymbol{\beta}_0^c,t)\cdot \mathrm{P}} 1)(\omega) = 0.$$

Thus $\sigma^2(\tau) = 0$ iff for $\lambda\!\!\!\lambda$-almost all t,

$$1 \in \operatorname{span}_{\nu(t)} \left\{ \frac{\rho_1(\boldsymbol{\alpha}_0^c,t)}{h_1(\boldsymbol{\beta}_0^c,t)}, \frac{\nabla_j\rho_1(\boldsymbol{\alpha}_0^c,t)}{h_1(\boldsymbol{\beta}_0^c,t)}, j=1,\dots,k_{\boldsymbol{\alpha}^c} \right\},$$

where $\nu(t) = I\{h_1(\boldsymbol{\beta}_0^c,t)\neq 0\}\cdot \mathrm{P}$ and $\operatorname{span}_{\nu(t)}$ denotes the span in $L_2(\nu(t))$. Hence, $\sigma^2(\tau)=0$ iff for $\lambda\!\!\!\lambda$-almost all t,

$$h_1(\boldsymbol{\beta}_0^c,t) \in \operatorname{span}_{\nu(t)} \left\{ \rho_1(\boldsymbol{\alpha}_0^c,t), \nabla_j\rho_1(\boldsymbol{\alpha}_0^c,t), j=1,\dots,k_{\boldsymbol{\alpha}^c} \right\}.$$

By the assumptions, we get $\sigma^2(\tau)=0$ iff

$$\text{for } \lambda\!\!\!\lambda\text{-almost all } t,\ h_1(\boldsymbol{\beta}_0^c,t) \in \operatorname{span}_{\mathrm{P}} \left\{ \rho_1(\boldsymbol{\alpha}_0^c,t), \nabla_j\rho_1(\boldsymbol{\alpha}_0^c,t), j=1,\dots,k_{\boldsymbol{\alpha}^c} \right\}, \tag{7.6}$$

where $\operatorname{span}_{\mathrm{P}}$ denotes the span in $L_2(\mathrm{P})$.

To use our test we need to ensure that (7.6) does not hold. Since $\boldsymbol{\theta}_0 = (\boldsymbol{\alpha}_0^{c\top},\boldsymbol{\beta}_0^{c\top})^\top$ is unknown, one needs to be careful when testing against alternatives for which there exists $\boldsymbol{\theta}_0$ such that (7.6) holds. Usually, there will be such $\boldsymbol{\theta}_0$: Most models include $\boldsymbol{\alpha}^c$ and $\boldsymbol{\beta}^c$ such that

$$h_i(\boldsymbol{\beta}^c,s) = \rho_i(\boldsymbol{\alpha}^c,s) = R_i(s) \quad \text{for all } i,s,$$

where $R_i(s)$ is the at-risk indicator. In the classical Cox model (2.1) one merely has to choose $\boldsymbol{\alpha}^c = \boldsymbol{0}$. Another example where (7.6) holds is the following: If $\boldsymbol{\rho}$ and $\boldsymbol{h}$ use the same type of model, like the Cox model, and share some covariates then $h_i = \rho_i$ if the other covariates do not have any influence.

One way to ensure that (7.6) does not hold is to first identify the set $\mathcal{D}$ of all $\boldsymbol{\theta}_0$ for which (7.6) holds and then to test the null hypothesis $\boldsymbol{\theta}_0 \in \mathcal{D}$. This requires special considerations for different combinations of $\boldsymbol{\rho}$ and $\boldsymbol{h}$. In Section 7.4, we propose a general test for (7.6), which can be used for many combinations of $\boldsymbol{\rho}$ and $\boldsymbol{h}$ without having to identify for which $\boldsymbol{\theta}_0$ the condition (7.6) holds true.

7.3 Distribution of the Test Statistic if the Variance Vanishes

In this section, we discuss informally the asymptotic behavior of $T(\boldsymbol{c}, \widehat{\boldsymbol{\theta}}, \tau)$ under the null hypothesis (7.1) if (7.6) holds true. That is we use the setup of Section 7.1 and assume that (7.5) holds true. We use the weights $\boldsymbol{c}$ given by (7.3) and thus the asymptotic variance of $T(\boldsymbol{c}, \widehat{\boldsymbol{\theta}}, \tau)$ is zero. The considerations will suggest that in this case $n^{1/2}T(\boldsymbol{c}, \widehat{\boldsymbol{\theta}}, \tau)$ converges to a weighted sum of χ^2-distributed random variables.

First, note that in this case $\boldsymbol{c}(\boldsymbol{\alpha}_0^c, \boldsymbol{\beta}_0^c, \cdot) = \mathbf{0}$. Let $\widehat{\boldsymbol{\theta}} = (\widehat{\boldsymbol{\alpha}^c}^\top, \widehat{\boldsymbol{\beta}^c}^\top)^\top$ be an estimator of $\boldsymbol{\theta} = (\boldsymbol{\alpha}^{c\top}, \boldsymbol{\beta}^{c\top})^\top$ and let $\boldsymbol{\theta}_0 = (\boldsymbol{\alpha}_0^{c\top}, \boldsymbol{\beta}_0^{c\top})^\top$ be the limit of $\widehat{\boldsymbol{\theta}}$. A Taylor expansion of $n^{1/2}T(\boldsymbol{c}, \widehat{\boldsymbol{\theta}}, \tau)$ around $\boldsymbol{\theta}_0$ yields that $n^{1/2}T(\boldsymbol{c}, \widehat{\boldsymbol{\theta}}, \tau)$ is equal to

$$
\begin{aligned}
& n^{\frac{1}{2}}(\widehat{\boldsymbol{\theta}} - \boldsymbol{\theta}_0)^\top n^{-\frac{1}{2}} \int_0^\tau \nabla_{\boldsymbol{\theta}} \boldsymbol{c}(\boldsymbol{\theta}_0, s)^\top \mathrm{d}\boldsymbol{M}(s) \\
& - n^{\frac{1}{2}}(\widehat{\boldsymbol{\theta}} - \boldsymbol{\theta}_0)^\top \left(\frac{1}{n} \int_0^\tau (\nabla_{\boldsymbol{\theta}} \boldsymbol{c})(\boldsymbol{\theta}_0, s)^\top (\nabla_{\boldsymbol{\alpha}^c} \boldsymbol{\rho})(\boldsymbol{\alpha}_0^c, s) \alpha_0^v(s) \, \mathrm{d}s \right) n^{\frac{1}{2}}(\widehat{\boldsymbol{\theta}} - \boldsymbol{\theta}_0) \\
& + n^{\frac{1}{2}}(\widehat{\boldsymbol{\theta}} - \boldsymbol{\theta}_0)^\top \left(\frac{1}{n} \sum_{i=1}^n \int_0^\tau \nabla_{\boldsymbol{\theta}}^\top \nabla_{\boldsymbol{\theta}} c_i(\boldsymbol{\theta}_0, s) \, \mathrm{d}M_i(s) \right) n^{\frac{1}{2}}(\widehat{\boldsymbol{\theta}} - \boldsymbol{\theta}_0) \\
& + R,
\end{aligned}
$$

where R denotes the remainder term of third order. Under suitable conditions, the last two terms converge stochastically to 0.

If it can be shown that $n^{1/2}(\widehat{\boldsymbol{\theta}} - \boldsymbol{\theta}_0)$ and $n^{-1/2} \int_0^\tau \nabla_{\boldsymbol{\theta}} \boldsymbol{c}(\boldsymbol{\theta}_0, s)^\top \mathrm{d}\boldsymbol{M}(s)$ converge jointly to a multivariate normal distribution then $n^{1/2}T(\boldsymbol{c}, \widehat{\boldsymbol{\theta}}, \tau)$ converges in distribution to a quadratic form of a multivariate normal distribution, which can be written as a weighted sum of χ^2-distributed random variables. If the estimators $\widehat{\boldsymbol{\alpha}^c}$ and $\widehat{\boldsymbol{\beta}^c}$ are both maximum partial likelihood estimators one can see the joint normality under i.i.d. assumptions as follows: The idea is to argue similar to Lemma 7.2 of Section 7.4. First, write $n^{-1/2} \int_0^\tau \nabla_{\boldsymbol{\theta}} \boldsymbol{c}(\boldsymbol{\theta}_0, s)^\top \mathrm{d}\boldsymbol{M}(s)$, $\widehat{\boldsymbol{\alpha}^c}$, and $\widehat{\boldsymbol{\beta}^c}$ as sums of i.i.d. random variables plus vanishing remainder terms. After that, apply the multivariate central limit theorem.

7.4 A Sequential Test

As in the previous sections, we work with an i.i.d. setup and consider the optimal test of the Cox-type model (7.1) directed against the Cox-type alternative (7.2) using the weights $\boldsymbol{c}$ given by (7.3). We have seen that in order to apply our test, we need to ensure that (7.6) does not hold. We propose to employ a test whose null hypothesis is (7.6) to verify this. We call this test 'preliminary test'. Only if we reject (7.6), we use our tests based on the standardized test statistic $T(\boldsymbol{c}, \widehat{\boldsymbol{\theta}}, \cdot)$, which we call 'test of fit'. Combining the preliminary test and the test of fit to a sequential test which rejects iff both tests reject, one arrives at a conservative test which can be used regardless of whether (7.6) holds or not. In particular, the sequential test is valid even if the null hypothesis and the competing model are nested or overlapping.

7.4.1 The Preliminary Test

Let $\boldsymbol{\eta} = (\boldsymbol{\rho}, \nabla_{\boldsymbol{\alpha}^c}\boldsymbol{\rho})$. For the preliminary test, we use the test statistic $H(\widehat{\boldsymbol{\theta}})$, where

$$H(\boldsymbol{\theta}) := \int_0^\tau \left\| Q^{\boldsymbol{\eta}(\boldsymbol{\alpha}^c,s)}_{\boldsymbol{R}(s)\cdot\mathbb{P}_n} \frac{\boldsymbol{h}(\boldsymbol{\beta}^c, s)}{\overline{\boldsymbol{h}}(\boldsymbol{\beta}^c, s)} \right\|^2_{\boldsymbol{R}(s)\cdot\mathbb{P}_n} \mathrm{d}s$$

and $\boldsymbol{R}(s)$ is the at-risk indicator defined by $R_i(s) = I\{\rho_i(\boldsymbol{\alpha}^c_0, s) > 0\}$. Thus, $H(\boldsymbol{\theta})$ is the integrated euclidean distance of $\boldsymbol{h}(\boldsymbol{\beta}^c, \cdot)$, suitably normed, to the space spanned by the columns of $\boldsymbol{\eta}(\boldsymbol{\alpha}^c, \cdot)$. We reject (7.6) for large values of $H(\widehat{\boldsymbol{\theta}})$. We divide $\boldsymbol{h}(\boldsymbol{\beta}^c, s)$ by $\overline{\boldsymbol{h}}(\boldsymbol{\beta}^c, s)$ to make the test statistic invariant with respect to certain transformations of $\boldsymbol{h}(\boldsymbol{\beta}^c, \cdot)$, see the following remark.

Remark 7.1. H and $V^{(1)}(\boldsymbol{c}) = T(\boldsymbol{c}, \boldsymbol{\theta}, \tau)/\sqrt{\hat{\sigma}^2(\boldsymbol{c}, \tau)}$ are invariant with respect to certain transformations of $\boldsymbol{\rho}$ and $\boldsymbol{h}$ that do not change the models: If $\boldsymbol{\rho}(\boldsymbol{\alpha}^c, s)$ is replaced by $\boldsymbol{\rho}(\boldsymbol{\alpha}^c, s)b(\boldsymbol{\theta}, s)$ for some (possibly random) $b : \boldsymbol{\mathcal{X}}_{\boldsymbol{\theta}} \times [0, \tau] \to (0, \infty)$ then H and $V^{(1)}(\boldsymbol{c})$ do not change. Indeed, H is unchanged since $\boldsymbol{\eta}(\boldsymbol{\alpha}^c, s)$ and $\boldsymbol{\eta}(\boldsymbol{\alpha}^c, s)b(\boldsymbol{\theta}, s)$ span the same space. For a similar reason, the weights $\boldsymbol{c}$ given in (7.3) and hence $V^{(1)}(\boldsymbol{c})$ do not change. Furthermore, H and $V^{(1)}(\boldsymbol{c})$ do not change if $\boldsymbol{h}(\boldsymbol{\beta}^c, s)$ is replaced by $\boldsymbol{h}(\boldsymbol{\beta}^c, s)b(\boldsymbol{\theta}, s)$.

Next, we show that H converges to a quadratic form of a multivariate normal distribution. We start with a lemma which assumes the following as given:

$$n^{\frac{1}{2}}(\widehat{\boldsymbol{\theta}} - \boldsymbol{\theta}_0) \xrightarrow{\mathrm{d}} N(\boldsymbol{0}, \boldsymbol{\Sigma}) \text{ and } \hat{\boldsymbol{\Sigma}} \xrightarrow{\mathrm{P}} \boldsymbol{\Sigma}, \tag{7.7}$$

where $\hat{\boldsymbol{\Sigma}}$ can be any consistent estimator of the asymptotic covariance matrix $\boldsymbol{\Sigma}$.

Lemma 7.1. *Suppose the Cox-type model* (7.5) *holds true. Suppose that* (ρ_i, h_i) *are i.i.d.,* $\rho_i(\boldsymbol{\alpha}^c, \cdot)$, $h_i(\boldsymbol{\beta}^c, \cdot)$ *are càglàd, locally bounded stochastic processes,* ρ_i *is three times continuously differentiable with respect to* $\boldsymbol{\alpha}^c$ *and* h_i *is twice continuously differentiable with respect to* $\boldsymbol{\beta}^c$*. Suppose that* $\mathrm{E}[\boldsymbol{\eta}_i(\boldsymbol{\alpha}^c, t)^\top \boldsymbol{\eta}_i(\boldsymbol{\alpha}^c, t)]$ *is invertible for all* $(\boldsymbol{\alpha}^c, t) \in K \times [0, \tau]$ *and continuous in* $(\boldsymbol{\alpha}^c, t)$ *on* $K \times [0, \tau]$*, where* K *is a convex, compact set containing* $\boldsymbol{\alpha}^c_0$ *in its interior. Suppose that* $\rho_i(\boldsymbol{\alpha}^c, t)$ *and its first, second, and third partial derivatives with respect to* $\boldsymbol{\alpha}^c$ *are uniformly bounded on* $K \times [0, \tau]$ *and suppose that* $h_i(\boldsymbol{\beta}^c, t)$ *and its first and second partial derivatives with respect to* $\boldsymbol{\beta}^c$ *are uniformly bounded on* $G \times [0, \tau]$*, where* G *is* *a convex, compact set containing $\boldsymbol{\beta}_0^c$ in its interior. Suppose that $R_i(s) = 0$ iff ($\rho_i(\boldsymbol{\alpha}^c, s) = 0$ and $h_i(\boldsymbol{\beta}^c, s) = 0$ for all $\boldsymbol{\alpha}^c \in K$, $\boldsymbol{\beta}^c \in G$). Furthermore, suppose that $\mathrm{E}\, h_1(\boldsymbol{\beta}_0^c, t)$ is bounded away from 0. If (7.6) and (7.7) hold true then*

$$H(\widehat{\boldsymbol{\theta}}) \xrightarrow{d} \boldsymbol{X}^\top \boldsymbol{A} \boldsymbol{X},$$

where $\boldsymbol{X} \sim N(\mathbf{0}, \boldsymbol{\Sigma})$ and $\boldsymbol{A}$ is some deterministic matrix. Furthermore,

$$\int_0^\tau \frac{1}{n} \boldsymbol{B}(\widehat{\boldsymbol{\theta}}, s)^\top \boldsymbol{B}(\widehat{\boldsymbol{\theta}}, s)\, \mathrm{d}s \xrightarrow{\mathrm{P}} \boldsymbol{A},$$

where

$$\boldsymbol{B}(\boldsymbol{\theta}, s) = \nabla_{\boldsymbol{\theta}} \left(\boldsymbol{Q}^{\boldsymbol{\eta}(\boldsymbol{\alpha}^c, s)}_{\boldsymbol{R}(s)\cdot \mathbb{P}_n} \frac{\boldsymbol{h}(\boldsymbol{\beta}^c, s)}{\overline{\boldsymbol{h}}(\boldsymbol{\beta}^c, s)} \right) \boldsymbol{R}(s).$$

Remark 7.2. If $\overline{\boldsymbol{\eta}\boldsymbol{\eta}}(\boldsymbol{\alpha}^c, s)$ is invertible then $\boldsymbol{B}(\boldsymbol{\theta}, s)$ has the following form, where we do not show the dependence on $\boldsymbol{\theta}$ and s:
For $j = 1, \ldots, k_{\boldsymbol{\alpha}^c}$, and $\nu = k_{\boldsymbol{\alpha}^c} + 1, \ldots, k_{\boldsymbol{\theta}}$,

$$\begin{aligned} B_j &= \Big[-(\nabla_j \boldsymbol{\eta})(\overline{\boldsymbol{\eta}\boldsymbol{\eta}})^{-1} \overline{\boldsymbol{\eta}\boldsymbol{h}} + \boldsymbol{\eta}(\overline{\boldsymbol{\eta}\boldsymbol{\eta}})^{-1} \left(\overline{(\nabla_j \boldsymbol{\eta})\boldsymbol{\eta}} + \overline{\boldsymbol{\eta}(\nabla_j \boldsymbol{\eta})} \right) (\overline{\boldsymbol{\eta}\boldsymbol{\eta}})^{-1} \overline{\boldsymbol{\eta}\boldsymbol{h}} \\ &\quad - \boldsymbol{\eta}(\overline{\boldsymbol{\eta}\boldsymbol{\eta}})^{-1} \overline{(\nabla_j \boldsymbol{\eta})\boldsymbol{h}} \Big] (\overline{\boldsymbol{h}})^{-1}, \\ B_\nu &= \left(\nabla_\nu \boldsymbol{h} - \boldsymbol{\eta}(\overline{\boldsymbol{\eta}\boldsymbol{\eta}})^{-1} \overline{\boldsymbol{\eta}(\nabla_\nu \boldsymbol{h})} \right) (\overline{\boldsymbol{h}})^{-1} - \left(\boldsymbol{h} - \boldsymbol{\eta}(\overline{\boldsymbol{\eta}\boldsymbol{\eta}})^{-1} \overline{\boldsymbol{\eta}\boldsymbol{h}} \right) (\overline{\boldsymbol{h}})^{-2} \overline{\nabla_\nu \boldsymbol{h}}. \end{aligned}$$

Proof of Lemma 7.1. Consider the event $A_n := \{\overline{\boldsymbol{\eta}\boldsymbol{\eta}}(\boldsymbol{\alpha}^c, t) \text{ invertible for all } t \in [0, \tau], \boldsymbol{\alpha}^c \in K\}$. By Lemma 3.3 and Theorem A.4, $\mathrm{P}(A_n) \to 1$. Hence, it suffices to show the assertions on A_n. From now on we work on A_n. By (7.6),

$$\boldsymbol{Q}^{\boldsymbol{\eta}(\boldsymbol{\alpha}_0^c, s)}_{\boldsymbol{R}(s)\cdot \mathbb{P}_n} \frac{\boldsymbol{h}(\boldsymbol{\beta}_0^c, s)}{\overline{\boldsymbol{h}}(\boldsymbol{\beta}_0^c, s)} = \mathbf{0}$$

for all $s \in [0, \tau]$. Hence,

$$\nabla_{\boldsymbol{\theta}} H(\boldsymbol{\theta}_0) = 2 \int_0^\tau \boldsymbol{B}(\boldsymbol{\theta}_0, s)^\top \left(\boldsymbol{Q}^{\boldsymbol{\eta}(\boldsymbol{\alpha}_0^c, s)}_{\boldsymbol{R}(s)\cdot \mathbb{P}_n} \frac{\boldsymbol{h}(\boldsymbol{\beta}_0^c, s)}{\overline{\boldsymbol{h}}(\boldsymbol{\beta}_0^c, s)} \right) \mathrm{d}s = \mathbf{0}$$

and furthermore,

$$\nabla_{\boldsymbol{\theta}}^\top \nabla_{\boldsymbol{\theta}} H(\boldsymbol{\theta}_0) = 2 \int_0^\tau \boldsymbol{B}(\boldsymbol{\theta}_0, s)^\top \boldsymbol{B}(\boldsymbol{\theta}_0, s)\, \mathrm{d}s.$$

By a Taylor expansion of $H(\widehat{\boldsymbol{\theta}})$ around $\boldsymbol{\theta}_0$,

$$H(\widehat{\boldsymbol{\theta}}) = \frac{1}{2} n^{\frac{1}{2}} (\widehat{\boldsymbol{\theta}} - \boldsymbol{\theta}_0)^\top \frac{1}{n} \nabla_{\boldsymbol{\theta}}^\top \nabla_{\boldsymbol{\theta}} H(\widetilde{\boldsymbol{\theta}}) n^{\frac{1}{2}} (\widehat{\boldsymbol{\theta}} - \boldsymbol{\theta}_0)$$

for some $\widetilde{\boldsymbol{\theta}}$ on the line segment between $\widehat{\boldsymbol{\theta}}$ and $\boldsymbol{\theta}_0$. It remains to show that $\frac{1}{2n} \nabla_{\boldsymbol{\theta}}^\top \nabla_{\boldsymbol{\theta}} H(\widetilde{\boldsymbol{\theta}})$ and $\frac{1}{2n} \nabla_{\boldsymbol{\theta}}^\top \nabla_{\boldsymbol{\theta}} H(\widehat{\boldsymbol{\theta}}) = \frac{1}{n} \int_0^\tau \boldsymbol{B}(\widehat{\boldsymbol{\theta}}, s)^\top \boldsymbol{B}(\widehat{\boldsymbol{\theta}}, s)\, \mathrm{d}s$ both converge stochastically to the same limit. By using a law of large numbers (Theorem A.5) and an explicit computation of $\frac{1}{n}\nabla_{\boldsymbol{\theta}}^{\top}\nabla_{\boldsymbol{\theta}}H(\boldsymbol{\theta})$, which is rather lengthy, this can be shown by the continuity of the limit in $\boldsymbol{\theta}$. To see the continuity of the limit in $\boldsymbol{\theta}$, one uses the dominated convergence theorem. We shall not write down the explicit formulas. □

The previous lemma shows that $H(\widehat{\boldsymbol{\theta}})$ converges in distribution to a quadratic form of normally distributed random vectors. The p-value of our preliminary test which rejects for large values of $H(\widehat{\boldsymbol{\theta}})$ can thus be approximated as follows: Generate an i.i.d. sample $\boldsymbol{X}_1, \ldots, \boldsymbol{X}_b$ from $N(\mathbf{0}, \hat{\boldsymbol{\Sigma}})$ for some $b \in \mathbb{N}$. The approximated p-value is

$$\frac{1}{b}\sum_{i=1}^{b} I\left\{ H(\widehat{\boldsymbol{\theta}}) < \boldsymbol{X}_i^{\top} \int_0^{\tau} \frac{1}{n}\boldsymbol{B}(\widehat{\boldsymbol{\theta}}, s)^{\top}\boldsymbol{B}(\widehat{\boldsymbol{\theta}}, s)\,\mathrm{d}s\boldsymbol{X}_i \right\}.$$

We recommend using $b \geq 1000$. A more elegant way of simulating the p-value can be derived by considering the eigenvalues of the quadratic form, see Johnson and Kotz (1970).

7.4.2 Joint Convergence of the Estimators of Two Different Cox Models

To apply Lemma 7.1, we still need the asymptotic distribution of $\widehat{\boldsymbol{\theta}} = (\widehat{\boldsymbol{\alpha}^c}^{\top}, \widehat{\boldsymbol{\beta}^c}^{\top})^{\top}$ under the null hypothesis, i.e. the joint asymptotic distribution of $\widehat{\boldsymbol{\alpha}^c}$ and $\widehat{\boldsymbol{\beta}^c}$ if (7.5) is satisfied. Suppose that both $\widehat{\boldsymbol{\alpha}^c}$ and $\widehat{\boldsymbol{\beta}^c}$ are maximum partial likelihood estimators, i.e. $\widehat{\boldsymbol{\alpha}^c}$ maximizes

$$\boldsymbol{X}(\boldsymbol{\alpha}^c) = \frac{1}{n}\sum_{i=1}^{n}\int_0^{\tau} \log(\rho_i(\boldsymbol{\alpha}^c, s))\,\mathrm{d}N_i(s) - \int_0^{\tau} \log(\overline{\boldsymbol{\rho}}(\boldsymbol{\alpha}^c, s))\,\mathrm{d}\overline{\boldsymbol{N}}(s)$$

and $\widehat{\boldsymbol{\beta}^c}$ maximizes

$$\boldsymbol{Y}(\boldsymbol{\beta}^c) = \frac{1}{n}\sum_{i=1}^{n}\int_0^{\tau} \log(h_i(\boldsymbol{\beta}^c, s))\,\mathrm{d}N_i(s) - \int_0^{\tau} \log(\overline{\boldsymbol{h}}(\boldsymbol{\beta}^c, s))\,\mathrm{d}\overline{\boldsymbol{N}}(s).$$

Since we work under the Cox-type model (7.5), the desired limit $\boldsymbol{\alpha}_0^c$ of $\widehat{\boldsymbol{\alpha}^c}$ is given by the model. Under regularity conditions, $\widehat{\boldsymbol{\beta}^c}$ converges to the maximizer $\boldsymbol{\beta}_0^c$ of

$$y(\boldsymbol{\beta}^c) = \int_0^{\tau} \left(\overrightarrow{\log(\boldsymbol{h})\boldsymbol{\rho}}(\boldsymbol{\alpha}_0^c, \boldsymbol{\beta}^c, s) - \log(\overrightarrow{\boldsymbol{h}}(\boldsymbol{\beta}^c, s))\overrightarrow{\boldsymbol{\rho}}(\boldsymbol{\alpha}_0^c, s) \right) \alpha_0^v(s)\,\mathrm{d}s.$$

Conditions for the stochastic convergence of $\widehat{\boldsymbol{\alpha}^c}$ and $\widehat{\boldsymbol{\beta}^c}$ can be given similar to the considerations in Andersen and Gill (1982), Prentice and Self (1983), and Gandy and Jensen (2005c). The next lemma assumes this stochastic convergence. It is an extension of results about the maximum partial likelihood estimator under misspecified models, see e.g. Lin and Wei (1989).

Lemma 7.2. *Suppose the Cox-type model (7.1) holds true. Suppose that (ρ_i, h_i) are i.i.d., ρ_i is twice continuously differentiable with respect to $\boldsymbol{\alpha}^c$ and h_i is twice continuously differentiable with respect to $\boldsymbol{\beta}^c$. Suppose that $\rho_i(\boldsymbol{\alpha}^c, t)$ and its first and second partial derivatives with respect to $\boldsymbol{\alpha}^c$ are uniformly bounded on $K \times [0, \tau]$ and suppose that $h_i(\boldsymbol{\beta}^c, t)$ and its first and second partial derivatives with respect to $\boldsymbol{\beta}^c$ are uniformly bounded on $G \times [0, \tau]$, where K is a convex, compact set containing $\boldsymbol{\alpha}_0^c$ in its interior and G is a convex, compact set containing $\boldsymbol{\beta}_0^c$ in its interior. Furthermore, suppose that $y(\boldsymbol{\beta}^c)$ is twice continuously differentiable and $\nabla_{\boldsymbol{\beta}^c}^\top \nabla_{\boldsymbol{\beta}^c} y(\boldsymbol{\beta}_0^c)$ is invertible. Suppose that $x(\boldsymbol{\alpha}^c)$ is twice continuously differentiable and $\nabla_{\boldsymbol{\alpha}^c}^\top \nabla_{\boldsymbol{\alpha}^c} x(\boldsymbol{\alpha}_0^c)$ is invertible, where*

$$x(\boldsymbol{\alpha}^c) = \int_0^\tau \left(\overrightarrow{\log(\boldsymbol{\rho})\boldsymbol{\lambda}}(\boldsymbol{\alpha}^c, s) - \log(\overrightarrow{\boldsymbol{\rho}}(\boldsymbol{\alpha}^c, s))\overrightarrow{\boldsymbol{\lambda}}(s) \right) \mathrm{d}s.$$

If $\overrightarrow{\boldsymbol{\rho}}$ and $\overrightarrow{\boldsymbol{h}}$ are bounded away from 0 and if $\widehat{\boldsymbol{\beta}^c} \xrightarrow{\mathrm{P}} \boldsymbol{\beta}_0^c$, $\widehat{\boldsymbol{\alpha}^c} \xrightarrow{\mathrm{P}} \boldsymbol{\alpha}_0^c$ then

$$n^{\frac{1}{2}}(\widehat{\boldsymbol{\theta}} - \boldsymbol{\theta}_0) \xrightarrow{d} N(0, \boldsymbol{\Sigma}),$$

for some matrix $\boldsymbol{\Sigma}$. Furthermore,

$$\hat{\boldsymbol{\Sigma}} := \left(\boldsymbol{A}(\widehat{\boldsymbol{\alpha}^c}, \widehat{\boldsymbol{\beta}^c})\right)^{-1} \overline{\boldsymbol{W}\boldsymbol{W}} \left(\boldsymbol{A}(\widehat{\boldsymbol{\alpha}^c}, \widehat{\boldsymbol{\beta}^c})\right)^{-1} \xrightarrow{\mathrm{P}} \boldsymbol{\Sigma},$$

where

$$\boldsymbol{A}(\boldsymbol{\alpha}^c, \boldsymbol{\beta}^c) = \begin{pmatrix} \nabla_{\boldsymbol{\alpha}^c}^\top \nabla_{\boldsymbol{\alpha}^c} \boldsymbol{X}(\boldsymbol{\alpha}^c) & \boldsymbol{0} \\ \boldsymbol{0} & \nabla_{\boldsymbol{\beta}^c}^\top \nabla_{\boldsymbol{\beta}^c} \boldsymbol{Y}(\boldsymbol{\beta}^c) \end{pmatrix}$$

and the ith row of $\boldsymbol{W} \in \mathbb{R}^{n \times (k_{\boldsymbol{\alpha}^c} + k_{\boldsymbol{\beta}^c})}$ is given by

$$\begin{aligned} \boldsymbol{W}_i = \Bigg(&\int_0^\tau \left(\frac{\nabla_{\boldsymbol{\alpha}^c} \rho_i(\widehat{\boldsymbol{\alpha}^c}, s)}{\rho_i(\widehat{\boldsymbol{\alpha}^c}, s)} - \frac{\overline{\nabla_{\boldsymbol{\alpha}^c} \boldsymbol{\rho}}(\widehat{\boldsymbol{\alpha}^c}, s)}{\overline{\boldsymbol{\rho}}(\widehat{\boldsymbol{\alpha}^c}, s)} \right) \mathrm{d}N_i(s), \\ &\int_0^\tau \left(\frac{\nabla_{\boldsymbol{\beta}^c} h_i(\widehat{\boldsymbol{\beta}^c}, s)}{h_i(\widehat{\boldsymbol{\beta}^c}, s)} - \frac{\overline{\nabla_{\boldsymbol{\beta}^c} \boldsymbol{h}}(\widehat{\boldsymbol{\beta}^c}, s)}{\overline{\boldsymbol{h}}(\widehat{\boldsymbol{\beta}^c}, s)} \right) \mathrm{d}N_i(s) \\ &- \int_0^\tau \left(\frac{(\nabla_{\boldsymbol{\beta}^c} h_i)(\widehat{\boldsymbol{\beta}^c}, s)}{\overline{\boldsymbol{h}}(\widehat{\boldsymbol{\beta}^c}, s)} - h_i(\widehat{\boldsymbol{\beta}^c}, s) \frac{\overline{\nabla_{\boldsymbol{\beta}^c} \boldsymbol{h}}(\widehat{\boldsymbol{\beta}^c}, s)}{(\overline{\boldsymbol{h}}(\widehat{\boldsymbol{\beta}^c}, s))^2} \right) \mathrm{d}\overline{\boldsymbol{N}}(s) \Bigg). \end{aligned}$$

Before we begin with the proof of the previous lemma, we give formulas for $\nabla_{\boldsymbol{\alpha}^c}^\top \nabla_{\boldsymbol{\alpha}^c} \boldsymbol{X}(\boldsymbol{\alpha}^c)$ and $\nabla_{\boldsymbol{\beta}^c}^\top \nabla_{\boldsymbol{\beta}^c} \boldsymbol{Y}(\boldsymbol{\beta}^c)$, which appear in the definition of $\boldsymbol{A}(\boldsymbol{\alpha}^c, \boldsymbol{\beta}^c)$. For $j, \mu = 1, \ldots, k_{\boldsymbol{\alpha}^c}$,

$$\begin{aligned} \nabla_\mu \nabla_j \boldsymbol{X}(\boldsymbol{\alpha}^c) = &\frac{1}{n} \int_0^\tau \left(\frac{\nabla_\mu \nabla_j \boldsymbol{\rho}(\boldsymbol{\alpha}^c, s)}{\boldsymbol{\rho}(\boldsymbol{\alpha}^c, s)} - \frac{\nabla_j \boldsymbol{\rho}(\boldsymbol{\alpha}^c, s) \nabla_\mu \boldsymbol{\rho}(\boldsymbol{\alpha}^c, s)}{\boldsymbol{\rho}(\boldsymbol{\alpha}^c, s)^2} \right)^\top \mathrm{d}\boldsymbol{N}(s) \\ &- \int_0^\tau \left(\frac{\overline{\nabla_\mu \nabla_j \boldsymbol{\rho}}(\boldsymbol{\alpha}^c, s)}{\overline{\boldsymbol{\rho}}(\boldsymbol{\alpha}^c, s)} - \frac{\overline{\nabla_j \boldsymbol{\rho}}(\boldsymbol{\alpha}^c, s) \overline{\nabla_\mu \boldsymbol{\rho}}(\boldsymbol{\alpha}^c, s)}{(\overline{\boldsymbol{\rho}}(\boldsymbol{\alpha}^c, s))^2} \right) \mathrm{d}\overline{\boldsymbol{N}}(s) \end{aligned}$$

and for $j, \mu = k_{\boldsymbol{\alpha}^c} + 1, \ldots, k_{\boldsymbol{\theta}}$,

$$\begin{aligned}\nabla_\mu \nabla_j \boldsymbol{Y}(\boldsymbol{\beta}^c) =& \frac{1}{n} \int_0^\tau \left(\frac{\nabla_\mu \nabla_j \boldsymbol{h}(\boldsymbol{\beta}^c, s)}{\boldsymbol{h}(\boldsymbol{\beta}^c, s)} - \frac{\nabla_j \boldsymbol{h}(\boldsymbol{\beta}^c, s) \nabla_\mu \boldsymbol{h}(\boldsymbol{\beta}^c, s)}{\boldsymbol{h}(\boldsymbol{\beta}^c, s)^2} \right)^\top \mathrm{d}\boldsymbol{N}(s) \\ &- \int_0^\tau \left(\frac{\overline{\nabla_\mu \nabla_j \boldsymbol{h}}(\boldsymbol{\beta}^c, s)}{\overline{\boldsymbol{h}}(\boldsymbol{\beta}^c, s)} - \frac{\overline{\nabla_j \boldsymbol{h}(\boldsymbol{\beta}^c, s) \nabla_\mu \boldsymbol{h}}(\boldsymbol{\beta}^c, s)}{(\overline{\boldsymbol{h}}(\boldsymbol{\beta}^c, s))^2} \right) \mathrm{d}\overline{\boldsymbol{N}}(s).\end{aligned}$$

Proof of Lemma 7.2. The score function for $\boldsymbol{\theta} = (\boldsymbol{\alpha}^{c\top}, \boldsymbol{\beta}^{c\top})^\top$ is given by

$$\boldsymbol{U}(\boldsymbol{\theta}) := n^{\frac{1}{2}} (\nabla_{\boldsymbol{\alpha}^c} \boldsymbol{X}(\boldsymbol{\alpha}^c), \nabla_{\boldsymbol{\beta}^c} \boldsymbol{Y}(\boldsymbol{\beta}^c))^\top.$$

Since $\boldsymbol{U}(\widehat{\boldsymbol{\theta}}) = \boldsymbol{0}$, a Taylor expansion of $\boldsymbol{U}(\widehat{\boldsymbol{\theta}})$ around $\boldsymbol{\theta}_0$ yields

$$-\boldsymbol{U}(\boldsymbol{\theta}_0) = \boldsymbol{A}(\widetilde{\boldsymbol{\theta}})(\widehat{\boldsymbol{\theta}} - \boldsymbol{\theta}_0)$$

for $\widetilde{\boldsymbol{\theta}}$ on the line segment between $\boldsymbol{\theta}_0$ and $\widehat{\boldsymbol{\theta}}$ and hence,

$$n^{\frac{1}{2}}(\widehat{\boldsymbol{\theta}} - \boldsymbol{\theta}_0) = -\boldsymbol{A}(\widetilde{\boldsymbol{\theta}})^{-1} n^{\frac{1}{2}} \boldsymbol{U}(\boldsymbol{\theta}_0).$$

The assumptions guarantee that $\boldsymbol{A}(\widetilde{\boldsymbol{\theta}})$ and $\boldsymbol{A}(\widehat{\boldsymbol{\theta}})$ converge stochastically to the same limit given by $\operatorname{diag}(\nabla_{\boldsymbol{\alpha}^c}^\top \nabla_{\boldsymbol{\alpha}^c} x(\boldsymbol{\alpha}_0^c), \nabla_{\boldsymbol{\beta}^c}^\top \nabla_{\boldsymbol{\beta}^c} y(\boldsymbol{\beta}_0^c))$, which is invertible by the assumptions. Similar to the proof of Theorem 2.1 in Lin and Wei (1989), we shall write $n^{1/2}\boldsymbol{U}(\boldsymbol{\theta}_0)$ as a sum of i.i.d. random variables plus a vanishing remainder. For ease of notation, we shall frequently drop the dependence on $\boldsymbol{\theta}_0$ and on the integration variable s.

$$\begin{aligned} n^{\frac{1}{2}} \nabla_{\boldsymbol{\beta}^c} \boldsymbol{Y}(\boldsymbol{\beta}_0^c) &= n^{-\frac{1}{2}} \sum_{i=1}^n \int_0^\tau \frac{\nabla_{\boldsymbol{\beta}^c} h_i}{h_i} \,\mathrm{d}N_i - n^{\frac{1}{2}} \int_0^\tau \frac{\overline{\nabla_{\boldsymbol{\beta}^c} \boldsymbol{h}}}{\overline{\boldsymbol{h}}} \,\mathrm{d}\overline{\boldsymbol{N}} \\ &= n^{-\frac{1}{2}} \sum_{i=1}^n \int_0^\tau \left(\frac{\nabla_{\boldsymbol{\beta}^c} h_i}{h_i} - \frac{\overrightarrow{\nabla_{\boldsymbol{\beta}^c} \boldsymbol{h}}}{\overrightarrow{\boldsymbol{h}}} \right) \mathrm{d}N_i - \boldsymbol{B} - \boldsymbol{C} - \boldsymbol{D}, \end{aligned}$$

where

$$\boldsymbol{B} = n^{\frac{1}{2}} \int_0^\tau \left(\frac{\overline{\nabla_{\boldsymbol{\beta}^c} \boldsymbol{h}}}{\overline{\boldsymbol{h}}} - \frac{\overrightarrow{\nabla_{\boldsymbol{\beta}^c} \boldsymbol{h}}}{\overrightarrow{\boldsymbol{h}}} \right) \left(\mathrm{d}\overline{\boldsymbol{N}} - \overline{\boldsymbol{\lambda}} \,\mathrm{d}s \right),$$

$$\boldsymbol{C} = n^{\frac{1}{2}} \int_0^\tau \left(\frac{\overline{\nabla_{\boldsymbol{\beta}^c} \boldsymbol{h}}}{\overline{\boldsymbol{h}}} - \frac{\overrightarrow{\nabla_{\boldsymbol{\beta}^c} \boldsymbol{h}}}{\overrightarrow{\boldsymbol{h}}} \right) \left(\overline{\boldsymbol{\lambda}} - \overrightarrow{\boldsymbol{\lambda}} \right) \mathrm{d}s, \text{ and}$$

$$\boldsymbol{D} = n^{\frac{1}{2}} \int_0^\tau \left(\frac{\overline{\nabla_{\boldsymbol{\beta}^c} \boldsymbol{h}}}{\overline{\boldsymbol{h}}} - \frac{\overrightarrow{\nabla_{\boldsymbol{\beta}^c} \boldsymbol{h}}}{\overrightarrow{\boldsymbol{h}}} \right) \overrightarrow{\boldsymbol{\lambda}} \,\mathrm{d}s.$$

$\boldsymbol{B}$ converges stochastically to $\boldsymbol{0}$ by Lenglart's inequality. To show that $\boldsymbol{C}$ vanishes asymptotically one can argue as follows: Rewriting $\boldsymbol{C}$ and using the Cauchy-Schwarz inequality we get

$$\begin{aligned}\|\boldsymbol{C}\|^2 \le & n^{\frac{1}{2}} \sup_{s\in[0,\tau]} |\overrightarrow{\boldsymbol{h}}|^{-1} \left(\int_0^\tau \left\|\overline{\nabla_{\boldsymbol{\beta}^c}\boldsymbol{h}} - \overrightarrow{\nabla_{\boldsymbol{\beta}^c}\boldsymbol{h}}\right\|^2 \mathrm{d}s\right)^{\frac{1}{2}} \left(\int_0^\tau \left|\overline{\boldsymbol{\lambda}} - \overrightarrow{\boldsymbol{\lambda}}\right|^2 \mathrm{d}s\right)^{\frac{1}{2}} \\ & + n^{\frac{1}{2}} \sup_{s\in[0,\tau]} \|\overline{\nabla_{\boldsymbol{\beta}^c}\boldsymbol{h}}\| \left(\int_0^\tau \left|\overline{\boldsymbol{h}}^{-1} - \overrightarrow{\boldsymbol{h}}^{-1}\right|^2 \mathrm{d}s\right)^{\frac{1}{2}} \left(\int_0^\tau \left|\overline{\boldsymbol{\lambda}} - \overrightarrow{\boldsymbol{\lambda}}\right|^2 \mathrm{d}s\right)^{\frac{1}{2}}.\end{aligned}$$

Using the central limit theorem for the Hilbert space $L_2(\lambda)$ (Theorem A.6) we get

$$\left(\int_0^\tau \left\|\overline{\nabla_{\boldsymbol{\beta}^c}\boldsymbol{h}} - \overrightarrow{\nabla_{\boldsymbol{\beta}^c}\boldsymbol{h}}\right\|^2 \mathrm{d}s\right)^{\frac{1}{2}} = O_{\mathrm{P}}(n^{-\frac{1}{2}}), \quad \left(\int_0^\tau \left|\overrightarrow{\boldsymbol{h}} - \overline{\boldsymbol{h}}\right|^2 \mathrm{d}s\right)^{\frac{1}{2}} = O_{\mathrm{P}}(n^{-\frac{1}{2}}),$$

$$\text{and} \quad \left(\int_0^\tau \left|\overline{\boldsymbol{\lambda}} - \overrightarrow{\boldsymbol{\lambda}}\right|^2 \mathrm{d}s\right)^{\frac{1}{2}} = O_{\mathrm{P}}(n^{-\frac{1}{2}}).$$

By e.g. Lemma 3.3 we also get

$$\left(\int_0^\tau \left|\overrightarrow{\boldsymbol{h}}^{-1} - \overline{\boldsymbol{h}}^{-1}\right|^2 \mathrm{d}s\right)^{\frac{1}{2}} = O_{\mathrm{P}}(n^{-\frac{1}{2}}).$$

Hence, $|\boldsymbol{C}| \xrightarrow{\mathrm{P}} 0$. Using a 2-dimensional Taylor expansion of the mapping $(a, b) \to (a/b)$, $\boldsymbol{D}$ can be rewritten as

$$\boldsymbol{D} = n^{\frac{1}{2}} \int_0^\tau \frac{1}{\overrightarrow{\boldsymbol{h}}} \left(\overline{\nabla_{\boldsymbol{\beta}^c}\boldsymbol{h}} - \overline{\boldsymbol{h}}\frac{\overrightarrow{\nabla_{\boldsymbol{\beta}^c}\boldsymbol{h}}}{\overrightarrow{\boldsymbol{h}}}\right) \overrightarrow{\boldsymbol{\lambda}}\,\mathrm{d}s + o_{\mathrm{P}}(1).$$

Hence,

$$n^{\frac{1}{2}} \nabla_{\boldsymbol{\beta}^c} \boldsymbol{Y}(\boldsymbol{\beta}_0^c) = n^{-\frac{1}{2}} \sum_{i=1}^n \boldsymbol{w}_i^{(\boldsymbol{\beta}^c)} + o_{\mathrm{P}}(1),$$

where

$$\boldsymbol{w}_i^{(\boldsymbol{\beta}^c)} = \int_0^\tau \left(\frac{\nabla_{\boldsymbol{\beta}^c} h_i}{h_i} - \frac{\overrightarrow{\nabla_{\boldsymbol{\beta}^c}\boldsymbol{h}}}{\overrightarrow{\boldsymbol{h}}}\right) \mathrm{d}N_i - \int_0^\tau \frac{1}{\overrightarrow{\boldsymbol{h}}} \left((\nabla_{\boldsymbol{\beta}^c} h_i) - h_i \frac{\overrightarrow{\nabla_{\boldsymbol{\beta}^c}\boldsymbol{h}}}{\overrightarrow{\boldsymbol{h}}}\right) \overrightarrow{\boldsymbol{\lambda}}\,\mathrm{d}s.$$

Using similar arguments, we can rewrite $n^{1/2}\nabla_{\boldsymbol{\alpha}^c}\boldsymbol{X}(\boldsymbol{\alpha}_0^c)$ as follows:

$$n^{\frac{1}{2}} \nabla_{\boldsymbol{\alpha}^c} \boldsymbol{X}(\boldsymbol{\alpha}_0^c) = n^{-\frac{1}{2}} \sum_{i=1}^n \boldsymbol{w}_i^{(\boldsymbol{\alpha}^c)} + o_{\mathrm{P}}(1),$$

where

$$\boldsymbol{w}_i^{(\boldsymbol{\alpha}^c)} = \int_0^\tau \left(\frac{\nabla_{\boldsymbol{\alpha}^c}\rho_i}{\rho_i} - \frac{\overrightarrow{\nabla_{\boldsymbol{\alpha}^c}\boldsymbol{\rho}}}{\overrightarrow{\boldsymbol{\rho}}}\right) \mathrm{d}N_i.$$

$\boldsymbol{w}_i^{(\boldsymbol{\alpha}^c)}$ has a simpler form than $\boldsymbol{w}_i^{(\boldsymbol{\beta}^c)}$ because (7.1) holds true. Let $\boldsymbol{w}_i := (\boldsymbol{w}_i^{(\boldsymbol{\alpha}^c)}, \boldsymbol{w}_i^{(\boldsymbol{\beta}^c)})$. The definition of $\boldsymbol{\beta}_0^c$ and (7.1) guarantee that $\boldsymbol{w}_i$ has mean zero. We want to apply the multivariate central limit theorem for $\boldsymbol{w}_i$. For this we need to show that $\boldsymbol{w}_i$ is square integrable. To see that $\boldsymbol{w}_i^{(\boldsymbol{\alpha}^c)}$ is square integrable, first we rewrite it as follows:

$$\int_0^T \left(\frac{\nabla_{\boldsymbol{\alpha}^c} \rho_i}{\rho_i} - \frac{\overrightarrow{\nabla_{\boldsymbol{\alpha}^c} \boldsymbol{\rho}}}{\overrightarrow{\boldsymbol{\rho}}} \right) (\mathrm{d}N_i - \lambda_i \, \mathrm{d}s) + \int_0^T \left(\frac{\nabla_{\boldsymbol{\alpha}^c} \rho_i}{\rho_i} - \frac{\overrightarrow{\nabla_{\boldsymbol{\alpha}^c} \boldsymbol{\rho}}}{\overrightarrow{\boldsymbol{\rho}}} \right) \lambda_i \, \mathrm{d}s.$$

To see that the first term is square integrable, we use Lenglart's inequality. The second term is square integrable by the assumptions. Similarly, it can be shown that $\boldsymbol{w}_i^{(\boldsymbol{\beta}^c)}$ is square integrable. Hence, by the multivariate central limit theorem,

$$n^{\frac{1}{2}} \boldsymbol{U}(\boldsymbol{\theta}_0) \xrightarrow{\mathrm{d}} N(\boldsymbol{0}, \mathrm{E}[\boldsymbol{w}_1^\top \boldsymbol{w}_1]).$$

By means of Lenglart's inequality and a law of large numbers (Theorem A.5), it can be shown that

$$\frac{1}{n} \boldsymbol{W}^\top \boldsymbol{W} \xrightarrow{\mathrm{P}} \mathrm{E}[\boldsymbol{w}_1^\top \boldsymbol{w}_1].$$

□

7.5 Using the Bootstrap

Another approach to deal with the problem of a vanishing variance is to use bootstrap methods. Bootstrap methods are computer-intensive resampling methods to obtain distributions of test statistics.

In our case, we want to obtain the distribution of a test statistic that is to be used as a goodness-of-fit test. If we use the classical bootstrap approach, i.e. we sample from the individuals with replacement then we approximate the distribution from which the sample was generated - irrespective of whether the null hypothesis holds true. This would guarantee the level of the test, but unfortunately the level would also be kept under alternatives, i.e. the power and the level of the test would agree. So the standard scheme of resampling from the individuals with replacement is not suitable for our case. In order to achieve a powerful test, one needs to sample from the distribution under the null hypothesis.

Again, we focus on the Cox model. In fact,we only consider the simple right censorship model described in Section 2.1: For each individual i, there are the time of the event T_i and the censoring time C_i. We observe only $X_i = C_i \wedge T_i$ and whether the event occurs via the indicator variable $\delta_i = I\{T_i = X_i\}$. Furthermore, for each individual we observe covariates $\boldsymbol{Z}_i$ which, for simplicity, are assumed to be constant over time.

We estimate the regression parameter $\boldsymbol{\alpha}^c$ by the maximum partial likelihood estimator $\widehat{\boldsymbol{\alpha}^c}$ and the integrated baseline $A^v(t) = \int_0^t \alpha^v(s)\,ds$ by the Breslow estimator $\hat{A}^v(t)$. The distribution of the censoring times C_i can be estimated by the Kaplan-Meier estimate

$$\hat{G}(t) = 1 - \prod_{i=1}^{n} \left(1 - \frac{I\{X_i \le t\}(1-\delta_i)}{\sum_{j=1}^{n} I(X_j \le X_i)}\right).$$

The estimated distribution function of the event time T given the row vector of covariates $\boldsymbol{Z}$ is

$$\hat{F}_{\boldsymbol{Z}}(t) = 1 - \exp(-\exp(\boldsymbol{Z}\widehat{\boldsymbol{\alpha}^c})\hat{A}^v(t)).$$

If we assume that censoring is independent of the covariates and the survival time then we may use several resampling schemes. We denote one bootstrap sample by $(X_i^*, \delta_i^*, \boldsymbol{Z}_i^*), i = 1, \ldots, n$.

Davison and Hinkley (1997, Algorithm 7.2) suggest the following resampling scheme conditional on the value of the covariates and conditional on the censoring distribution. We shall call this algorithm conditional bootstrap.
For $i = 1, \ldots, n$, independently do the following:

1. Generate T_i^* from the estimated event time distribution $\hat{F}_{\boldsymbol{Z}_i}(t)$.

2. If $\delta_i = 0$ then set $C_i^* = X_i$. If $\delta_i = 1$ then generate C_i^* from the conditional censoring distribution given that $C_i > X_i$, that is generate from $\frac{\hat{G}(t)-\hat{G}(X_i)}{1-\hat{G}(X_i)}$, $t \ge X_i$.

3. Set $\boldsymbol{Z}_i^* = \boldsymbol{Z}_i$, $X_i^* = C_i^* \wedge T_i^*$, and $\delta_i^* = I\{X_i^* = T_i^*\}$.

We also consider the following approach that does not condition on covariates and censoring distribution. We shall call this algorithm unconditional bootstrap.

1. Draw an i.i.d. sample $\boldsymbol{Z}_1^*, \ldots, \boldsymbol{Z}_n^*$ from the set $\{\boldsymbol{Z}_1, \ldots, \boldsymbol{Z}_n\}$ with replacement.

2. For $i = 1, \ldots, n$, generate independent random variables T_i^* from the estimated event time distribution $\hat{F}_{\boldsymbol{Z}_i^*}(t)$.

3. For $i = 1, \ldots, n$, generate independent random variables C_i^* from the estimated censoring distribution $\hat{G}(t)$.

4. Set $X_i^* = C_i^* \wedge T_i^*$ and $\delta_i^* = I\{X_i^* = T_i^*\}$, $i = 1, \ldots, n$.

The previous approaches can be adapted to several events per individual as long as the censoring is independent.

Suppose we want to use a one-sided test based on $V(\boldsymbol{c}) = T(\boldsymbol{c}, \tau)/\sqrt{\hat{\sigma}^2(\boldsymbol{c}, \tau)}$ that rejects for large values. For each bootstrap sample, we can compute the appropriate weight $\boldsymbol{c}^*$ and the test statistic $V^*(\boldsymbol{c}^*)$. Now we draw m independent bootstrap samples $V_1^*(\boldsymbol{c}_1^*), \ldots, V_m^*(\boldsymbol{c}_m^*)$. The p-value based on these bootstrap samples is

$$\frac{1}{m}\sum_{j=1}^{m} I\{V(\boldsymbol{c}) \leq V_j^*(\boldsymbol{c}_j^*)\}.$$

Note that we work with the standardized test statistic. This approach is sometimes called prepivoting.

Both resampling schemes are similar concerning the amount of computation needed. In Subsection 8.3.3, we try both schemes in a simulation study. The results of the simulation study do not lead to a clear recommendation for one of the two resampling schemes.

Chapter 8

Special Cases and Simulation Results

The goal of this chapter is twofold: Firstly, we give explicit formulas to test the fit of several special models. We suggest several choices of weights, focusing particularly on weights chosen via a competing Cox-type model. Besides giving the optimal weights, we also consider ad hoc weights based on unweighted orthogonal projections, since they are computationally simpler. As in the previous chapters, those weights are often called least squares weights.

Secondly, we investigate the behavior of the methods presented in this thesis using simulation studies. For this, the methods have been implemented by the author in the statistical programming language R.

In Section 8.1, we consider tests for the Aalen model as null hypothesis, mainly using the setup of Chapter 3. In Section 8.2, we discuss the semiparametric restriction of the Aalen model given in (2.7). Similarly to the Aalen model, we do not need to estimate the parameters of the model to apply our methods, since the parameters act linearly on the intensity. In Section 8.3, we consider Cox-type models as null hypothesis. In Section 8.4, we consider parametric models. In particular, we shall consider parametric Cox models.

8.1 Checking Aalen Models

In this section, we present simulation studies using Aalen's model (2.6) as null hypothesis, i.e. we test the model

$$\lambda_i(t) = \boldsymbol{Y}_i(t)\boldsymbol{\alpha}(t).$$

In particular, we will consider tests that are sensitive against Cox's model (2.1).

8.1.1 The Weights

In Chapter 3, for testing against the Cox model $\lambda_i(t) = \lambda_0(t)\exp(\boldsymbol{Z}_i(t)\boldsymbol{\beta})R_i(t)$, we suggested the ad hoc weights $\boldsymbol{c}(\widehat{\boldsymbol{\beta}}, t)$, where

$$\boldsymbol{c}(\boldsymbol{\beta}, t) = \boldsymbol{Q}_{\mathbb{P}_n}^{\boldsymbol{Y}(t)}\boldsymbol{\rho}(\boldsymbol{\beta}, t)$$

with

$$\rho_i(\boldsymbol{\beta}, t) = \exp(\boldsymbol{Z}_i(t)\boldsymbol{\beta})R_i(t).$$

If $\overline{\boldsymbol{Y}\boldsymbol{Y}}(t)$ is invertible then

$$\boldsymbol{c}(\boldsymbol{\beta}, t) = \boldsymbol{\rho}(\boldsymbol{\beta}, t) - \boldsymbol{Y}(t)\left(\overline{\boldsymbol{Y}\boldsymbol{Y}}(t)\right)^{-1}\overline{\boldsymbol{Y}\boldsymbol{\rho}}(\boldsymbol{\beta}, t). \tag{8.1}$$

We also derived optimal weights against fixed alternatives in Subsection 3.6.1. Using the estimates derived in that section, we get the weights

$$\boldsymbol{c}(t) = \boldsymbol{Q}_{\boldsymbol{h}(t)\cdot\mathbb{P}_n}^{\tilde{\boldsymbol{B}}(t)}\mathbf{1},$$

where $\tilde{\boldsymbol{B}}(t) = \operatorname{diag}(\boldsymbol{h}^{-1}(t))\boldsymbol{Y}(t)$ with $h_i(t) = \lambda_0(t)\rho_i(\boldsymbol{\beta}_0, t)$. Since $\lambda_0(t)$ is a scalar value,

$$\boldsymbol{c}(t) = \boldsymbol{Q}_{\boldsymbol{h}(t)\cdot\mathbb{P}_n}^{\tilde{\boldsymbol{B}}(t)}\mathbf{1} = \boldsymbol{Q}_{\boldsymbol{\rho}(\boldsymbol{\beta}_0,t)\cdot\mathbb{P}_n}^{\boldsymbol{B}(t)}\mathbf{1},$$

where $\boldsymbol{B}(t) = \operatorname{diag}(\boldsymbol{\rho}^{-1}(\boldsymbol{\beta}_0, t))\boldsymbol{Y}(t)$. Instead of $\boldsymbol{\beta}_0$, we plug in the maximum partial likelihood estimator $\widehat{\boldsymbol{\beta}}$ of $\boldsymbol{\beta}$. If $\overline{\boldsymbol{Y}\boldsymbol{\rho}^{-1}\boldsymbol{Y}}(\widehat{\boldsymbol{\beta}}, t)$ is invertible then

$$\boldsymbol{Q}_{\boldsymbol{\rho}(\widehat{\boldsymbol{\beta}},t)\cdot\mathbb{P}_n}^{\boldsymbol{B}(t)}\mathbf{1} = \mathbf{1} - \operatorname{diag}(\boldsymbol{\rho}^{-1}(\widehat{\boldsymbol{\beta}}, t))\boldsymbol{Y}(t)\left(\overline{\boldsymbol{Y}\boldsymbol{\rho}^{-1}\boldsymbol{Y}}(\widehat{\boldsymbol{\beta}}, t)\right)^{-1}\overline{\boldsymbol{Y}}(t). \tag{8.2}$$

8.1.2 Simulation Results

Our simulation study uses a setup which was also considered in McKeague and Utikal (1991). As covariates, we take independent random variables $x_i, i = 1, \ldots, n$, that are uniformly distributed on $[0, 1]$. The simulation is for classical survival analysis, i.e. we have $\lambda_i(t) = 0$ if $N_i(t-) = 1$. We assume independent right censoring with i.i.d. random variables $C_i, i = 1, \ldots, n$, following an exponential distribution with parameter chosen such that 27% of the observations before τ are censored. Let $R_i(t) := I\{C_i > t, N_i(t-) = 0\}$. In our simulations we have $\boldsymbol{Y}_i(t) = (1, x_i)R_i(t)$, i.e. we consider checks for the Aalen model

$$\lambda_i(t) = R_i(t)\alpha_1(t) + R_i(t)x_i\alpha_2(t).$$

First, we consider $\boldsymbol{c}$ given by the ad hoc choice (8.1) to make our tests powerful against a Cox model. As covariates $\boldsymbol{Z}$ for the Cox model we use $\boldsymbol{Z}_i(t) = (x_i)$. Table 8.1 gives levels and powers at the asymptotic 5% level. The test statistics $V^{(1)}$, $V^{(2)}$, and $V^{(3)}$ are as given in Section 3.4. We also display some

Table 8.1: Observed levels and powers for tests of the Aalen model $\lambda_i(t) = R_i(t)\alpha_1(t) + R_i(t)x_i\alpha_2(t)$ using the Cox model $\lambda_i(t) = \lambda_0(t)\exp(\beta x_i)R_i(t)$ as competing model with weights given by (8.1). The asymptotic level is 5%. The number of replications is 10000. We used $\tau = 2$. The test based on $V^{(1)}$ is one-sided. Results from McKeague and Utikal (1991, table 1(b)) are also displayed (1000 replications).

true λ_i	$\lambda_i(t) = (1+x_i)R_i(t)$				$\lambda_i(t) = 1/2\exp(2x_i)R_i(t)$			
statistic	$V^{(1)}$	$V^{(2)}$	$V^{(3)}$	McKU	$V^{(1)}$	$V^{(2)}$	$V^{(3)}$	McKU
n	observed level				observed power			
75	0.0350	0.0272	0.0341		0.1693	0.0262	0.0538	
150	0.0430	0.0399	0.0431		0.2996	0.0838	0.1354	
300	0.0452	0.0420	0.0442	0.212	0.5198	0.2286	0.3084	0.243
600	0.0490	0.0472	0.0459		0.8060	0.5320	0.6276	
1200	0.0468	0.0485	0.0486	0.106	0.9734	0.8847	0.9256	0.579

Table 8.2: Observed levels and powers for tests using the optimal weights given by (8.2). The setup is as in Table 8.1.

true λ_i	$\lambda_i(t) = (1+x_i)R_i(t)$			$\lambda_i(t) = 1/2\exp(2x_i)R_i(t)$		
statistic	$V^{(1)}$	$V^{(2)}$	$V^{(3)}$	$V^{(1)}$	$V^{(2)}$	$V^{(3)}$
n	observed level			observed power		
75	0.0452	0.0374	0.0414	0.2067	0.0583	0.0896
150	0.0451	0.0387	0.0431	0.3646	0.1458	0.1944
300	0.0449	0.0427	0.0469	0.6003	0.3286	0.4063
600	0.0465	0.0462	0.0468	0.8651	0.6601	0.7345
1200	0.0478	0.0468	0.0461	0.9900	0.9449	0.9665

results from the goodness-of-fit test suggested by McKeague and Utikal (1991, table 1(b)). In the simulation where the true model is an Aalen model most observed levels are close to or below the nominal level of 5%. The tests are conservative for small n. If $\lambda_i(t) = 1/2\exp(2x_i)R_i(t)$ then the power increases as n increases with best results for the one-sided test based on $V^{(1)}$.

In a simulation using the optimal weights given by (8.2) instead of the ad hoc weights given by (8.1), the results are as given in Table 8.2. The power of the tests simulated under the competing Cox model against which the test is directed increases.

We also consider the alternative $\lambda_i(t) = \min(x_i, 1 - x_i)R_i(t)$. To detect this alternative we choose $d_i = I\{x_i \notin [0.25, 0.75]\}$ and use the weight $\boldsymbol{c}(t) =$

$\boldsymbol{Q}_{\mathbb{P}_n}^{\boldsymbol{Y}(t)}\boldsymbol{d}(t)$. Results of the simulations are in Table 8.3. The test is sensitive against this alternative for small sample sizes. In the simulations where the null hypothesis (2.6) holds true, the observed level is close to or below 5% again.

Table 8.3: Observed levels and powers for the test of the Aalen model $\lambda_i(t) = R_i(t)\alpha_1(t) + R_i(t)x_i\alpha_2(t)$ using $d_i = I\{x_i \notin [0.25, 0.75]\}$ and $\tau = 10$ with asymptotic level 5%. The number of replications was 10000. The test based on $V^{(1)}$ is two-sided. Results from McKeague and Utikal (1991, table 1(b)) are also displayed (1000 replications).

true $\lambda_i(t)$	$\lambda_i(t) = (1 + x_i)R_i(t)$			$\lambda_i(t) = \min(x_i, 1 - x_i)R_i(t)$			
statistic	$V^{(1)}$	$V^{(2)}$	$V^{(3)}$	$V^{(1)}$	$V^{(2)}$	$V^{(3)}$	McKU
n	observed level			observed power			
50	0.0417	0.0187	0.0269	0.8746	0.7170	0.8194	
100	0.0455	0.0287	0.0346	0.9959	0.9828	0.9919	
180	0.0455	0.0348	0.0397	1.0000	1.0000	1.0000	0.912

Comparing these results to those of McKeague and Utikal (1991) we see that our tests are much better at attaining the prescribed level in the simulations in which the null hypothesis (2.6) holds true. Furthermore, we get a greater power against the stated alternatives. Of course, the greater power is not surprising since the test of McKeague and Utikal (1991) is an omnibus test and our test was designed to detect these specific alternatives.

In order to direct the test against the Cox model we need to estimate the regression parameter of the Cox model. To assess how much power is lost due to this estimation, we conducted some simulation studies which indicate that the loss is small. For example in the setup of Table 8.1 with $\lambda_i(t) = 1/2\exp(2x_i)R_i(t)$ and a sample size of $n = 300$ we used $\boldsymbol{c}(t) = \boldsymbol{Q}_{\mathbb{P}_n}^{\boldsymbol{Y}(t)}\boldsymbol{d}$, where $d_i = \exp(2x_i)$, and got an empirical rejection rate of 0.5267 for the one-sided test based on $V^{(1)}$, the simulation from Table 8.1 with the estimated parameter resulted in an empirical rejection rate of 0.5198.

8.2 Checking Semiparametric Additive Risk Models

Using the notation of Chapter 6, the intensity under the semiparametric restriction of the Aalen model (2.7) can be written as

$$\lambda_i(t) = f_i(\boldsymbol{\alpha}^v(t), \boldsymbol{\alpha}^c, t) := \boldsymbol{Y}_i^v(t)\boldsymbol{\alpha}^v(t) + \boldsymbol{Y}_i^c(t)\boldsymbol{\alpha}^c$$

for some $\boldsymbol{\alpha}^v \in \mathrm{bm}(\boldsymbol{\mathcal{X}}_{\boldsymbol{\alpha}^v})$, $\boldsymbol{\alpha}^c \in \boldsymbol{\mathcal{X}}_{\boldsymbol{\alpha}^c}$. Since $\nabla_{\boldsymbol{\alpha}^v} f_i(\boldsymbol{\alpha}(t), t) = \boldsymbol{Y}_i^v(t)$ and $\nabla_{\boldsymbol{\alpha}^c} f_i(\boldsymbol{\alpha}(t), t) = \boldsymbol{Y}_i^c(t)$, condition (SP1a) can be rewritten as

$$\int_0^\tau \sum_{i=1}^n c_i(\boldsymbol{\psi}(s), \widehat{\boldsymbol{\eta}}(\boldsymbol{\psi}), s)\boldsymbol{Y}_i^c(s)\,\mathrm{d}s = \boldsymbol{0}, \quad \forall\, \boldsymbol{\psi} \in \boldsymbol{\Psi} \tag{8.3}$$

and (SP1b) as

$$\sum_{i=1}^n c_i(\boldsymbol{\psi}(t), \widehat{\boldsymbol{\eta}}(\boldsymbol{\psi}), t)\boldsymbol{Y}_i^v(t) = \boldsymbol{0}, \quad \forall\, \boldsymbol{\psi} \in \boldsymbol{\Psi}, t \in [0, \tau]. \tag{8.4}$$

If the weights satisfy (8.3) and (8.4) then

$$\int_0^\tau \sum_{i=1}^n c_i(\boldsymbol{\psi}(s), \widehat{\boldsymbol{\eta}}(\boldsymbol{\psi}), s) f_i(\boldsymbol{\alpha}(s), s)\,\mathrm{d}s = \boldsymbol{0}, \quad \forall \boldsymbol{\psi} = (\boldsymbol{\alpha}, \boldsymbol{\gamma}) \in \boldsymbol{\Psi}$$

and hence we may rewrite the test statistic $T(\boldsymbol{c}, \tau)$ as

$$n^{-\frac{1}{2}} \int_0^\tau \boldsymbol{c}(\boldsymbol{\psi}(s), \widehat{\boldsymbol{\eta}}(\boldsymbol{\psi}), s)^\top \mathrm{d}\boldsymbol{N}(s).$$

Furthermore, since (8.3) and (8.4) only depend on $\boldsymbol{\alpha}$ through $\boldsymbol{c}$, we may choose $\boldsymbol{c}$ and $\widehat{\boldsymbol{\eta}}$ that do not depend on $\boldsymbol{\alpha} = (\boldsymbol{\alpha}^v, \boldsymbol{\alpha}^c)$.

8.2.1 Least Squares Weights

We want to use the projection approach of Section 6.3 in which we project arbitrary weights orthogonal to

$$U = \{\text{columns of } \boldsymbol{Y}^c(\cdot)\} + \{\boldsymbol{y}(\cdot)g(\cdot) : \boldsymbol{y} \text{ column of } \boldsymbol{Y}^v \text{ and } g \in \mathrm{bm}([0, \tau])\}.$$

As sketched in Section 6.3, if $\boldsymbol{A} := \int_0^\tau \tilde{\boldsymbol{Y}^c}(s)^\top \tilde{\boldsymbol{Y}^c}(s)\,\mathrm{d}s$ is invertible, where $\tilde{\boldsymbol{Y}^c}(t) := \boldsymbol{Q}_{\mathbb{P}_n}^{\boldsymbol{Y}^v(t)} \boldsymbol{Y}^c(t)$, then for $\boldsymbol{a} \in L_2(\mathbb{P}_n \otimes \lambda)$,

$$\boldsymbol{Q}^U_{\mathbb{P}_n \otimes \lambda}(\boldsymbol{a})(t) = \boldsymbol{Q}_{\mathbb{P}_n}^{\boldsymbol{Y}^v(t)} \boldsymbol{a}(t) - \tilde{\boldsymbol{Y}^c}(t)\boldsymbol{A}^{-1} \int_0^\tau \tilde{\boldsymbol{Y}^c}(s)^\top \boldsymbol{Q}_{\mathbb{P}_n}^{\boldsymbol{Y}^v(s)} \boldsymbol{a}(s)\,\mathrm{d}s.$$

For example, to test against a Cox model, we may use

$$\boldsymbol{c}(\lambda_0, \boldsymbol{\beta}, \widehat{\boldsymbol{\eta}}(\lambda_0, \boldsymbol{\beta})) = \boldsymbol{Q}_{\mathbb{P}_n}^{\boldsymbol{Y}^v(s)} \boldsymbol{\rho}(\boldsymbol{\beta}, s)\lambda_0(s) - \tilde{\boldsymbol{Y}^c}(s)\widehat{\boldsymbol{\eta}}(\lambda_0, \boldsymbol{\beta}), \tag{8.5}$$

where

$$\widehat{\boldsymbol{\eta}}(\lambda_0, \boldsymbol{\beta}) = \boldsymbol{A}^{-1} \int_0^\tau \tilde{\boldsymbol{Y}^c}(s)^\top \boldsymbol{Q}_{\mathbb{P}_n}^{\boldsymbol{Y}^v(s)} \boldsymbol{\rho}(\boldsymbol{\beta}, s)\lambda_0(s)\,\mathrm{d}s.$$

Here, we need an estimator of the baseline and we shall use the smoothing approach of Section 4.2.

If one wants to test against the full Aalen model

$$\lambda_i(t) = \boldsymbol{Y}_i^v(t)\boldsymbol{\alpha}^v(t) + \boldsymbol{Y}_i^c(t)\boldsymbol{\alpha}^c(t), \tag{8.6}$$

where now $\boldsymbol{\alpha}^c$ may depend on time, then the problem of an asymptotically vanishing variance arises. Here, one cannot use an estimate of the intensity under (8.6) and use its orthogonal project to U as weights. Furthermore, using the weights $\boldsymbol{Q}^U_{\mathbb{P}_n\otimes\lambda}\boldsymbol{y^c}$, where $\boldsymbol{y^c}$ is a column of $\boldsymbol{Y}^c$, leads to $\sigma^2 = 0$ as well. A simple approach to solve this particular problem is to use $\boldsymbol{c} = \boldsymbol{Q}^U_{\mathbb{P}_n\otimes\lambda}\boldsymbol{d}$, where $\boldsymbol{d}(t) = \boldsymbol{y^c}(t)I\{t \leq s\}$ for some $0 < s < \tau$.

8.2.2 Optimal Weights

As optimal weights against a fixed alternative, Proposition 6.2 suggests to use an element of the equivalence class

$$\boldsymbol{c}(\boldsymbol{\alpha}(\cdot),\cdot) = \boldsymbol{Q}^V_{\boldsymbol{h}\cdot(\mathbb{P}_n\otimes\lambda)}\frac{\boldsymbol{h}(\cdot) - (\boldsymbol{Y}^v(\cdot)\boldsymbol{\alpha}^v(\cdot) + \boldsymbol{Y}^c(\cdot)\boldsymbol{\alpha}^c)}{\boldsymbol{h}(\cdot)} = \boldsymbol{Q}^V_{\boldsymbol{h}\cdot(\mathbb{P}_n\otimes\lambda)}\boldsymbol{1},$$

where $V = V_v + \left\{\frac{\widetilde{\boldsymbol{f}}(\cdot)}{\boldsymbol{h}(\cdot)} : \widetilde{\boldsymbol{f}} \text{ column of } \boldsymbol{Y}^c\right\}$ and

$$V_v = \left\{\frac{\widetilde{\boldsymbol{f}}(\cdot)}{\boldsymbol{h}(\cdot)}g(\cdot) : \widetilde{\boldsymbol{f}} \text{ column of } \boldsymbol{Y}^v \text{ and } g \in \text{bm}([0,\tau])\right\}.$$

Note that because of the linear structure of the intensity, V and $\boldsymbol{c}$ do not depend on $\boldsymbol{\alpha}$. If $\boldsymbol{A} := \int_0^\tau \overline{\widetilde{\boldsymbol{Y}^c}\boldsymbol{h}\widetilde{\boldsymbol{Y}^c}}(s)\,\mathrm{d}s$ is invertible, where

$$\widetilde{\boldsymbol{Y}^v}(t) := \text{diag}(\boldsymbol{h}^{-1}(t))\boldsymbol{Y}^v(t) \quad \text{and} \quad \widetilde{\boldsymbol{Y}^c}(t) := \boldsymbol{Q}^{\widetilde{\boldsymbol{Y}^v}(t)}_{\boldsymbol{h}(t)\cdot\mathbb{P}_n}\,\text{diag}(\boldsymbol{h}^{-1}(t))\boldsymbol{Y}^c(t),$$

then

$$\boldsymbol{c}(t) = \boldsymbol{Q}^{\widetilde{\boldsymbol{Y}^v}(t)}_{\boldsymbol{h}(t)\cdot\mathbb{P}_n}\left(\boldsymbol{1} - \widetilde{\boldsymbol{Y}^c}(t)\boldsymbol{A}^{-1}\int_0^\tau \overline{\widetilde{\boldsymbol{Y}^c}\boldsymbol{h}}(s)\,\mathrm{d}s\right). \tag{8.7}$$

To test against a competing Cox model, one may want to use

$$h_i(t) = \widehat{\lambda}_0(t)\rho_i(\widehat{\boldsymbol{\beta}},t), \tag{8.8}$$

where $\widehat{\boldsymbol{\beta}}$ is the maximum partial likelihood estimator and $\widehat{\lambda}_0$ is a smoothed version of the Breslow estimator using the approach of Section 4.2.

8.2.3 Simulation Results

The simulation is based on the same setup as in Subsection 8.1.2. We use $\boldsymbol{Y}_i^c(t) = (R_i(t)x_i)$ and $\boldsymbol{Y}_i^v(t) = (R_i(t))$, i.e. the null hypothesis is

$$\lambda_i(t) = R_i(t)\alpha_1^v(t) + R_i(t)x_i\alpha_1^c. \tag{8.9}$$

We will consider three true intensities, the first two of them, $(1+x_i)R_i(t)$ and $1/2\exp(2x_i)R_i(t)$, have been used in Subsection 8.1.2. The censoring in these two cases is 27% and we take $\tau = 2$. The third true intensity we use is $(1+10t)\exp(x_i)R_i(t)$. In this case 28% of the values are censored and $\tau = 1/2$. Our simulations are based on 10000 replications. For the tests we use a nominal size of 5%.

The first simulations use the weights given by (8.5), i.e. we test against the Cox model with the ad hoc weights. The estimator $\widehat{\lambda}_0$ of the baseline is the smoothed Breslow estimator using the smoother given in Example 4.1 and the bandwidth given by (4.7). We observed the levels/powers displayed in Table 8.4. The last two rows were added to verify that the observed level actually converges to 5% under the null hypothesis.

Table 8.5 contains the tests using the optimal weights against a Cox model given in (8.7) and (8.8). using the same smoothed Breslow estimator $\widehat{\lambda}_0$.

Next, we consider the following parametric version of Cox's model as competing model:

$$\lambda_0(t) = (1+at)\exp(\beta x_i)R_i(t), \tag{8.10}$$

where a and β are the unknown parameters. We will use the maximum partial likelihood estimators $\hat{a}$ and $\hat{\beta}$ defined as maximizer of the partial likelihood given by (2.9), which we computed numerically. As weights, we shall use the same choice as for the tests with a competing semiparametric Cox model with $\hat{\lambda}_0(t) = 1 + \hat{a}t$. Table 8.6 contains the corresponding simulation results.

For comparison purposes, Table 8.7 gives the simulation results of testing the full Aalen model

$$\lambda_i(t) = R_i(t)\alpha_1(t) + R_i(t)x_i\alpha_2(t), \tag{8.11}$$

using the one-sided test based on $V^{(1)}$ and the ad hoc choice of weights against a Cox model given in Subsection 3.5.3.

Table 8.4: Test of the null hypothesis (8.9) using the competing Cox model $\lambda_i(t) = \lambda_0(t)\exp(\beta x_i)R_i(t)$ with the weights given in (8.5).

true $\lambda_i(t)$	$(1+x_i)R_i(t)$	$1/2\exp(2x_i)R_i(t)$	$(1+10t)\exp(x_i)R_i(t)$
n			
75	0.0508	0.1835	0.1995
150	0.0571	0.3334	0.4433
300	0.0639	0.5707	0.7697
600	0.0623	0.8405	0.9723
1200	0.0638	0.9871	0.9999
2400	0.0586		
4800	0.0541		

Table 8.5: Test of the null hypothesis (8.9) using the one-sided test based on $V^{(1)}$ and the optimal weights given by (8.8) and (8.7).

true $\lambda_i(t)$	$(1+x_i)R_i(t)$	$1/2\exp(2x_i)R_i(t)$	$(1+10t)\exp(x_i)R_i(t)$
n			
75	0.0594	0.2478	0.2676
150	0.0612	0.4166	0.5659
300	0.0670	0.6641	0.8821
600	0.0630	0.9083	0.9962
1200	0.0623	0.9930	1.0000

Table 8.6: Test of the null hypothesis (8.9) using the competing parametric Cox model (8.10) with the weights given in (8.5) using $\hat{\lambda}_0(t) = 1 + \hat{a}t$.

true $\lambda_i(t)$	$(1+x_i)R_i(t)$	$1/2\exp(2x_i)R_i(t)$	$(1+10t)\exp(x_i)R_i(t)$
n			
75	0.0556	0.1708	0.2189
150	0.0532	0.3085	0.4542
300	0.0532	0.4968	0.7789
600	0.0578	0.7731	0.9756
1200	0.0534	0.9583	0.9999

From the simulations one can draw the following conclusions. Under the null hypothesis, the tests using the semiparametric approach are slightly liberal. For the simulations with true intensity $\lambda_i(t) = 1/2\exp(2x_i)R_i(t)$, the power of the tests using the ad hoc weights are in the same range. In the case of the true intensity $\lambda_i(t) = (1+10t)\exp(x_i)R_i(t)$, we get a considerable improvement with the tests that assume time-independent influence of the second covariate. Of course, the tests using the optimal weights have the greatest power.

Table 8.7: Simulation results for tests of the full Aalen model (8.11) using the one-sided test based on $V^{(1)}$ and the weights given in Subsection 3.5.3.

true $\lambda_i(t)$	$(1+x_i)R_i(t)$	$1/2\exp(2x_i)R_i(t)$	$(1+10t)\exp(x_i)R_i(t)$
n			
75	0.0350	0.1693	0.1601
150	0.0430	0.2996	0.2913
300	0.0452	0.5198	0.5101
600	0.0490	0.8060	0.7973
1200	0.0468	0.9734	0.9722

8.3 Checking Cox Models

In this section, we consider checks of a Cox-type model (2.5), i.e. the model given by the intensity

$$\lambda_i(t) = f_i(\alpha^v(t), \boldsymbol{\alpha}^c) := \alpha^v(t)\rho_i(\boldsymbol{\alpha}^c, t), \tag{8.12}$$

where the baseline $\alpha^v(t)$ and the regression parameter $\boldsymbol{\alpha}^c \in \boldsymbol{\mathcal{X}}_{\boldsymbol{\alpha}^c} \subset \mathbb{R}^{k_{\alpha^c}}$ are unknown, and where the nonnegative stochastic processes ρ_i are observable.

In Section 7.1, we already considered testing this model based on the non-parametric approach of Chapter 5.

8.3.1 Checks Using the Semiparametric Approach

In this section, we use checks based on the approach of Chapter 6. In this setup, for the generalized Cox model (8.12), we have $\nabla_{\boldsymbol{\alpha}^v} f_i(\alpha^v(t), \boldsymbol{\alpha}^c, t) = \rho_i(\boldsymbol{\alpha}^c, t)$ and $\nabla_{\boldsymbol{\alpha}^c} f_i(\alpha^v(t), \boldsymbol{\alpha}^c, t) = \alpha^v(t)(\nabla_{\boldsymbol{\alpha}^c}\rho_i)(\boldsymbol{\alpha}^c, t)$. Again, for weights satisfying (SP1b), the test statistic simplifies to

$$n^{-\frac{1}{2}} \int_0^{\mathcal{T}} \boldsymbol{c}(s)^\top \mathrm{d}\boldsymbol{N}(s).$$

Furthermore, we may rewrite the optimal weights against the fixed alternative $\lambda_i(t) = h_i(t)$ as follows:

$$\boldsymbol{c}(\alpha^v(t), \boldsymbol{\alpha}^c, \boldsymbol{\eta}, t) = \boldsymbol{Q}_{\boldsymbol{h}(t)\cdot\mathbb{P}_n}^{\boldsymbol{\rho}(\boldsymbol{\alpha}^c,t)/\boldsymbol{h}(t)} \left(\mathbf{1} - \alpha^v(t)(\nabla_{\boldsymbol{\alpha}^c}\boldsymbol{\rho})(\boldsymbol{\alpha}^c, t)\boldsymbol{\eta}\right)$$

and

$$\widehat{\boldsymbol{\eta}}(\alpha^v, \boldsymbol{\alpha}^c) = \left(\int_0^{\mathcal{T}} \overline{\tilde{\boldsymbol{B}}^c \boldsymbol{h} \tilde{\boldsymbol{B}}^c}(\boldsymbol{\alpha}^c, s)\alpha^v(s)^2 \,\mathrm{d}s\right)^{-1} \int_0^{\mathcal{T}} \overline{\tilde{\boldsymbol{B}}^c \boldsymbol{h}}(\boldsymbol{\alpha}^c, s)\alpha^v(s)\,\mathrm{d}s,$$

where

$$\boldsymbol{Q}_{\boldsymbol{h}(t)\cdot\mathbb{P}_n}^{\boldsymbol{\rho}(\boldsymbol{\alpha}^c,t)/\boldsymbol{h}(t)} = \boldsymbol{I} - \frac{\boldsymbol{\rho}(\boldsymbol{\alpha}^c, t)}{\boldsymbol{h}(t)} \left(\overline{\boldsymbol{\rho}\boldsymbol{h}^{-1}\boldsymbol{\rho}}(\boldsymbol{\alpha}^c, t)\right)^{-1} \frac{1}{n}\boldsymbol{\rho}(\boldsymbol{\alpha}^c, t)^\top$$

and

$$\tilde{\boldsymbol{B}}^c(\boldsymbol{\alpha}^c, t) = \boldsymbol{Q}_{\boldsymbol{h}(t)\cdot\mathbb{P}_n}^{\boldsymbol{\rho}(\boldsymbol{\alpha}^c,t)/\boldsymbol{h}(t)} \frac{(\nabla_{\boldsymbol{\alpha}^c}\boldsymbol{\rho})(\boldsymbol{\alpha}^c, t)}{\boldsymbol{h}(t)}.$$

In particular, if we consider a competing Cox-type model given by

$$\lambda_i(t) = \gamma^v(t)h_i(\boldsymbol{\gamma}^c, t),$$

then the above simplifies to

$$\boldsymbol{c}(\alpha^v(t), \boldsymbol{\alpha}^c, \gamma^v(t), \boldsymbol{\gamma}^c, \boldsymbol{\eta}, t) = \boldsymbol{Q}_{\boldsymbol{h}(\boldsymbol{\gamma}^c,t)\cdot\mathbb{P}_n}^{\boldsymbol{\rho}(\boldsymbol{\alpha}^c,t)/\boldsymbol{h}(\boldsymbol{\gamma}^c,t)} \left(\mathbf{1} - \alpha^v(t)(\nabla_{\boldsymbol{\alpha}^c}\boldsymbol{\rho})(\boldsymbol{\alpha}^c, t)\boldsymbol{\eta}\right)$$

and

$$\widehat{\eta}(\alpha^v, \boldsymbol{\alpha}^c, \gamma^v, \boldsymbol{\gamma}^{\mathbf{c}}) = \left(\int_0^\tau \overline{\tilde{\boldsymbol{B}}^c \boldsymbol{h} \tilde{\boldsymbol{B}}^c}(\boldsymbol{\alpha}^c, \boldsymbol{\gamma}^{\mathbf{c}}, s) \frac{\alpha^v(s)^2}{\gamma^v(s)} \,\mathrm{d}s \right)^{-1} \int_0^\tau \overline{\tilde{\boldsymbol{B}}^c \boldsymbol{h}}(\boldsymbol{\alpha}^c, s) \alpha^v(s) \,\mathrm{d}s,$$

where

$$\boldsymbol{Q}_{\boldsymbol{h}(\boldsymbol{\gamma}^{\mathbf{c}},t)\cdot\mathbb{P}_n}^{\boldsymbol{\rho}(\boldsymbol{\alpha}^c,t)/\boldsymbol{h}(\boldsymbol{\gamma}^{\mathbf{c}},t)} = \boldsymbol{I} - \frac{\boldsymbol{\rho}(\boldsymbol{\alpha}^c, t)}{\boldsymbol{h}(\boldsymbol{\gamma}^{\mathbf{c}}, t)} \left(\overline{\boldsymbol{\rho} \boldsymbol{h}^{-1} \boldsymbol{\rho}}(\boldsymbol{\alpha}^c, \boldsymbol{\gamma}^{\mathbf{c}}, t) \right)^{-1} \frac{1}{n} \boldsymbol{\rho}(\boldsymbol{\alpha}^c, t)^\top$$

and $\tilde{\boldsymbol{B}}^c(\boldsymbol{\alpha}^c, \boldsymbol{\gamma}^{\mathbf{c}}, t) = \boldsymbol{Q}_{\boldsymbol{h}(\boldsymbol{\gamma}^{\mathbf{c}},t)\cdot\mathbb{P}_n}^{\boldsymbol{\rho}(\boldsymbol{\alpha}^c,t)/\boldsymbol{h}(\boldsymbol{\gamma}^{\mathbf{c}},t)} \dfrac{(\nabla_{\boldsymbol{\alpha}^c} \boldsymbol{\rho})(\boldsymbol{\alpha}^c, t)}{\boldsymbol{h}(\boldsymbol{\gamma}^{\mathbf{c}}, t)}.$

8.3.2 Using an Alternative Estimator of the Variance

Under suitable conditions, another consistent estimator for the variance σ^2 in Theorem 5.1 for the particular case of the Cox model (8.12) is given by

$$\hat{\sigma}^2_{(2)}(\boldsymbol{c}, t) := \int_0^t \overline{\boldsymbol{c}\boldsymbol{\rho}\boldsymbol{c}}(\widehat{\boldsymbol{\theta}}, s) \frac{\mathrm{d}\overline{\boldsymbol{N}}(s)}{\overline{\boldsymbol{\rho}}(\widehat{\boldsymbol{\alpha}^c}, s)}.$$

If the fixed alternative $\lambda_i(t) = h_i(t)$ holds then under suitable conditions,

$$\hat{\sigma}^2_{(2)}(\boldsymbol{c}, t) \xrightarrow{\mathrm{P}} \int_0^t \overline{\boldsymbol{c}\boldsymbol{\rho}\vec{\boldsymbol{c}}}(\boldsymbol{\theta}_0, s) \frac{\vec{\boldsymbol{h}}(s)}{\vec{\boldsymbol{\rho}}(\boldsymbol{\alpha}_0^c, s)} \,\mathrm{d}s.$$

Hence, we might also consider a test based on

$$\tilde{V}^{(1)}(\boldsymbol{c}) := T(\boldsymbol{c}, \widehat{\boldsymbol{\theta}}, \tau) / \sqrt{\hat{\sigma}^2_{(2)}(\boldsymbol{c}, \tau)}.$$

An argument similar to the derivation of optimal tests using the nonparametric approach of Chapter 5 (which we also used in Chapter 7) against the completely known fixed alternative $\lambda_i(t) = h_i(t)$ leads to the following optimal choice for $\boldsymbol{c}$ in the sense of approximate Bahadur efficiency:

$$\boldsymbol{c}^*_{(2)}(\boldsymbol{\alpha}^c, s) := \boldsymbol{Q}_{\boldsymbol{\rho}(\boldsymbol{\alpha}^c,s)\cdot\mathbb{P}_n}^{\boldsymbol{B}(\boldsymbol{\alpha}^c,s)} \frac{\boldsymbol{h}(s)}{\boldsymbol{\rho}(\boldsymbol{\alpha}^c, s)} \frac{\overline{\boldsymbol{\rho}}(\boldsymbol{\alpha}^c, s)}{\overline{\boldsymbol{h}}(s)},$$

where $\boldsymbol{B}(\boldsymbol{\alpha}^c, s) = \left(\mathbf{1}, \dfrac{\nabla_1 \boldsymbol{\rho}(\boldsymbol{\alpha}^c, s)}{\boldsymbol{\rho}(\boldsymbol{\alpha}^c, s)}, \dots, \dfrac{\nabla_{k_{\boldsymbol{\alpha}^c}} \boldsymbol{\rho}(\boldsymbol{\alpha}^c, s)}{\boldsymbol{\rho}(\boldsymbol{\alpha}^c, s)} \right)$. In this case,

$$\left(\frac{1}{n} \int_0^\tau \left\| \boldsymbol{c}^*_{(2)}(\boldsymbol{\alpha}_0^c, s) \right\|^2_{\boldsymbol{\rho}(\boldsymbol{\alpha}_0^c,s)\cdot\mathbb{P}_n} \frac{\overline{\boldsymbol{h}}(s)}{\overline{\boldsymbol{\rho}}(\boldsymbol{\alpha}_0^c, s)} \,\mathrm{d}s \right)^{\frac{1}{2}}$$

converges to the approximate Bahadur slope.

8.3.3 Simulation Results

We start with some simulations for the simpler tests suggested in Section 7.1. In Table 8.8, we present a simulation study for the special case with at most one event per individual. We use independent censoring and assume that the lifetimes and the covariates of the individuals are i.i.d. All tests except one are one-sided tests based on $V^{(1)} = T(\boldsymbol{c}, \tau)/\sqrt{\hat{\sigma}^2(\boldsymbol{c}, \tau)}$, where the optimal weights $\boldsymbol{c}$ are given by (7.3). The tests use an asymptotic level of 0.05.

Setup 1 considers the case of nested Cox models. Simulation a) illustrates that under the null hypothesis the preliminary tests attains the prescribed level, whereas - as expected - the test of fit does not meet the prescribed level. Simulation b) demonstrates that under the alternative, both the preliminary and the test of fit reject. Simulation b) was modeled after a simulation used by Lin and Wei (1991) and Marzec and Marzec (1997a) to allow comparisons. The test by Lin Wei, Cox's test, Schoenfeld's test, and Wei's test all have powers of less than 0.25 for $n = 100$. Only some of the tests suggested by Marzec and Marzec (1997a) come close with the best test achieving a power of 0.971 for $n = 100$.

Setup 2 shows how the test behaves when the alternative is separated but close to the alternative. In Simulation a) the test of fit is slightly liberal. We also did Simulation a) for $n = 400, 800, 1600$ and got rejection rates of the preliminary test of $0.078, 0.321, 0.879$. This indicates that our test of fit seems to be applicable in this setup, only the preliminary test is slow in picking this up. Simulation b) is identical to Simulation a), except that now we used the test based on $\tilde{V}^{(1)}(\boldsymbol{c}^*_{(2)})$ from Subsection 8.3.2, i.e. we use an alternative estimator for the variance. Now, the test of fit is slightly conservative. It is no general conclusion that $\tilde{V}^{(1)}(\boldsymbol{c}^*_{(2)})$ leads to a conservative test and $V^{(1)}$ leads to a liberal test; other simulations, not reported here, suggest that in some cases the test based on $\tilde{V}^{(1)}(\boldsymbol{c}^*_{(2)})$ is liberal and the test based on $V^{(1)}$ is conservative. Simulation c) is modeled after Lin and Wei (1991) and Marzec and Marzec (1997a) to allow comparisons. The test by Lin Wei, Cox's test, Schoenfeld's test, and Wei's test all have powers of less than 0.25 for $n = 100$. The test by Lin and Wei (1991) has a power of 0.447 for $n = 200$, Cox's test, Schoenfeld's test, and Wei's test all have powers of less than 0.15 for $n = 200$ and the best test from Marzec and Marzec (1997a) achieves a power of 0.577 for $n = 200$.

Setup 3 illustrates the case where the null hypothesis and the alternative share a common submodel. The sequential procedure achieves the asymptotic level in Simulation a). In Simulation b), the power increases with n.

Summarizing, our tests seem to have a good power, even if the sequential procedure is used. The test of fit may be slightly liberal, but the sequential procedure is conservative.

Table 8.8: Observed rejection rates. The true intensity is $\lambda_i(t) = R_i(t)g_i$, where g_i is given in the third column. The censoring distribution is $U(0,\tau)$, where τ is given in the second column. Simulations used 1000 replications. All but one test are based on $V^{(1)}$ using the optimal weights given by (7.3).

	τ	$g_i \setminus n =$	preliminary test 50	100	200	test of fit 50	100	200	sequential test 50	100	200
Setup 1: $\lambda_i(t) = \lambda_0(t)R_i(t)\exp(aZ_{i1})$ *vs.* $\lambda_i(t) = \lambda_0(t)R_i(t)\exp(aZ_{i1} + bZ_{i4})$											
a)	5	$\exp(0.2Z_{i1})$	0.055	0.057	0.035	0.122	0.122	0.103	0.053	0.056	0.035
b)	7	$\exp(0.2Z_{i1} + 0.5Z_{i4})$	0.823	0.990	1.000	0.892	0.995	1.000	0.822	0.990	1.000
Setup 2: $\lambda_i(t) = \lambda_0(t)R_i(t)\exp(aZ_{i5})$ *vs.* $\lambda_i(t) = \lambda_0(t)R_i(t)(1 + aZ_{i5})$											
a)	4	$\exp(0.2Z_{i5})$	0.009	0.008	0.026	0.082	0.078	0.064	0.008	0.006	0.011
b)	4	$\exp(0.2Z_{i5})$ †	0.011	0.021	0.029	0.045	0.054	0.037	0.005	0.006	0.006
c)	5	$1 + 0.5Z_{i5}$	0.183	0.581	0.911	0.521	0.628	0.774	0.143	0.407	0.696
Setup 3: $\lambda_i(t) = \lambda_0(t)R_i(t)\exp(aZ_{i1} + bZ_{i2})$ *vs.* $\lambda_i(t) = \lambda_0(t)R_i(t)\exp(aZ_{i1} + bZ_{i3})$											
a)	4	$\exp(0.2Z_{i1})$	0.037	0.054	0.044	0.099	0.121	0.093	0.031	0.053	0.042
b)	4	$\exp(0.2Z_{i1} + 0.2Z_{i3})$	0.214	0.393	0.671	0.320	0.488	0.758	0.200	0.379	0.667

$Z_{i1}, Z_{i2}, Z_{i3} \sim N(0,1)$, $\mathrm{P}(Z_{i4} = 1) = \mathrm{P}(Z_{i4} = -1) = 1$,
Z_{i5} is a standard normal random variable truncated at ± 1.96.
†: Test based on $\tilde{V}^{(1)}(\boldsymbol{c}^*_{(2)})$ from Subsection 8.3.2.

We also redid some of the simulations of Table 8.8 using the optimal weights based on the semiparametric approach. Based on 10000 replications, for the simulation 2a), we got rejection rates of 0.0741 for $n = 50$, 0.0575 for $n = 100$, and 0.0535 for $n = 200$. For the simulation 2c), we got rejection rates of 0.5018 for $n = 50$, 0.6199 for $n = 100$, and 0.7872 for $n = 200$. Even though the power did not increase under the alternative, the observed level under the null hypothesis was closer to the nominal level. This can be attributed to the less restrictive weights of the semiparametric orthogonality conditions.

We also did some bootstrap simulations corresponding to the two bootstrap procedures suggested in Section 7.5. Since simulating a bootstrap procedure requires much computing time, we only considered the case $n = 50$ and used $m = 500$ bootstrap replications. We also only did 500 repetitions of the simulations. Still, the result in Table 8.9 give some insight. No clear suggestion as to which bootstrap procedure to use can be made - none of them is always better with respect to being more conservative or being more powerful. But indeed, both bootstrap procedures seem to work well in the simulations under the null hypothesis, even though in some of the simulations the bootstrap procedures are slightly liberal. Distinguishing between the cases in which the hypotheses are nested, overlapping, or separated does not seem to be necessary for the bootstrap procedures. There is no clear picture whether it is better, with respect to the power of the tests, to use one of the bootstrap procedures or the sequential procedure. In Setup 1b) the power using the bootstrap is comparable to using the sequential procedure, in Setup 2c) it is greater, and in Setup 3b) it is less.

Table 8.9: Simulation results for the setups of Table 8.8 using the bootstrap procedures described in Section 7.5 using $n = 50$.

Setup:	1a)	1b)	2a)	2b)	2c)	3a)	3b)
Unconditional Bootstrap:	0.070	0.830	0.048	0.072	0.250	0.046	0.152
Conditional Boostrap:	0.048	0.790	0.056	0.056	0.216	0.060	0.170

8.4 Checking Parametric Models

Thus far, the test statistic always simplified to an integral with respect to the counting process. For parametric models this is usually not the case. Consider the following parametric Cox model

$$\lambda_i(t) = (1 + at)\exp(\boldsymbol{Z}_i(t)\boldsymbol{\beta})R_i(t),$$

where a is a real-valued parameter, $\boldsymbol{\beta}$ is a vector-valued parameter, the observable covariates $\boldsymbol{Z}_i$ are row vectors of predictable stochastic processes, and R_i are the at-risk indicators. We do not have time-dependent parameters in this model. (SP1b) consists of the conditions

$$\int_0^\tau \sum_{i=1}^n c_i(s)\exp(\boldsymbol{Z}_i(s)\boldsymbol{\beta})R_i(s)\,\mathrm{d}s = 0$$

and

$$\int_0^\tau \sum_{i=1}^n c_i(s)\boldsymbol{Z}_i(s)\exp(\boldsymbol{Z}_i(s)\boldsymbol{\beta})R_i(s)\,\mathrm{d}s = \mathbf{0}.$$

So in this case, the test statistic simplifies to

$$T(\boldsymbol{c}, \tau) = \sum_{i=1}^n \int_0^\tau c_i(s)\left(\mathrm{d}N_i(s) - \exp(\boldsymbol{Z}_i(s)\boldsymbol{\beta})R_i(s)\,\mathrm{d}s\right).$$

Even though we do not present simulation studies with a parametric null hypothesis, we give the form of optimal weights against fixed alternatives. Consider a general parametric model of type

$$\lambda_i(s) = f_i(\boldsymbol{\alpha}^c, s),$$

where $\boldsymbol{\alpha}^c$ is a finite-dimensional unknown parameter and f_i is observable. Then the optimal weights against a fixed alternative with intensity $h_i(t)$ is given by

$$\begin{aligned} \boldsymbol{c}(\boldsymbol{\alpha}^c, t) =& \boldsymbol{Q}^{\boldsymbol{B}(\boldsymbol{\alpha}^c,\cdot)}_{\boldsymbol{h}\cdot(\mathbb{P}_n\otimes\lambda)} \left(\frac{\boldsymbol{h}(\cdot) - \boldsymbol{f}(\boldsymbol{\alpha}^c,\cdot)}{\boldsymbol{h}(\cdot)} \right)(t) \\ =& \mathbf{1} - \frac{\boldsymbol{f}(\boldsymbol{\alpha}^c, t)}{\boldsymbol{h}(t)} - \frac{\nabla_{\boldsymbol{\alpha}^c}\boldsymbol{f}(\boldsymbol{\alpha}^c, t)}{\boldsymbol{h}(t)} \boldsymbol{D}^{-1}(\boldsymbol{\alpha}^c)\boldsymbol{A}(\boldsymbol{\alpha}^c), \end{aligned}$$

where $\boldsymbol{B}(\boldsymbol{\alpha}^c, t) := \operatorname{diag}(\boldsymbol{h}^{-1}(t))\nabla_{\boldsymbol{\alpha}^c}\boldsymbol{f}(\boldsymbol{\alpha}^c, t)$,

$$\boldsymbol{A}(\boldsymbol{\alpha}^c) = \int_0^T \left(\overline{\nabla_{\boldsymbol{\alpha}^c}\boldsymbol{f}}(\boldsymbol{\alpha}^c, s) - \overline{(\nabla_{\boldsymbol{\alpha}^c}\boldsymbol{f})\boldsymbol{f}\boldsymbol{h}^{-1}}(\boldsymbol{\alpha}^c, s) \right) \mathrm{d}s,$$

and

$$\boldsymbol{D}(\boldsymbol{\alpha}^c) = \int_0^T \overline{(\nabla_{\boldsymbol{\alpha}^c}\boldsymbol{f})\boldsymbol{h}^{-1}(\nabla_{\boldsymbol{\alpha}^c}\boldsymbol{f})}(\boldsymbol{\alpha}^c, s)\,\mathrm{d}s.$$

Of course, the above assumes that $\boldsymbol{D}(\boldsymbol{\alpha}^c)$ is invertible. As estimators, one can plug in the maximum partial likelihood estimators based on (2.9).

With some parametric models, one has to be careful, though. Consider a simple Weibull distribution. If the shape parameter is negative then the intensity is unbounded. If it is positive, then the intensity is zero at time 0 even though the individual is at risk at time 0. Using such a model as null hypothesis or as alternative with the optimal weights may lead to problems since then the integrals $\boldsymbol{A}(\boldsymbol{\alpha}^c)$ and $\boldsymbol{D}(\boldsymbol{\alpha}^c)$ need not be finite.

Chapter 9

Applications to Real Datasets

We applied our tests to three real datasets. The first from software reliability is chosen to illustrate the applicability of the methods outside classical survival analysis where we may have multiple events per point process. The second example considers some models for the PBC dataset given in Fleming and Harrington (1991). The last example is an application to the well-known Stanford heart transplant data (Miller and Halpern, 1982).

9.1 A Dataset from Software Reliability

We start by considering a dataset from software reliability used in Gandy and Jensen (2004). It contains bug reports of 73 open source software projects. Two different sets of covariates are considered in the aforementioned paper and Aalen models with those covariates are fitted. The first set of covariates includes only the current size of the source code of the projects as covariate. The second set of covariates includes the size of the source code of the projects a fixed time ago, changes in the size of source code since then and the number of recent bug reports. Note that in both cases, no baseline in the form of a covariate identically equal to 1 was included. Checking this using $\boldsymbol{d}(t) = \mathbf{1}$ as suggested in Subsection 3.5.2, we get the p-values of Table 9.1 for our tests applied to the two different sets of covariates, where $V^{(1)}$, $V^{(2)}$, and $V^{(3)}$ are as in Section 3.4. In the dataset with one covariate all three tests suggest a bad fit of the model that could possibly be improved by including a baseline. In the case of three covariates the hypothesis that the Aalen model is the correct one is supported. This agrees with the conclusion of Gandy and Jensen (2004).

Table 9.1: p-values for the software reliability datasets from Gandy and Jensen (2004) using $d_i(t) = 1$. The test based on $V^{(1)}$ is two-sided.

test statistic	$V^{(1)}$	$V^{(2)}$	$V^{(3)}$
one covariate	0.0100	0.0022	0.0155
three covariates	0.7805	0.3655	0.5914

9.2 PBC Data

Our next example considers the PBC dataset presented in Fleming and Harrington (1991), where it is analyzed at length using Cox's model. It contains data about the survival of 312 patients with primary biliary cirrhosis (PBC). We use the corrections of the dataset given by Fleming and Harrington (1991, p. 188).

We discuss the fit of several Cox models, Table 9.2 gives an overview over the covariates in these models. A more detailed description will follow.

Table 9.2: Covariates of the Cox models used for the PBC dataset.

Model	Covariates
4.4.2(b)	age, albumin, bilirubin, edema, hepatomegaly, prothrombin
4.4.3(c)	age, edema, log(albumin), log(bilirubin), log(prothrombin)
A	edema, log(albumin), log(bilirubin), log(prothrombin), $I\{\text{age} > 50\}$, $I\{\text{age} > 60\}$
B	edema, log(albumin), log(bilirubin), log(prothrombin), $I\{\text{age} > 50\}$, $I\{\text{age} > 60\}$, log(copper), log(SGOT)

We use the methods given in Section 7.1 and the sequential procedure suggested in Section 7.4. The results are displayed in Table 9.3. Using the alternative test statistic from Subsection 8.3.2 we arrive at the same conclusions.

Fleming and Harrington (1991) first develop a Cox model (model 4.4.2(b)) which includes the covariates age, albumin, bilirubin, edema, hepatomegaly, and prothrombin time. Hereby, albumin is the amount of a certain protein in the blood, bilirubin is the level of a liver bile pigment, edema is an indicator for the presence of a certain swelling caused by excess fluids, hepatomegaly is an indicator for the presence of a swelling or enlargement of the liver, and prothrombin time is the amount of time it takes a blood sample to begin coagulation in a certain laboratory test. To improve the fit, Fleming and Harrington (1991) add some transformed covariates and use model selection techniques to arrive at a Cox model (model 4.4.3(c)) which uses the following covariates: age, edema, log(albumin), log(bilirubin), and log(prothrombin time). They claim that model 4.4.2(b) does not have a good fit. Our test can be used to formally check this. As can be seen in Table 9.3, model 4.4.2(b) is clearly rejected.

Next, we consider a modification of model 4.4.3(c), called model A, in which we replace the covariate age by two covariates indicating whether the individual is older than 50 or 60 years. Note that model A and model 4.4.3(c) are overlapping models. Using model 4.4.3(c) as null hypothesis and directing against model A indicates that the fit of model 4.4.3(c) is not perfect. Reversing the roles and using model A as null hypothesis, model A is not rejected.

Using further covariates (the amount of copper in the urine per day and the level of the enzyme SGOT) the models can be improved. This is not totally surprising, since Fleming and Harrington (1991) deliberately built models that did not include these covariates. Consider a Cox model, called model B, where we include the covariates edema, log(albumin), log(bilirubin), log(prothrombin time), log(urine copper), log(SGOT), and indicators for persons older than 50 and older than 60. Using this model as competing model to the models 4.4.3(c) and A, we get a clear rejection. Note that model A and model B are nested, whereas model 4.4.3(c) and model B are overlapping.

We also computed p-values of our test of fit based on the two bootstrap procedures of Section 7.5 using $m = 1000$ bootstrap replications. Furthermore, we also used the tests of Subsection 8.3.1 that use the information that $\boldsymbol{\beta}$ does not depend on time. In Table 9.3, these tests are described as 'semiparametric test of fit'. The results reinforce our conclusions except for maybe the rejection of model 4.4.3(c) with the competing model A.

Table 9.3: Comparison of several models from the PBC dataset. The table contains the p-values. The p-values of the preliminary test are based on 10000 replications.

null hypothesis competing model	4.4.2(b) 4.4.3(c)	4.4.3(c) A	A 4.4.3(c)	4.4.3(c) B	A B
preliminary test	$< 10^{-4}$	0.0335	$< 10^{-4}$	0.0144	0.0008
test of fit	$< 10^{-4}$	0.0178	0.4373	$< 10^{-4}$	$< 10^{-4}$
sequential test	$< 10^{-4}$	0.0335	0.4373	0.0144	0.0008
test of fit †	0.000	0.043	0.429	0.000	0.004
test of fit *	0.001	0.049	0.465	0.001	0.002
semiparametric test of fit	$< 10^{-4}$	0.0441	0.2648	$< 10^{-4}$	$< 10^{-4}$
semiparametric test of fit †	0.000	0.090	0.223	0.001	0.000
semiparametric test of fit *	0.000	0.101	0.234	0.001	0.000

†: p-value based on unconditional bootstrap

*: p-value based on conditional bootstrap

In Grønnesby and Borgan (1996), the dataset is analyzed with Aalen's model using the covariates baseline, bilirubin, edema dichotomized (the values 0 and 0.5

are pooled together), albumin (zero for the highest half then linear), prothrombin time (zero for the lowest half then linear), age, interaction of age and prothrombin time. Grønnesby and Borgan (1996) investigate the fit of the linear model and the model 4.4.3(c) of Fleming and Harrington (1991) and conclude that "the fit of both models is acceptable". Their formal goodness-of-fit test for the linear model yields a p-value of .075 for Aalen's model and .197 for the Cox model. Furthermore they mention that the linear model "suffers from [...] negative estimated intensities".

Using the just mentioned coding of covariates for an Aalen model, our one-sided test of Aalen's model with the competing Cox model 4.4.3(c) based on $V^{(1)}$ and the least squares weights leads to a p-value of .034 rejecting Aalen's model at the 5% level. Using the optimal weights the picture is far clearer, leading to a p-value of less than 10^{-5}. This clearly indicates that the model suggested by Grønnesby and Borgan (1996) lacks fit.

9.3 Stanford Heart Transplant Data

The last example uses the Stanford heart transplant dataset given by Miller and Halpern (1982). It contains information about the survival of $n = 184$ patients that received heart transplants. Two covariates are given: age at time of transplant (denoted by a_i) and a donor-recipient mismatch score (denoted by b_i). The covariate b_i was only recorded for $n = 157$ patients, and at the beginning we only consider those patients.

Among the models fitted by Miller and Halpern (1982) are several Cox models. The mismatch score b_i is not significant in a Cox model with covariates $\boldsymbol{Z}_i = (a_i, b_i)$. Based on a graphical method, Miller and Halpern (1982) state that the fit of the model with covariates $\boldsymbol{Z}_i = (a_i)$ is "not ideal". To improve the fit, they consider a model based on the covariates $\boldsymbol{Z}_i = (a_i, a_i^2)$ for which they find no lack of fit. These findings are supported using formal tests by Lin et al. (1993) and Marzec and Marzec (1997a).

In the sequel, we fit and test Aalen's additive model. Let $R_i(t)$ denote the at-risk indicator of the i-th patient. If we use the model $\boldsymbol{Y}_i(t) = (1, a_i)R_i(t)$ our two-sided least squares test based on $V^{(1)}$ using $d_i(t) = b_i$ is not significant (p-value .401) whereas the test using $d_i(t) = a_i^2$ is significant (p-value .004). Our one-sided test based on $V^{(1)}$ directed against the Cox model with $\boldsymbol{Z}_i = (a_i)$ based on least squares weights is significant as well (p-value .0167). If we include a_i^2 as suggested by our test, and consider the model

$$\lambda_i(t) = R_i(t)(\alpha_1^v(t) + a_i\alpha_2^v(t) + a_i^2\alpha_3^v(t)), \tag{9.1}$$

our test against the Cox model with covariates $\boldsymbol{Z}_i = (a_i, a_i^2)$ using the one-sided test based on $V^{(1)}$ and the least squares weights is not significant (p-value .153). Using the optimal weights, the test is significant (p-value .030), suggesting that (9.1) does not fit perfectly.

From now on, we no longer use the covariate b_i and thus can use all $n = 184$ patients in the study. If we test (9.1) against the Cox model with $\boldsymbol{Z}_i = (a_i, a_i^2)$ using all patients, we still do not get a rejection with the least squares weights (p-value .114), and with the optimal weights the test is still significant (p-value .032).

Next, we want to check the following submodel of (9.1):

$$\lambda_i(t) = R_i(t)(\alpha_1^v(t) + a_i\alpha_1^c + a_i^2\alpha_2^c). \tag{9.2}$$

We expect that this model fits worse than (9.2) and hope that our tests lead to this conclusion. We test against the same Cox model as before using the nonparametric estimator for the baseline with bandwidth given by (4.7). With the least squares weights proposed in (8.2.1), the p-value is .043, and with the optimal weights proposed in (8.2.2). the p-value is .008. We may conclude that the model (9.2) has a bad fit, leaving the Cox model suggested by Miller and Halpern (1982).

Chapter 10

Some Remarks

In this chapter, we give some further remarks concerning our approach. We only sketch the ideas. In Section 10.1 we show that with a modified variance estimator, the need to estimate the unknown model parameters does not affect the power of the test. In Section 10.2, we mention how our test can be modified to direct it against several competing models. In Section 10.3, we sketch some possible extensions of our approach. In Section 10.4, we mention a relation to estimating equations in semiparametric models.

10.1 Plugging in Estimators does not Lower the Power

Usually, some parameters from the null hypothesis are unknown. If they do not act affine-linearly on the intensity we have to estimate them. To guarantee that these estimated parameters do no affect the asymptotic distribution of the test statistic T we introduced the orthogonality conditions. If we use an alternative estimator for the variance, then - under certain conditions - the need to estimate the parameters and the orthogonality conditions do not affect the approximate Bahadur slope of the test, i.e. we do not loose any power. We demonstrate this in the parametric setup of Section 8.4. As null hypothesis we use the model

$$\lambda_i(s) = f_i(\boldsymbol{\alpha}^c, s),$$

where $\boldsymbol{\alpha}^c$ is an unknown finite-dimensional parameter and the observable $f_i(\boldsymbol{\alpha}^c, s)$ is a predictable stochastic process.

Under suitable conditions, an alternative consistent estimator for the variance of $T(\boldsymbol{c}, \widehat{\boldsymbol{\alpha}^c}, \tau)$ is given by

$$\hat{\sigma}^2_{(2)}(\boldsymbol{c}) := \int_0^\tau \overline{\boldsymbol{c}\boldsymbol{f}\boldsymbol{c}}(\widehat{\boldsymbol{\alpha}^c}, s)\,\mathrm{d}s,$$

where $\widehat{\boldsymbol{\alpha}^c}$ is some estimator for $\boldsymbol{\alpha}^c$. In contrast to the usual estimator of the variance, $1/n \int_0^\tau \boldsymbol{c}(\widehat{\boldsymbol{\alpha}^c}, s)^\top \operatorname{diag}(\mathrm{d}\boldsymbol{N}(s))\boldsymbol{c}(\widehat{\boldsymbol{\alpha}^c}, s)$, the estimator $\hat{\sigma}^2_{(2)}(\boldsymbol{c})$ uses explicitly the structure of the null hypothesis.

Similarly to Section 6.4, one can derive the following optimal weights in the sense of approximate Bahadur efficiency:

$$\begin{aligned}\boldsymbol{c}(\boldsymbol{\alpha}^c, t) =& \boldsymbol{Q}^{\boldsymbol{B}(\boldsymbol{\alpha}^c,\cdot)}_{\boldsymbol{f}(\boldsymbol{\alpha}^c,\cdot)\cdot(\mathbb{P}_n\otimes\lambda\!\!\lambda)} \left(\frac{\boldsymbol{h}(\cdot) - \boldsymbol{f}(\boldsymbol{\alpha}^c, \cdot)}{\boldsymbol{f}(\boldsymbol{\alpha}^c, \cdot)}\right)(t) \\ =& \frac{\boldsymbol{h}(t) - \boldsymbol{f}(\boldsymbol{\alpha}^c, t)}{\boldsymbol{f}(\boldsymbol{\alpha}^c, t)} - \boldsymbol{B}(\boldsymbol{\alpha}^c, t)\boldsymbol{D}^{-1}(\boldsymbol{\alpha}^c)\boldsymbol{A}(\boldsymbol{\alpha}^c)\end{aligned}$$

where $\boldsymbol{B}(\boldsymbol{\alpha}^c, t) = \operatorname{diag}(\boldsymbol{f}(\boldsymbol{\alpha}^c, \cdot)^{-1})\nabla_{\boldsymbol{\alpha}^c}\boldsymbol{f}(\boldsymbol{\alpha}^c, \cdot)$, $\nabla_{\boldsymbol{\alpha}^c}\boldsymbol{f}(\boldsymbol{\alpha}^c, t) = \frac{\partial}{\partial \boldsymbol{\alpha}^c}\boldsymbol{f}(\boldsymbol{\alpha}^c, t)$,

$$\boldsymbol{A}(\boldsymbol{\alpha}^c) = \int_0^\tau \left(\overline{(\nabla_{\boldsymbol{\alpha}^c}\boldsymbol{f})(\boldsymbol{f}^{-1})\boldsymbol{h}}(\boldsymbol{\alpha}^c, s) - \overline{\nabla_{\boldsymbol{\alpha}^c}\boldsymbol{f}}(\boldsymbol{\alpha}^c, s)\right) \mathrm{d}s,$$

and

$$\boldsymbol{D}(\boldsymbol{\alpha}^c) = \int_0^\tau \overline{(\nabla_{\boldsymbol{\alpha}^c}\boldsymbol{f})(\boldsymbol{f}^{-1})(\nabla_{\boldsymbol{\alpha}^c}\boldsymbol{f})}(\boldsymbol{\alpha}^c, s)\, \mathrm{d}s.$$

The main difference to Section 8.4 is that now the orthogonal projections are weighted by $\boldsymbol{f}$ and not by $\boldsymbol{h}$. Now the approximate Bahadur slope is given by

$$\left\| \boldsymbol{Q}^{\boldsymbol{B}_1(\boldsymbol{\alpha}^c_0,\cdot)}_{f_1(\boldsymbol{\alpha}^c_0,\cdot)\cdot(\mathrm{P}\otimes\lambda\!\!\lambda)} \left(\frac{h_1(\cdot) - f_1(\boldsymbol{\alpha}^c_0, \cdot)}{f_1(\boldsymbol{\alpha}^c_0, \cdot)}\right) \right\|^2_{f_1(\boldsymbol{\alpha}^c_0,\cdot)\cdot(\mathrm{P}\otimes\lambda\!\!\lambda)}.$$

We estimate $\boldsymbol{\alpha}^c$ by the maximum partial likelihood estimator $\widehat{\boldsymbol{\alpha}^c}$ defined as the maximizer of the log partial likelihood C given in (2.9). Hence, $(\nabla_{\boldsymbol{\alpha}^c}C)(\widehat{\boldsymbol{\alpha}^c}) = \boldsymbol{0}$, where the score function $(\nabla_{\boldsymbol{\alpha}^c}C)(\boldsymbol{\alpha}^c)$ is given by

$$(\nabla_{\boldsymbol{\alpha}^c}C)(\boldsymbol{\alpha}^c) = \int_0^\tau \sum_{i=1}^n \frac{\nabla_{\boldsymbol{\alpha}^c} f_i(\boldsymbol{\alpha}^c, t)}{f_i(\boldsymbol{\alpha}^c, t)}\, \mathrm{d}N_i(t) - \int_0^\tau \sum_{i=1}^n \nabla_{\boldsymbol{\alpha}^c} f_i(\boldsymbol{\alpha}^c, t)\, \mathrm{d}t.$$

Under suitable conditions, it can be shown that if $\lambda_i(t) = h_i(t)$ then

$$\frac{1}{n}(\nabla_{\boldsymbol{\alpha}^c}C)(\boldsymbol{\alpha}^c) - \boldsymbol{A}(\boldsymbol{\alpha}^c) \xrightarrow{\mathrm{P}} \boldsymbol{0},$$

uniformly in a neighborhood of $\boldsymbol{\alpha}^c_0$. Furthermore,

$$\boldsymbol{A}(\boldsymbol{\alpha}^c) \xrightarrow{\mathrm{P}} \left(< \frac{\nabla_j f_1(\boldsymbol{\alpha}^c, \cdot)}{f_1(\boldsymbol{\alpha}^c, \cdot)}, \frac{h(\cdot) - f_1(\boldsymbol{\alpha}^c, \cdot)}{f_1(\boldsymbol{\alpha}^c, \cdot)} >_{f_1(\boldsymbol{\alpha}^c,\cdot)\cdot(\mathrm{P}\otimes\lambda\!\!\lambda)} \right)_{j=1,\ldots,k_{\boldsymbol{\alpha}^c}}.$$

Thus if the right hand side is continuous in $\boldsymbol{\alpha}^c$ and if $\widehat{\boldsymbol{\alpha}^c} \xrightarrow{\mathrm{P}} \boldsymbol{\alpha}^c_0$, then for $j = 1, \ldots, k_{\boldsymbol{\alpha}^c}$,

$$< \frac{\nabla_j f_1(\boldsymbol{\alpha}^c_0, \cdot)}{f_1(\boldsymbol{\alpha}^c_0, \cdot)}, \frac{h(\cdot) - f_1(\boldsymbol{\alpha}^c_0, \cdot)}{f_1(\boldsymbol{\alpha}^c_0, \cdot)} >_{f_1(\boldsymbol{\alpha}^c_0,\cdot)\cdot(\mathrm{P}\otimes\lambda\!\!\lambda)} = 0.$$

Hence,

$$Q^{B_1(\alpha_0^c,\cdot)}_{f_1(\alpha_0^c,\cdot)\cdot(\mathrm{P}\otimes\lambda)}\left(\frac{h_1(\cdot)-f_1(\alpha_0^c,\cdot)}{f_1(\alpha_0^c,\cdot)}\right)=\frac{h_1(\cdot)-f_1(\alpha_0^c,\cdot)}{f_1(\alpha_0^c,\cdot)}.$$

And thus the approximate Bahadur slope is equal to

$$\left\|\frac{h_1(\cdot)-f_1(\alpha_0^c,\cdot)}{f_1(\alpha_0^c,\cdot)}\right\|^2_{f_1(\alpha_0^c,\cdot)\cdot(\mathrm{P}\otimes\lambda)}.$$

This is precisely the slope we would get if the parameter α_0^c was known. Thus if we direct against a fixed alternative which happens to be true, then we do not loose any efficiency by estimating α^c. Of course this depends on the specific estimator for α^c and on the specific estimator for the variance of T.

10.2 Using Several Competing Models

We confined ourselves to a single competing model. To check a model against ν competing models with a test of level α, one can proceed as follows: Use the test for each competing model separately and reject if the p-value of any of the tests exceeds α/ν. Of course, this is only a very crude approach.

To improve it, we would have to consider the joint distribution of our test statistics. In fact, instead of using an n-dimensional vector of stochastic processes as weights, one can use an $n\times p$-matrix C of predictable stochastic processes as weights. To direct against different alternatives we may use the columns of C. The following p-variate vector of stochastic processes can be used as basis for tests:

$$\widetilde{T}(C,t)=n^{-\frac{1}{2}}\int_0^t C(s)^\top\left(\mathrm{d}N(s)-\hat{\lambda}(s)\,\mathrm{d}s\right).$$

Indeed, the asymptotic distribution of $\widetilde{T}(C,t)$ can be derived similarly to Chapter 3, Chapter 5, and Chapter 6. Under orthogonality conditions, $\widetilde{T}(C,t)$ converges to a mean zero p-variate Gaussian martingale. One can construct a tests that rejects for large values of

$$\widetilde{T}(C,\tau)^\top\left(\widehat{\Sigma}(\tau)\right)^{-1}\widetilde{T}(C,\tau).$$

where $\widehat{\Sigma}(\tau)$ is some consistent estimator of the asymptotic covariance matrix of $\widetilde{T}(C,\tau)$ - which we assume to be invertible.

10.3 Possible Extensions

In our setup, we observe n counting processes $N_1,\dots,N_n$. One can interpret this as a marked point process $\tilde{N}$ on $[0,\tau]$ where the marks are in the set $\{1,\dots,n\}$

and indicate which individual experiences the event. We could replace the mark space $\{1, \ldots, n\}$ by some general mark space, say X. Now all sums $\sum_{i=1}^{n}$ can be replaced by integrals over the mark space X.

An example of this are spatio-temporal models, where events are observed in time and space. Here, the mark space is $X = \mathbb{R}^2$ or $X = \mathbb{R}^3$. Asymptotics can be derived if the area/volume of an observation window tends to infinity.

The approach used in this thesis is not confined to survival analysis or point processes. In classical statistic, usually one has n independent observations. For each of these observations, one can compute some residual. After that one can use a weighted sum of these residuals as test statistics - with weights restricted such that the derivative of the test statistic with respect to the model parameters is zero. The one can expect that the (asymptotic) distribution of the weighted residuals does not depend on which estimator is used for the model parameters.

10.4 Relation to Estimation in Semiparametric Models

In our approach, we want to test the fit of the model and orthogonalize our test statistic in a special way such that it does not depend on which estimator of the model parameters is used. This is related to an approach used for estimation purposes in general semiparametric models. The approach is as follows:

In a semiparametric model, usually there is a finite-dimensional parameter θ which is the focus of interest and there is another, possibly infinite-dimensional, parameter η one is less interested in. The parameter η is usually called a 'nuisance' parameter. For example, in the semiparametric Cox model (2.1), the regression parameter $\boldsymbol{\beta}$ is the parameter of interest and the baseline λ_0 is the nuisance parameter.

Often, estimation of θ is based on the score function, which is defined as the derivative of the (log)-likelihood with respect to θ. In general, this score function depends on the unknown nuisance parameter η. Plugging in an estimator $\widehat{\eta}$ for η into the score function for θ usually leads to an estimator $\widehat{\theta}$, whose distribution depends on the distribution of the estimator $\widehat{\eta}$.

Asymptotically, one can avoid this by modifying the score function for θ suitably. Indeed, this can be accomplished by a certain orthogonal projection. This has the advantage that the asymptotic distribution of the estimator of θ does not depend on which estimator $\widehat{\eta}$ of η is used. Basically, one can use any estimator $\widehat{\eta}$ of η that converges at an appropriate rate.

Chapter 11

Conclusions

We introduced model checks based on weighted martingale residuals for a large class of models from survival analysis. A particular feature is that we may direct the tests to be sensitive against certain 'competing models' by choosing the weights suitably. We would like to stress that the tests we propose are model checks. Since it may be the case that neither the null hypothesis nor the competing model is appropriate, a rejection of the null hypothesis is no conclusive evidence for the competing model.

A key idea of the test is that the weights are required to be orthogonal to the derivative of the intensity of the counting process with respect to the model parameters. This has two advantages. First, the asymptotic distribution of the test statistic does not depend on the distribution of the estimators of the model parameters. Therefore, we can use any 'off-the-shelf' estimator from the literature that is consistent at a certain rate. The second advantage is that the asymptotic distribution is relatively simple, allowing us to choose optimal weights against local and fixed alternatives.

The weights are n-variate stochastic processes with the essential requirement that they should be predictable. For some tests, as for example for the test against a Cox model, we have to insert estimates of parameters which destroy the property of predictability of the weights. Nevertheless we want that plugging in the parameter does not change the asymptotic distribution of our test statistic. If the parameter is finite-dimensional then by Taylor expansions, it can be shown that this is the case. However, in some cases we need to plug in estimated functions. Now we need estimators that besides converging uniformly also have the property that the total variation of the difference to the limiting value converges to zero. We propose a modified kernel smoothing algorithm to satisfy these requirements. In the kernel smoothing approach the choice of the bandwidth is critical and our ad hoc choice could conceivably be improved. The condition on the total variation is needed because the estimated functions are not assumed to be predictable.

Model checks based on martingale residuals have been previously considered in the literature. Most of these checks use a Cox model as null hypothesis with ad hoc weights for the martingale residuals. Even for the Cox model, our approach of choosing weights such that they satisfy certain orthogonality conditions with respect to all parameters - both the finite and infinite dimensional parameters - seems to be new. For other models, the field of model checks is still underdeveloped.

In Chapters 7-9, we mainly focused on the additive Aalen model (2.6), a semiparametric restriction of it (2.7), and on Cox-type models (2.5). The results from Chapter 5 and Chapter 6 can be used for other (semiparametric) models from survival analysis as well, most notably the combined Cox-Aalen models (2.11) and (2.12).

We derived the asymptotic distribution of our test statistic and showed how to choose optimal weights. The optimal weights are essentially weighted orthogonal projections of the difference between the intensity of the alternative and the intensity under the null hypothesis. We also provided ad hoc weights based on unweighted orthogonal projections.

One has to be careful with some weights - the asymptotic variance of our test statistic may converge to zero and thus we cannot use the asymptotic normality of our test statistic to construct tests. In particular, this may be the case if we are dealing with nested hypotheses. For a particular test of Cox-type models we suggested two approaches to deal with this problem: a sequential procedure and a bootstrap procedure. In the sequential procedure we test whether the asymptotic variance is zero or not before applying our test of fit. The sequential procedure is asymptotically conservative.

We conducted extensive simulation studies that reveal that the tests can be used for practical sample sizes. Furthermore, not surprisingly, our directed test has greater power than other goodness-of-fit tests. Of course, no general suggestions for the sample size to attain a certain power can be made, because this depends on the alternative to be detected. The simulations show that the optimal weights lead to some improvement in the power compared with ad hoc weights.

We applied some of our methods to three different datasets: a dataset from software reliability and two datasets from medical studies. We have seen that our methods are useful in practice: formal conclusions can be drawn that have not been possible with previous methods.

Summarizing, we suggested and analyzed model checks that can be used for many standard models in survival analysis. Simulation studies and applications to real datasets showed that the tests are promising.

Appendix A

Some Tools From Probability Theory

In this chapter, we collect some results mainly from the theory of stochastic processes. We only give proofs for some non-standard results.

A.1 Counting Processes

Counting processes are basic to survival analysis. We shall introduce them briefly. An adapted, càdlàg stochastic process N is called a *counting process* if $N(0) = 0$ and if the paths of N are step functions, whose discontinuities are upward jumps of size 1.

If N_1 and N_2 are counting processes, we say that they "*have no common jumps*", if $\Delta N_1 \Delta N_2$ is indistinguishable from 0, where $\Delta N_1(t) = N_1(t) - N_1(t-)$ is the size of the jump of N_1 at time t.

If $\boldsymbol{N} = (N_1, \ldots, N_k)^\top$ is a k-variate process then it is a k-variate counting process, if each N_i is a counting process and if for $i \neq j$, the processes N_i and N_j do not have common jumps.

A.2 Predictable Covariation

In this section, we recall the predictable covariation of a local martingale.

Theorem A.1. *If X, Y are local square integrable martingales then there exists a predictable process $\langle X, Y\rangle$ of finite variation, which is unique up to indistinguishability, such that $XY - \langle X, Y\rangle$ is a local square integrable martingale.*

If $X = Y$, we abbreviate $\langle X, X\rangle$ by $\langle X\rangle$.

Lemma A.1. *Let N be a counting process on $[0,\tau]$, $0<\tau<\infty$, which satisfies $\mathrm{E}\,N(\tau)<\infty$. If N admits a continuous compensator A, i.e. $M=N-A$ is a local martingale, then the following holds true.*

(i) A, N, M are locally bounded and locally square-integrable.

(ii) $\langle M\rangle = A$

The following theorem can be found in Fleming and Harrington (1991, Theorem 2.4.3).

Theorem A.2. *Suppose that H_1 and H_2 are locally bounded predictable processes and N_1 and N_2 are counting processes with compensators A_1 and A_2. Let $M_i = N_i - A_i, i=1,2$. Then*

$$\left\langle \int_0^{\cdot} H_1(s)\,\mathrm{d}M_1(s), \int_0^{\cdot} H_2(s)\,\mathrm{d}M_2(s)\right\rangle(t) = \int_0^t H_1(s)H_2(s)\,\mathrm{d}\left\langle M_1, M_2\right\rangle(s).$$

A.3 Rebolledo's Theorem

This section contains a central limit theorem for martingales. For each $n\in\mathbb{N}$, $i\in\{1,\ldots,r\}$, let M_i^n be a local square integrable càdlàg martingale defined on $[0,\tau]$, $0<\tau<\infty$, and let $\boldsymbol{M}^n=(M_1^n,\ldots,M_r^n)^\top$.

Suppose that for each $\epsilon>0$, $\boldsymbol{M}^{n,\epsilon}=(M_1^{n,\epsilon},\ldots,M_r^{n,\epsilon})^\top$ is a vector of local square integrable càdlàg martingales that contains all jumps of components of $\boldsymbol{M}^n$ larger than ϵ, i.e. $|\Delta M_i^n(t)-\Delta M_i^{n,\epsilon}(t)|\le\epsilon$ for all $t\in[0,\tau], i\in\{1,\ldots r\}$. Of course, we could choose $\boldsymbol{M}^{n,\epsilon}=\boldsymbol{M}^n$, but in general, this choice will not satisfy the conditions needed for Theorem A.3.

Let $\boldsymbol{V}:[0,\tau]\to\mathbb{R}^{r\times r}$ be a continuous function that has positive semidefinite increments, i.e. for all $s,t\in[0,\tau], s\le t$, $\boldsymbol{V}(t)-\boldsymbol{V}(s)$ is positive semidefinite.

Let $\boldsymbol{M}^\infty$ be an r-variate continuous Gaussian martingale with $\langle\boldsymbol{M}^\infty\rangle=\boldsymbol{V}$. Such a process can be constructed by starting with an r-dimensional Brownian motion and multiplying it with a suitable decomposition of $\boldsymbol{V}$.

The following theorem is due to Rebolledo (1980). We give the version from Andersen et al. (1993, p. 83f). Note that as always all limits are as n tends to infinity.

Theorem A.3. *If*

$$\langle\boldsymbol{M}^n\rangle(t)\xrightarrow{\mathrm{P}}\boldsymbol{V}(t)\qquad\forall t\in[0,\tau]\ \textit{and}$$
$$\langle M_i^{n,\epsilon}\rangle(t)\xrightarrow{\mathrm{P}}0\qquad\forall t\in[0,\tau],\epsilon>0,i\in\{1,\ldots,r\},$$

then in $(D[0,\tau])^r$,

$$\boldsymbol{M}^n\xrightarrow{d}\boldsymbol{M}^\infty.$$

Furthermore, uniformly in $t \in [0, \tau]$,

$$[\boldsymbol{M}^n](t) \xrightarrow{\mathrm{P}} \boldsymbol{V}(t).$$

In the previous theorem, $[\boldsymbol{M}^n] = ([M_i^n, M_j^n])_{i,j=1,\dots,r}$ is the quadratic covariation of $\boldsymbol{M}^n$. The quadratic covariation of two semimartingales X and Y is defined by $[X,Y](t) = X(t)Y(t) - X(0)Y(0) - \int_0^t Y(s-)\,\mathrm{d}X(s) - \int_0^t X(s-)\,\mathrm{d}Y(s)$. In our applications, usually $M^n(t) = \int_0^t \boldsymbol{c}(s)^\top(\,\mathrm{d}\boldsymbol{N}(s) - \boldsymbol{\lambda}(s)\,\mathrm{d}s)$, where $\boldsymbol{c}$ is an n-variate vector of predictable processes, $\boldsymbol{N}$ is an n-variate counting process and $\boldsymbol{\lambda}$ is the intensity of $\boldsymbol{N}$. Then $[M^n](t) = \sum_{i=1}^n \int_0^t c_i(s)^2\,\mathrm{d}N_i(s)$. We usually use $M^{n,\epsilon}(t) = \sum_{i=1}^n \int_0^t c_i(s) I\{c_i(s) > \epsilon\}\,\mathrm{d}(N_i(s) - \Lambda_i(s))$. Then $\langle M^{n,\epsilon}\rangle(t) = \int_0^t c_i^2(s) I\{c_i(s) > \epsilon\}\,\mathrm{d}\Lambda_i(s)$. Thus the second condition of Theorem A.3 is satisfied if $\sup_{\substack{i=1\dots n\\ s\in[0,\tau]}} |c_i(s)| \xrightarrow{\mathrm{P}} 0$.

A.4 Lenglart's Inequality

The inequality we present in this section is due to Lenglart (1977). The following version is from Jacod and Shiryaev (1987).

Lemma A.2. *Let X be a càdlàg, adapted process and A an increasing, predictable process such that for every bounded stopping time T,*

$$\mathrm{E}\,|X(T)| \le \mathrm{E}\,|A(T)|.$$

Then, for all stopping times T and all $\epsilon, \eta > 0$,

$$\mathrm{P}\left(\sup_{s\le T}|X(s)| \ge \epsilon\right) \le \frac{\eta}{\epsilon} + \mathrm{P}\left(A(T) \ge \eta\right)$$

We use this inequality in the following context: If M is a square integrable local martingale and T a bounded stopping time then $EM(T)^2 \le E\langle M\rangle(T)$. Hence, for all stopping times T and all $\epsilon, \eta > 0$,

$$\mathrm{P}\left(\sup_{s\le T}|M(s)| \ge \epsilon\right) \le \frac{\eta}{\epsilon^2} + \mathrm{P}\left(\langle M\rangle(T) \ge \eta\right).$$

The following is essentially an application of Lenglart's inequality.

Lemma A.3. *Suppose $\boldsymbol{N}$, $\boldsymbol{\lambda}$, and $\boldsymbol{M}$ are as given in Section 2.2. If $e_i^{(n)} : \Omega \times [0,\tau] \to \mathbb{R}, 1 \le i \le n,\ n \in \mathbb{N}$, are functions, $\sup_{\substack{i=1\dots n\\ t\in[0,\tau]}} |e_i^{(n)}(t)| \xrightarrow{\mathrm{P}} 0$, and $\overline{\boldsymbol{\lambda}}$ converges uip on $[0,\tau]$ then*

$$\frac{1}{n}\int_0^t \boldsymbol{e}^{(n)}(s)^\top\,\mathrm{d}\boldsymbol{N}(s) \xrightarrow{\mathrm{P}} 0 \text{ and } \frac{1}{n}\int_0^t \boldsymbol{e}^{(n)}(s)^\top\,\mathrm{d}\boldsymbol{M}(s) \xrightarrow{\mathrm{P}} 0 \text{ uniformly in } t \in [0,\tau].$$

We implicitly assume measurability of $\sup_{\substack{i=1\dots n\\ t\in[0,\tau]}} |e_i^{(n)}(t)|$ and the existence of the integrals.

Proof. Since

$$\begin{aligned}\langle \overline{\boldsymbol{M}} \rangle (\tau) &= \left\langle \frac{1}{n}\sum_{i=1}^{n}\left(N_i(\cdot) - \int_0^{\cdot} \lambda_i(s)\,\mathrm{d}s\right)\right\rangle (\tau) = \frac{1}{n^2}\sum_{i=1}^{n}\int_0^{\tau} \lambda_i(s)\,\mathrm{d}s \\ &= \frac{1}{n}\int_0^{\tau} \overline{\boldsymbol{\lambda}}(s)\,\mathrm{d}s \xrightarrow{\mathrm{P}} 0,\end{aligned}$$

Lenglart's inequality implies $\overline{\boldsymbol{M}} \to 0$ uip on $[0,\tau]$. Therefore,

$$\frac{1}{n}\sum_{i=1}^{n}\int_0^t |\,\mathrm{d}N_i(s)| = \overline{\boldsymbol{N}}(t) = \overline{\boldsymbol{M}}(t) + \int_0^t \overline{\boldsymbol{\lambda}}(s)\,\mathrm{d}s \xrightarrow{\mathrm{P}} \int_0^t \vec{\overline{\boldsymbol{\lambda}}}(s)\,\mathrm{d}s$$

uniformly in $t \in [0,\tau]$. Hence, dropping the dependence of $\boldsymbol{e}$ on n,

$$\begin{aligned}\left|\frac{1}{n}\int_0^t \boldsymbol{e}(s)^\top \mathrm{d}\boldsymbol{N}(s)\right| &\le \frac{1}{n}\sum_{i=1}^{n}\left|\int_0^t e_i(s)\,\mathrm{d}N_i(s)\right| \\ &\le \frac{1}{n}\sum_{i=1}^{n}\int_0^t |\,\mathrm{d}N_i(s)| \sup_{\substack{i=1\dots n\\ s\in[0,\tau]}} |e_i(s)| \xrightarrow{\mathrm{P}} 0\end{aligned}$$

uniformly in $t \in [0,\tau]$. The second statement can be shown similarly. □

A.5 Limit Theorems in the i.i.d. Case

The following strong law of large numbers is due to Rao (1963), where, for some $0 < \tau < \infty$, the symbol $D[0,\tau]$ is defined as the space of all càdlàg functions from $[0,\tau]$ into $\mathbb{R}$.

Theorem A.4. *Suppose $X_1, X_2, \dots$ are i.i.d. $D[0,\tau]$-valued random variables. If $\mathrm{E}\left[\sup_{t\in[0,\tau]} |X_1(t)|\right] < \infty$ then*

$$\sup_{t\in[0,\tau]}\left|\frac{1}{n}\sum_{i=1}^{n} X_i(t) - \mathrm{E}[X_1(t)]\right| \to 0 \quad \textit{almost surely.}$$

The following extension to càdlàg processes in Banach spaces can be found in Andersen and Gill (1982, Theorem III.I,Appendix III). Let B be a separable Banach space with norm $\|\cdot\|$. The space of càdlàg functions from $[0,\tau]$ into B is denoted by $D_B[0,\tau]$. We will only use $B = C(K)$, where K is some compact euclidean space and $C(K)$ is the space of all continuous functions $K \to \mathbb{R}$ endowed with the supremum norm.

Theorem A.5. *Suppose $X_1, X_2, \ldots$ are i.i.d. elements of $D_B[0,\tau]$. If*

$$\mathrm{E} \sup_{t\in[0,\tau]} \|X_1(t)\| < \infty$$

then

$$\sup_{t\in[0,\tau]} \left\| \frac{1}{n}\sum_{i=1}^{n} X_i(t) - \mathrm{E}[X_1(t)] \right\| \to 0 \quad \textit{almost surely.}$$

Remark A.1. The laws of large numbers we discussed are for càdlàg processes. But, as suggested in Andersen and Gill (1982), one merely has to reverse the time scale to use them for càglàd processes.

Lemma A.4. *Let $X_1, X_2, \ldots$ be i.i.d. random variables with $\mathrm{E}\,|X_1|^p < \infty$ for some $p > 0$. Then*

$$n^{-1/p} \sup_{i=1,\ldots,n} |X_i| \xrightarrow{\mathrm{P}} 0 \quad (n \to \infty).$$

Proof. Let $Y_i := |X_i|^p$. For all $\epsilon > 0$, using Bernoulli's inequality and dominated convergence,

$$\begin{aligned}
\mathrm{P}(\frac{1}{n} \sup_{i=1,\ldots,n} Y_i > \epsilon) &= 1 - \mathrm{P}(Y_i \le n\epsilon, i = 1,\ldots,n) = 1 - (1 - \mathrm{P}(Y_1 > n\epsilon))^n \\
&\le 1 - (1 - n\mathrm{P}(Y_1 > n\epsilon)) = n\mathrm{P}(Y_1 > n\epsilon) \\
&\le \epsilon^{-1} \int Y_1 I\{Y_1 > n\epsilon\}\, \mathrm{dP} \to 0 \quad (n \to \infty).
\end{aligned}$$

Hence,

$$n^{-1/p} \sup_{i=1,\ldots,n} |X_i| = (n^{-1} \sup_{i=1,\ldots,n} |Y_i|)^{1/p} \xrightarrow{\mathrm{P}} 0 \quad (n \to \infty).$$

□

The following is a central limit theorem in Hilbert spaces given in Ledoux and Talagrand (1991, Corollary 10.9) and in van der Vaart and Wellner (1996, p. 92).

Theorem A.6. *Let H be a separable Hilbert space and let $X, X_1, X_2, \ldots$ be i.i.d. stochastic elements in H (equipped with its Borel-σ algebra). If $\mathrm{E}(f(X)) = 0$ for all f in the dual space H', and $\mathrm{E}\,\|X\|^2 < \infty$ then $n^{-1/2}\sum_{i=1}^{n} X_i$ converges in distribution in H.*

A.6 Stochastic Processes Indexed by Time and Other Parameters

In this section we collect some results concerning stochastic processes that do not only depend on $t \in [0, \tau]$ but also on further parameters.

In the next lemma, we implicitly assume that appropriate measurability conditions are satisfied.

Lemma A.5. *Let $\boldsymbol{\mathcal{X}}_{\boldsymbol{\beta}} \subset \mathbb{R}^{k_\beta}$ be an open set, $a_n, a : \Omega \times \boldsymbol{\mathcal{X}}_{\boldsymbol{\beta}} \times [0, \tau] \to \mathbb{R}$, $n \in \mathbb{N}$ be functions. Suppose $\hat{\boldsymbol{\beta}} \xrightarrow{\mathrm{P}} \boldsymbol{\beta}_0 \in \boldsymbol{\mathcal{X}}_{\boldsymbol{\beta}}$, and suppose there exist an open neighborhood $\mathcal{C} \subset \boldsymbol{\mathcal{X}}_{\boldsymbol{\beta}}$ of $\boldsymbol{\beta}_0$ such that $a_n \xrightarrow{\mathrm{P}} a$ uniformly on $\mathcal{C} \times [0, \tau]$ and $a(\cdot, t) : \boldsymbol{\mathcal{X}}_{\boldsymbol{\beta}} \to \mathbb{R}$ is continuous at $\boldsymbol{\beta}_0$ uniformly in $t \in [0, \tau]$.*
Then $a_n(\hat{\boldsymbol{\beta}}, t) \xrightarrow{\mathrm{P}} a(\boldsymbol{\beta}_0, t)$ uniformly in $t \in [0, \tau]$.

Proof. For all $t \in [0, \tau]$,

$$\left|a_n(\hat{\boldsymbol{\beta}}, t) - a(\boldsymbol{\beta}_0, t)\right| \leq \left|a_n(\hat{\boldsymbol{\beta}}, t) - a(\hat{\boldsymbol{\beta}}, t)\right| + \left|a(\hat{\boldsymbol{\beta}}, t) - a(\boldsymbol{\beta}_0, t)\right|.$$

The continuity of a shows that the second term converges to 0 uniformly on $[0, \tau]$. Since $\mathcal{C}$ is an open neighborhood of $\boldsymbol{\beta}_0$, $\mathrm{P}(\hat{\boldsymbol{\beta}} \in \mathcal{C}) \to 1$ and hence the convergence of a_n implies $\left|a_n(\hat{\boldsymbol{\beta}}, t) - a(\hat{\boldsymbol{\beta}}, t)\right| \xrightarrow{\mathrm{P}} 0$ uniformly in $t \in [0, \tau]$. □

Lemma A.6. *Let $A \subset \mathbb{R}^k$ be a measurable set. For all $a \in A$, let $X_a : \Omega \times [0, \tau] \to \mathbb{R}$ be a predictable stochastic process such that for each $(\omega, t) \in \Omega \times [0, \tau]$, the mapping $X_\cdot(\omega, t) : A \to \mathbb{R}$ is continuous. Let $\alpha : \Omega \times [0, \tau] \to A$ be a predictable process. Then the process $(\omega, t) \mapsto X_{\alpha(\omega,t)}(\omega, t)$ is predictable.*

Proof. Recall that a stochastic process is called predictable if it is measurable with respect to the predictable σ-algebra on $\Omega \times [0, \tau]$. There are predictable processes $\beta_k : \Omega \times [0, \tau]$, $k \in \mathbb{N}$, of the form

$$\beta_k(\omega, t) = \sum_{i=1}^{k} x_k(i) I\{(\omega, t) \in B_k(i)\},$$

such that $\beta_k \to \alpha$ pointwise as $k \to \infty$, where $x_k(i) \in A$, $i = 1, \ldots, k$, and $\{B_k(i), i = 1, \ldots, k\}$ is a partition of $\Omega \times [0, \tau]$ into predictable sets. Then

$$X_{\beta_k(t)}(\omega, t) = \sum_{i=1}^{k} X_{x_k(i)}(\omega, t) I\{(\omega, t) \in B_k(i)\}.$$

Therefore, $X_{\beta_k(t)}(\omega, t)$ is predictable as a sum/product of predictable processes. By the continuity of $X_\cdot(\omega, t)$, we have for all (ω, t) that

$$X_{\alpha(t)}(\omega, t) = \lim_{k \to \infty} X_{\beta_k(t)}(\omega, t).$$

Hence, $X_{\alpha(t)}(\omega, t)$ is predictable. □

Remark A.2. The preceding lemma is still true if we replace predictable by progressive.

The following lemma shows that certain real valued stochastic processes that depend on a parameter $\boldsymbol{a} \in A \subset \mathbb{R}^\nu$, where A is compact, can be considered stochastic processes that take values in the Banach space $C(A)$ of continuous functions $\boldsymbol{A} \to \mathbb{R}$ equipped with the supremum-norm and the corresponding Borel-σ-algebra. This property is part of the conditions (LS1), (LSW1), and (SPW1). It allows the application of the strong law of large numbers for càdlàg processes in Banach spaces (Theorem A.5) to get uniform convergence with respect to parameters and time. An example for these stochastic processes is the intensity of the nonparametric model (5.1).

Lemma A.7. *Let $f : R^l \times A \to \mathbb{R}$ be a continuous function, where $A \subset \mathbb{R}^\nu$ is a compact set. Suppose that $\boldsymbol{X}$ is an l-variate vector of stochastic processes. Let $x(\omega, t)$ denote the mapping $\Omega \times [0, \tau] \to C(A)$ given by $x(\omega, t)(\boldsymbol{a}) = f(\boldsymbol{X}(\omega, t), \boldsymbol{a})$. Then $x(t)$ is a stochastic process with values in $C(A)$. Furthermore, we have the following:*

(i) If $\boldsymbol{X}$ is càglàd then x is càglàd.

(ii) If $\boldsymbol{X}$ is locally bounded then x is locally bounded and, in particular, $\sup_{a\in A} |x(\cdot)(a)|$ is a locally bounded real valued stochastic process.

Proof. First, we show that the mapping $g : R^l \to C(A), g(\boldsymbol{z}) = f(\boldsymbol{z}, \cdot)$ is continuous. Let $\boldsymbol{z} \in R^l$ and $\epsilon > 0$. Choose any $\delta > 0$. Then f is uniformly continuous on the compact set $\bar{U}(\boldsymbol{z}, \delta) \times A$, where $\bar{U}(\boldsymbol{z}, \delta) := \{\boldsymbol{y} \in \mathbb{R}^l : |\boldsymbol{z} - \boldsymbol{y}| \leq \delta\}$. Hence, there is $\tilde{\delta} > 0$ such that for $\boldsymbol{y}_1, \boldsymbol{y}_2 \in \bar{U}(\boldsymbol{z}, \delta)$, $\boldsymbol{\theta}_1, \boldsymbol{\theta}_2 \in A$,

$$|\boldsymbol{y}_1 - \boldsymbol{y}_2| < \tilde{\delta}, |\boldsymbol{\theta}_1 - \boldsymbol{\theta}_2| < \tilde{\delta} \text{ implies } |f(\boldsymbol{y}_1, \boldsymbol{\theta}_1) - f(\boldsymbol{y}_2, \boldsymbol{\theta}_2)| < \epsilon.$$

Hence, for all $\boldsymbol{y} \in R^l$,

$$|\boldsymbol{y} - \boldsymbol{z}| < \min(\tilde{\delta}, \delta) \text{ implies } \sup_{\boldsymbol{\theta}\in A} |f(\boldsymbol{y}, \boldsymbol{\theta}) - f(\boldsymbol{z}, \boldsymbol{\theta})| < \epsilon,$$

showing that g is indeed continuous. Since $x(t) = g(\boldsymbol{X}(t))$, the properties of g and $\boldsymbol{X}$ imply that $x(t)$ is measurable for each t, i.e. it is a stochastic process. Furthermore, if $\boldsymbol{X}$ is càglàd then the continuity of g implies that x is càglàd. Next, we show (ii). Suppose that $\boldsymbol{X}$ is locally bounded. Let $\nu_i, i \in \mathbb{N}$ be a localizing sequence for $\boldsymbol{X}$. Then there exists $K_i < \infty$ such that $\|\boldsymbol{X}(t \wedge \nu_i)\| < K_i$. Let $B_i := \{\boldsymbol{y} \in \mathbb{R}^l : \|\boldsymbol{y}\| < K_i\}$. Since $B_i \times A$ is compact, $f(B_i \times A)$ is compact and hence bounded. Clearly,

$$\sup_{a\in A} |x(a, t \wedge \nu_i)| \leq \max_{b\in B_i, a\in A} |f(b, a)| < \infty$$

and hence ν_i is a localizing sequence for $x(t)$. Since the norm is continuous on $C(A)$, $\sup_{a \in A} |x(t)(a)|$ is a random variable for each t and thus a stochastic process. It is a locally bounded process since x is a locally bounded process.

$\square$

Appendix B

Orthogonal Projections

In this chapter we collect some results about orthogonal projections. Some of the notation used can be found in Section 3.2.

Lemma B.1. *Suppose H is a vector space over $\mathbb{R}$ with scalar product $<\cdot,\cdot>$. If $y_1,\dots,y_k \in H$ are such that the matrix $\boldsymbol{A} := (<y_\eta, y_\xi>)_{\eta,\xi=1,\dots,k}$ is invertible then*

$$Q : H \to H, x \mapsto x - (y_1,\dots,y_k)\,\boldsymbol{A}^{-1}(<y_1,x>,\dots,<y_k,x>)^\top$$

is the orthogonal projection onto the space orthogonal to G, where G is the space spanned by $y_1,\dots,y_k$.

Proof. For $x \in H$ and $i \in \{1,\dots,k\}$,

$$\begin{aligned}<y_i, Qx> &= <y_i, x> - (<y_i,y_1>,\dots,<y_i,y_k>)\,\boldsymbol{A}^{-1}(<y_1,x>,\dots,<y_k,x>)^\top \\ &= <y_i,x> - <y_i,x> = 0.\end{aligned}$$

Hence Q maps into $G^\perp$. Clearly $P : H \to H, x \mapsto x - Qx$ maps into G. Since P and Q are linear, Q is the orthogonal projection onto $G^\perp$, see e.g. Rudin (1974, p. 84). □

The following results are needed for the projections in Chapter 6. Let $(A, \mathcal{A}, \mu)$ be a measure space with $\mu(A) < \infty$. Consider the product space $C = (A, \mathcal{A}, \mu) \otimes ([0,\tau], \mathcal{B}[0,\tau], \lambda\!\!\lambda)$, where $0 < \tau < \infty$, $\mathcal{B}[0,\tau]$ is the Borel-σ-algebra on $[0,\tau]$, and $\lambda\!\!\lambda$ is Lebesgue measure. Let $w : C \to [0,\infty)$ be measurable and define the measure $\nu = w \cdot (\mu \otimes \lambda\!\!\lambda)$. Let $x_1,\dots,x_k \in L_2(\nu)$. Define the set

$$U = \left\{ \sum_{j=1}^{k} x_j(u,s) g_j(s) \in L_2(\nu) : g_1,\dots,g_k \text{ measurable} \right\}.$$

For each $s \in [0,\tau]$, let $\boldsymbol{Q}^{X(\cdot,s)}_{h(\cdot,s)\cdot\mu}$ be the projection onto $X(\cdot,s)^\perp$, with respect to the measure $h(\cdot,s)\cdot\mu$. where $X(\cdot,s) = \{x_1(\cdot,s),\dots,x_k(\cdot,s)\}$.

Lemma B.2. *$Q : L_2(\nu) \to L_2(\nu)$ defined by*

$$Q(\boldsymbol{x})(u,s) = (\boldsymbol{Q}^{X(\cdot,s)}_{h(\cdot,s)\cdot\mu}\boldsymbol{x}(\cdot,s))(u)$$

is the projection onto $U^\perp$ with respect to ν.

Proof. Clearly, Q maps into U and for each $\boldsymbol{x} \in U$, $Q\boldsymbol{x} = \boldsymbol{x}$. To show that Q is an orthogonal projection we need that Q is a linear bounded operator that satisfies $QQ = Q$ and $< Q\boldsymbol{x}, \boldsymbol{y} >_\nu = < \boldsymbol{x}, Q\boldsymbol{y} >_\nu$ for all $\boldsymbol{x}, \boldsymbol{y} \in L_2(\nu)$. The linearity of $\boldsymbol{Q}^v(t)$ for each $t \in [0,\tau]$ implies the linearity of Q. Let $\boldsymbol{x}, \boldsymbol{y} \in L_2(\nu)$. Since $\|Q\boldsymbol{x}\|^2_\nu = \int_0^\tau \|(Q\boldsymbol{x})(s)\|^2_{h(\cdot,s)\cdot\mu}\,\mathrm{d}s = \int_0^\tau \|\boldsymbol{Q}^v(s)\boldsymbol{x}(s)\|^2_{h(\cdot,s)\cdot\mu}\,\mathrm{d}s \le \int_0^\tau \|\boldsymbol{x}(s)\|^2_{h(\cdot,s)\cdot\mu}\,\mathrm{d}s = \|\boldsymbol{x}\|^2_\nu$, the mapping Q is a bounded operator. Since

$$\begin{aligned}((QQ)\boldsymbol{x})(u,s) &= (Q(Q\boldsymbol{x}))(u,s) = \boldsymbol{Q}^v(s)((Q\boldsymbol{x})(\cdot,s))(u)\\ &= \boldsymbol{Q}^v(s)(\boldsymbol{Q}^v(s)\boldsymbol{x}(\cdot,s))(u) = (\boldsymbol{Q}^v(s)\boldsymbol{Q}^v(s))\boldsymbol{x}(\cdot,s)(u)\\ &= \boldsymbol{Q}^v(s)\boldsymbol{x}(\cdot,s)(u) = (Q\boldsymbol{x})(s,u),\end{aligned}$$

we have $QQ = Q$. Furthermore,

$$\begin{aligned}<Q\boldsymbol{x},\boldsymbol{y}>_\nu &= \int_0^\tau <\boldsymbol{Q}^v(s)\boldsymbol{x}(\cdot,s),\boldsymbol{y}(\cdot,s)>_{h(\cdot,s)\cdot\mu}\,\mathrm{d}s\\ &= \int_0^\tau <\boldsymbol{x}(\cdot,s),\boldsymbol{Q}^v(s)\boldsymbol{y}(\cdot,s)>_{h(\cdot,s)\cdot\mu}\,\mathrm{d}s = <\boldsymbol{x},Q\boldsymbol{y}>_\nu .\end{aligned}$$

□

Lemma B.3. *Let H be a Hilbert space and suppose that $V, W \subset H$. Let $\tilde{V} = \boldsymbol{Q}^W V = \{\boldsymbol{Q}^W v : v \in V\}$, where $\boldsymbol{Q}^W$ is the orthogonal projection onto $W^\perp$. Then $\boldsymbol{Q}^{V+W} = \boldsymbol{Q}^{\tilde{V}}\boldsymbol{Q}^W = \boldsymbol{Q}^W\boldsymbol{Q}^{\tilde{V}}$ is the orthogonal projection onto $(V+W)^\perp$, where $\boldsymbol{Q}^{\tilde{V}}$ is the orthogonal projection onto $\tilde{V}^\perp$.*

Proof. Let $x = y + z \in \tilde{V} + W$, where $y \in \tilde{V}$ and $z \in W$. Since $\tilde{V} \perp W$, we have $y \in W^\perp$ and hence $\boldsymbol{Q}^{\tilde{V}}\boldsymbol{Q}^W y = \boldsymbol{Q}^{\tilde{V}} y = 0$. Furthermore, $\boldsymbol{Q}^{\tilde{V}}\boldsymbol{Q}^W z = \boldsymbol{Q}^{\tilde{V}} 0 = 0$ and hence $\boldsymbol{Q}^{\tilde{V}}\boldsymbol{Q}^W x = 0$. As a concatenation of two projections, $\boldsymbol{Q}^{\tilde{V}}\boldsymbol{Q}^W$ is clearly linear and bounded (and thus continuous). Hence, $\boldsymbol{Q}^{\tilde{V}}\boldsymbol{Q}^W\boldsymbol{x} = 0$ for all $\boldsymbol{x}$ in the closure of the span of $\tilde{V} + W$. For $x \in (\tilde{V}+W)^\perp$, we have $x \in \tilde{V}^\perp$ and $x \in W^\perp$, and hence $\boldsymbol{Q}^{\tilde{V}}\boldsymbol{Q}^W x$. Thus $\boldsymbol{Q}^{V+W}$ is the orthogonal projection onto $(\tilde{V}+W)^\perp$. It remains to show $(\tilde{V}+W)^\perp = (V+W)^\perp$. Let $u \in (V+W)^\perp$. Then for $\tilde{x} \in \tilde{V}$ and $y \in W$ there exists $x \in V$ such that $\tilde{x} = \boldsymbol{Q}^W x$. Hence,

$$<u,\tilde{x}+y>=<u,\boldsymbol{Q}^W x> +0 =<\boldsymbol{Q}^W u, x>=<u,x>= 0$$

and thus $u \in (\tilde{V}+W)^\perp$. If $u \in (\tilde{V}+W)^\perp$ then for $x \in V$ and $y \in W$ we have $\boldsymbol{Q}^W x \in \tilde{V}$ and hence

$$<u,x+y>=<\boldsymbol{Q}^W u,x> +0 =<u,\boldsymbol{Q}^W x>= 0.$$

Thus $u \in (V+W)^\perp$. □

List of Symbols and Abbreviations

List of Conditions

Bibliography

For each reference, the pages on which it is cited are indicated.

Aalen, O. O. (1977). Weak convergence of stochastic integrals related to counting processes. *Zeitschrift für Wahrscheinlichkeitstheorie und verwandte Gebiete*, 38:261–277. Cited on page 11.

Aalen, O. O. (1978). Nonparametric inference for a family of counting processes. *Ann. Statist.*, 6(4):701–726. Cited on page 11.

Aalen, O. O. (1980). A model for nonparametric regression analysis of counting processes. In Klonecki, W., Kozek, A., and Rosinski, J., editors, *Mathematical statistics and probability theory*, volume 2 of *Lecture Notes in Statist.*, pages 1–25. Springer, New York. Cited on pages 2, 11, and 17.

Aalen, O. O. (1989). A linear regression model for the analysis of life times. *Statist. Med.*, 8:907–925. Cited on page 18.

Aalen, O. O. (1993). Further results on the non-parametric linear regression model in survival analysis. *Statist. Med.*, 12:1569–1588. Cited on pages 18 and 27.

Andersen, P. K. (1982). Testing goodness of fit of Cox's regression and life model. *Biometrics*, 38:67–77. Cited on page 24.

Andersen, P. K., Borgan, Ø., Gill, R. D., and Keiding, N. (1993). *Statistical Models Based on Counting Processes.* Springer, New York. Cited on pages 3, 11, 18, 19, 21, 28, 30, and 150.

Andersen, P. K. and Gill, R. D. (1982). Cox's regression model for counting processes: a large sample study. *Ann. Statist.*, 10(4):1100–1120. Cited on pages 16, 17, 56, 57, 59, 116, 152, and 153.

Arjas, E. (1988). A graphical method for assessing goodness of fit in Cox's proportional hazards model. *J. Amer. Statist. Assoc.*, 83(401):204–212. Cited on pages 22 and 30.

Bagdonavicius, V. and Nikulin, M. (1998). *Additive and multiplicative semiparametric models in accelerated life testing and survival analysis.* Queen's Papers in Pure and Applied Mathematics. 108. Kingston, Ont.: Queen's Univ. Cited on page 11.

Bahadur, R. R. (1960). Stochastic comparison of tests. *Ann. Math. Statist.*, 31:276–295. Cited on page 39.

Barlow, W. E. and Prentice, R. L. (1988). Residuals for relative risk regression. *Biometrika*, 75:65–74. Cited on page 22.

Bauer, H. (1992). *Maß- und Integrationstheorie.* Walter de Gruyter, Berlin, second edition. Cited on pages 49 and 77.

Billingsley, P. (1999). *Convergence of Probability Measures.* Wiley, New York, 2nd edition. Cited on page 34.

Borodin, A. N. and Salminen, P. (2002). *Handbook of Brownian motion—facts and formulae.* Birkhäuser, Basel, 2nd edition. Cited on page 35.

Burke, M. D. and Yuen, K. C. (1995). Goodness-of-fit tests for the Cox model via bootstrap method. *J. Statist. Plann. Inference*, 47(3):237–256. Cited on pages 25 and 27.

Cai, Z. and Sun, Y. (2003). Local linear estimation for time-dependent coefficients in Cox's regression models. *Scand. J. Statist.*, 30(1):93–111. Cited on page 20.

Cox, D. R. (1972). Regression models and life-tables. *J. Roy. Statist. Soc. Ser. B*, 34(2):187–220. Cited on pages 1, 2, 16, and 24.

Cox, D. R. and Oakes, D. (1984). *Analysis of survival data.* Monographs on Statistics and Applied Probability. Chapman & Hall, London. Cited on page 11.

Dabrowska, D. M. (1997). Smoothed Cox regression. *Ann. Statistist.*, 25:1510–1540. Cited on page 20.

Davison, A. C. and Hinkley, D. V. (1997). *Bootstrap methods and their application*, volume 1 of *Cambridge Series in Statistical and Probabilistic Mathematics.* Cambridge University Press, Cambridge. Cited on page 121.

Elandt-Johnson, R. C. and Johnson, N. L. (1999). *Survival models and data analysis.* Wiley. Cited on page 11.

Fine, J. P. (2002). Comparing nonnested Cox models. *Biometrika*, 89(3):635–647. Cited on pages 26, 37, 56, and 60.

Fleming, T. R. and Harrington, D. P. (1991). *Counting Processes and Survival Analysis*. Wiley, New York. Cited on pages 8, 11, 21, 62, 137, 138, 139, 140, and 150.

Gandy, A. (2002). A nonparametric additive risk model with applications in software reliability. Diplomarbeit, Universität Ulm. Cited on page 43.

Gandy, A. and Jensen, U. (2004). A nonparametric approach to software reliability. *Appl. Stoch. Models Bus. Ind.*, 20:3–15. Cited on pages 8, 9, 137, and 138.

Gandy, A. and Jensen, U. (2005a). Checking a semi-parametric additive risk model. Accepted for publication in *Lifetime Data Anal.* Cited on page 9.

Gandy, A. and Jensen, U. (2005b). Model checks for Cox-type regression models based on optimally weighted martingale residuals. Submitted. Cited on page 9.

Gandy, A. and Jensen, U. (2005c). On goodness of fit tests for Aalen's additive risk model. *Scand. J. Statist.*, 32:425–445. Cited on pages 9 and 116.

Grambsch, P. M. and Therneau, T. M. (1994). Proportional hazards tests and diagnostics based on weighted residuals. *Biometrika*, 81:515–526. Correction: Volume 82 page 668. Cited on pages 20 and 25.

Grønnesby, J. K. and Borgan, Ø. (1996). A method for checking regression models in survival analysis based on the risk score. *Lifetime Data Anal.*, 2(4):315–328. Cited on pages 23, 27, 139, and 140.

Grund, B. and Polzehl, J. (2001). Semiparametric lack-of-fit tests in an additive hazards regression model. *Statistics and Computing*, 11(4):323–335. Cited on page 28.

Hall, W. and Wellner, J. (1980). Confidence bands for a survival curve from censored data. *Biometrika*, 67:133–144. Cited on page 35.

Hjort, N. L. (1990). Goodness of fit tests in models for life history data based on cumulative hazard rates. *Ann. Statist.*, 18(3):1221–1258. Cited on page 28.

Hjort, N. L. (1992). On inference in parametric survival data. *Internat. Statist. Rev.*, 60(3):355–387. Cited on pages 19, 37, 54, and 56.

Hosmer, D. W. and Lemeshow, S. (1999). *Applied survival analysis. Regression modeling of time to event data.* Wiley. Cited on page 11.

Hougaard, P. (2000). *Analysis of multivariate survival data.* Springer. Cited on pages 3 and 11.

Huffer, F. W. and McKeague, I. W. (1991). Weighted least squares estimation of Aalen's additive risk model. *J. Amer. Statist. Assoc.*, 86(413):114–129. Cited on page 18.

Jacod, J. and Shiryaev, A. N. (1987). *Limit Theorems for Stochastic Processes.* Springer, Berlin Heidelberg. Cited on page 151.

Johnson, N. L. and Kotz, S. (1970). *Distributions in statistics. Continuous univariate distributions. 2.* Houghton Mifflin Co., Boston, Mass. Cited on page 116.

Khmaladze, E. V. (1981). Martingale approach in the theory of goodness-of-fit tests. *Theory Probab. Appl.*, 26:240–257. Cited on page 28.

Kim, J., Song, M. S., and Lee, S. (1998). Goodness–of–fit tests for the additive risk model with ($p > 2$)-dimensional time-invariant covariates. *Lifetime Data Anal.*, 4:405–416. Cited on page 27.

Klein, J. P. and Moeschberger, M. L. (2003). *Survival analysis. Techniques for censored and truncated data. 2nd ed.* Springer. Cited on page 11.

Kleinbaum, D. G. (1996). *Survival analysis. A self-learning text.* Springer. Cited on page 11.

Kosorok, M. R., Lee, B. L., and Fine, J. P. (2004). Robust inference for univariate proportional hazards frailty regression models. *Ann. Statist.*, 32(4):1448–1491. Cited on page 20.

Kraus, D. (2004). Goodness-of-fit inference for the Cox-Aalen additive-multiplicative regression model. *Statist. Probab. Lett.*, 70(4):285–298. Cited on page 30.

Kvaløy, J. T. and Neef, L. R. (2004). Tests for the proportional intensity assumption based on the score process. *Lifetime Data Anal.*, 10(2):139–157. Cited on page 23.

Ledoux, M. and Talagrand, M. (1991). *Probability in Banach Spaces.* Springer. Cited on page 153.

Lee, E. T. and Wang, J. W. (2003). *Statistical methods for survival data analysis. 3rd ed.* Wiley. Cited on page 11.

Lenglart, E. (1977). Relation de domination entre deux processus. *Ann. Inst. H. Poincaré Sect. B (N.S.)*, 13(2):171–179. Cited on page 151.

León, L. F. and Tsai, C.-L. (2004). Functional form diagnostics for Cox's proportional hazards model. *Biometrics*, 60:75–84. Cited on page 26.

Lin, D. Y. (1991). Goodness-of-fit analysis for the Cox regression model based on a class of parameter estimators. *J. Amer. Statist. Assoc.*, 86(415):725–728. Cited on page 26.

Lin, D. Y. and Spiekerman, C. F. (1996). Model checking techniques for parametric regression with censored data. *Scand. J. Statist.*, 23(2):157–177. Cited on pages 28 and 29.

Lin, D. Y. and Wei, L. J. (1989). The robust inference for the Cox proportional hazards model. *J. Amer. Statist. Assoc.*, 84(408):1074–1078. Cited on pages 37, 56, 60, 116, and 118.

Lin, D. Y. and Wei, L. J. (1991). Goodness-of-fit tests for the general Cox regression model. *Statistica Sinica*, 1:1–17. Cited on pages 25 and 133.

Lin, D. Y., Wei, L. J., and Ying, Z. (1993). Checking the Cox model with cumulative sums of martingale-based residuals. *Biometrika*, 80(3):557–572. Cited on pages 23, 27, 29, and 140.

Lin, D. Y. and Ying, Z. (1994). Semiparametric analysis of the additive risk model. *Biometrika*, 81(1):61–71. Cited on pages 18 and 27.

Lin, D. Y. and Ying, Z. (1995). Semiparametric analysis of general additive-multiplicative hazard models for counting processes. *Ann. Statist.*, 23(5):1712–1734. Cited on pages 3 and 19.

Martinussen, T. and Scheike, T. H. (2002). A flexible additive multiplicative hazard model. *Biometrika*, 89(2):283–298. Cited on pages 3, 19, and 20.

Martinussen, T., Scheike, T. H., and Skovgaard, I. M. (2002). Efficient estimation of fixed and time-varying covariate effects in multiplicative intensity models. *Scand. J. Statist.*, 29(1):57–74. Cited on pages 3, 20, and 30.

Marzec, L. and Marzec, P. (1993). Goodness of fit inference based on stratification in Cox's regression model. *Scand. J. Statist.*, 20(3):227–238. Cited on page 22.

Marzec, L. and Marzec, P. (1997a). Generalized martingale-residual processes for goodness-of-fit inference in Cox's type regression models. *Ann. Statist.*, 25(2):683–714. Cited on pages 24, 133, and 140.

Marzec, L. and Marzec, P. (1997b). On fitting Cox's regression model with time-dependent coefficients. *Biometrika*, 84(4):901–908. Cited on page 30.

Marzec, L. and Marzec, P. (1998). Testing based on sampled data for proportional hazards model. *Statist. Probab. Lett.*, 37(2):303–313. Cited on page 26.

May, S. and Hosmer, D. W. (1998). A simplified method of calculating an overall goodness-of-fit test for the Cox proportional hazards model. *Lifetime Data Analysis*, 4:393–403. Cited on page 23.

May, S. and Hosmer, D. W. (2004). A cautionary note on the use of the Grønnesby and Borgan goodness-of-fit test for the Cox proportional hazards model. *Lifetime Data Analysis*, 10:283–291. Cited on page 23.

McKeague, I. W. (1988). Asymptotic theory for weighted least squares estimators in Aalen's additive risk model. *Contemp. Math.*, 80:139–152. Cited on pages 18 and 43.

McKeague, I. W. and Sasieni, P. D. (1994). A partly parametric additive risk model. *Biometrika*, 81(3):501–514. Cited on pages 3 and 18.

McKeague, I. W. and Sun, Y. (1996). Towards an omnibus distribution-free goodness-of-fit test for the Cox model. *Statistica Sinica*, 6:579–588. Cited on page 25.

McKeague, I. W. and Utikal, K. J. (1991). Goodness-of-fit tests for additive hazards and proportional hazards models. *Scand. J. Statist.*, 18:177–195. Cited on pages 25, 27, 124, 125, and 126.

Miller, R. and Halpern, J. (1982). Regression with censored data. *Biometrika*, 69(3):521–531. Cited on pages 8, 137, 140, and 141.

Miller, R. G. (1998). *Survival analysis.* Wiley. Notes by Gail Gong. Problem solutions by Alvaro Munoz. Reprint. Cited on page 11.

Moreau, T., O'Quigley, J., and Mesbah, M. (1985). A global goodness-of-fit statistic for the proportional hazards model. *J. Roy. Statist. Soc. Ser. C*, 34(3):212–218. Cited on page 24.

Murphy, S. A. (1995). Asymptotic theory for the frailty model. *Ann. Statist.*, 23(1):182–198. Cited on page 20.

Murphy, S. A. and Sen, P. K. (1991). Time-dependent coefficients in a Cox-type regression model. *Stochastic. Process. Appl.*, 39:153–180. Cited on pages 3 and 20.

Nagelkerke, N., Oosting, J., and Hart, A. (1984). A simple test for goodness of fit of Cox's proportional hazards model. *Biometrics*, 40:483–486. Cited on page 26.

Ng'andu, N. H. (1997). An empirical comparison of statistical tests for assessing the proportional hazards assumption of Cox's model. *Statist. Med.*, 16:611–626. Cited on page 27.

Nielsen, G. G., Gill, R. D., Andersen, P. K., and Sørensen, T. I. A. (1992). A counting process approach to maximum likelihood estimation in frailty models. *Scand. J. Statist.*, 19(1):25–43. Cited on page 20.

Nikitin, Y. (1995). *Asymptotic Efficiency of Nonparametric Tests.* Cambridge University Press. Cited on page 39.

Parmar, M. K. and Machin, D. (1995). *Survival analysis. A practical approach.* Wiley. Cited on page 11.

Parzen, M. and Lipsitz, S. R. (1999). A global goodness-of-fit statistic for Cox regression models. *Biometrics*, 55(2):580–584. Cited on page 26.

Pons, O. (2000). Nonparametric estimation in a varying-coefficient Cox model. *Math. Meth. Statist.*, pages 376–398. Cited on page 20.

Prentice, R. L. and Self, S. G. (1983). Asymptotic distribution theory for Cox-type regression models with general relative risk form. *Ann. Statist.*, 11(3):804–813. Cited on pages 16, 17, and 116.

Rao, R. R. (1963). A law of large numbers for D[0,1]-valued random variables. *Theory of Probability and its Applications*, 8:70–74. Cited on page 152.

Rebolledo, R. (1980). Central limit theorems for local martingales. *Z. Wahrsch. Verw. Gebiete*, 51(3):269–286. Cited on page 150.

Rudin, W. (1974). *Real and complex analysis.* McGraw-Hill, New York, 2nd edition. Cited on page 157.

Sasieni, P. (1993). Some new estimators for Cox regression. *Ann. Statist.*, 21(4):1721–1759. Cited on pages 17, 37, 56, and 60.

Scheike, T. H. (2002). The additive nonparametric and semiparametric Aalen model as the rate function for a counting process. *Lifetime Data Anal.*, 8(3):247–262. Cited on page 28.

Scheike, T. H. and Martinussen, T. (2004). On estimation and tests of time-varying effects in the proportional hazards model. *Scand. J. Statist.*, 31(1):51–62. Cited on pages 20 and 30.

Scheike, T. H. and Zhang, M.-J. (2002). An additive multiplicative Cox-Aalen regression model. *Scand. J. Statist.*, 29:75–88. Cited on pages 3 and 19.

Scheike, T. H. and Zhang, M.-J. (2003). Extensions and applications of the Cox-Aalen survival model. *Biometrics*, 59(4):1036–1045. Cited on pages 20 and 30.

Schoenfeld, D. (1980). Goodness-of-fit tests for the proportional hazards regression model. *Biometrika*, 67:145–154. Cited on page 21.

Smith, P. J. (2002). *Analysis of failure and survival data.* Chapman and Hall/CRC. Cited on page 11.

Song, H. H. and Lee, S. (2000). Comparison of goodness of fit tests for the Cox proportional hazards model. *Commun.Statist.-Simula.*, 29(1):187–206. Cited on page 27.

Song, M.-U., Jeong, D.-M., and Song, J.-K. (1996). Checking the additive risk model with martingale residuals. *J. Korean Statist. Soc.*, 25(3):433–444. Cited on page 27.

Spiekerman, C. F. and Lin, D. (1996). Checking the marginal Cox model for correlated failure time data. *Biometrika*, 83(1):143–156. Cited on page 23.

Stute, W., González Manteiga, W., and Sánchez Sellero, C. (2000). Nonparametric model checks in censored regression. *Comm. Statist. Theory Methods*, 29(7):1611–1629. Cited on pages 29 and 30.

Tableman, M. and Kim, J. S. (2004). *Survival analysis using S. Analysis of time-to-event data. With a contribution from Stephen Portnoy.* Chapman and Hall/CRC. Cited on page 11.

Therneau, T. and Grambsch, P. (2000). *Modeling survival data: Extending the Cox model.* Springer. Cited on pages 3, 11, and 21.

Therneau, T. M., Grambsch, P. M., and Fleming, T. R. (1990). Martingale-based residuals for survival models. *Biometrika*, 77(1):147–160. Cited on pages 22, 25, and 27.

van der Vaart, A. W. and Wellner, J. A. (1996). *Weak convergence and empirical processes.* Springer Series in Statistics. Springer-Verlag, New York. With applications to statistics. Cited on page 153.

Verweij, P. J., van Houwelingen, H. D., and Stijnen, T. (1998). A godness-of-fit test for Cox's proportional hazards model based on martingale residuals. *Biometrics*, 54(4):1516–1526. Cited on page 24.

Vuong, Q. H. (1989). Likelihood ratio tests for model selection and nonnested hypotheses. *Econometrica*, 57(2):307–333. Cited on page 7.

Wei, L. J. (1984). Testing goodness of fit for proportional hazards model with censored observations. *J. Amer. Statist. Assoc.*, 79(387):649–652. Cited on page 23.

Winnett, A. and Sasieni, P. (2003). Iterated residuals and time-varying covariate effects in Cox regression. *J. Roy. Statist. Soc. Ser. B*, 65:473–488. Cited on page 20.

Yuen, K. C. and Burke, M. D. (1997). A test of fit for a semiparametric additive risk model. *Biometrika*, 84(3):631–639. Cited on page 27.

Zucker, D. M. and Karr, A. F. (1990). Nonparametric survival analysis with time-dependent covariate effects: A penalized partial likelihood approach. *Ann. Statist.*, 18:329–353. Cited on page 20.

Zusammenfassung

In dieser Arbeit werden gerichtete Anpassungstests für Regressionsmodelle aus der Lebensdaueranalyse (Survival Analysis) vorgeschlagen und untersucht. In der Lebensdaueranalyse versucht man das Auftreten gewisser Ereignisse zu analysieren. Hierzu werden mehrere Individuen über einen Zeitraum beobachtet, denen diese Ereignisse zugeordnet sind. Anders formuliert, beobachtet man für jedes Individuum einen Zählprozess, welcher die Anzahl der Ereignisse bis zu einem gewissen Zeitpunkt angibt. Das klassische Beispiel ist eine klinische Studie, wobei die Individuen Patienten sind und die Ereignisse Tod oder Rückfall sein können. Regressionsmodelle werden eingesetzt, um das Auftreten der Ereignisse mit Einflussgrößen in Verbindung zu setzen. Das bekannteste Modell ist ein Modell mit proportionalen Ausfallraten, welches auf Cox zurückgeht. Häufig sind die Regressionsmodelle semiparametrische Modelle, d.h. sie beinhalten sowohl unbekannte endlichdimensionale Parameter als auch unbekannte Funktionen. Regressionsmodelle werden typischerweise über die Intensität der Zählprozesse der Ereignisse definiert. Die Differenz zwischen dem Zählprozess und der integrierten Intensität ist ein Martingal.

Als Ausgangspunkt für Modellüberprüfungen wird häufig die Differenz zwischen den Zählprozessen und den geschätzten integrierten Intensitäten verwendet. Diese Differenzen werden Martingalresiduen genannt. Die Teststatistiken, die in dieser Arbeit betrachtet werden, sind die Summe gewichteter Integraltransformationen der Martingalresiduen.

Eine Kernidee der Arbeit ist, dass durch geschickte Wahl der Gewichte die asymptotische Verteilung der Teststatistik nicht von den gewählten Schätzern der Modellparameter abhängt, solange diese gewisse Konsistenzraten aufweisen. Man kann somit zwischen verschiedenen Standardschätzern aus der Literatur auswählen, ohne dass sich die asympotische Verteilung der Teststatistik ändert. Außerdem ergibt sich eine beträchtliche Vereinfachung der asymptotischen Verteilung. Bisherige Ansätze betrachten Teststatistiken, die nur unabhängig von der Verteilung der Schätzer der unendlichdimensionalen Parameter der Modelle sind.

Eine weitere Kernidee der Arbeit besteht darin, die Gewichte in der Teststatistik so zu wählen, dass die Macht des Tests gegen gewisse Alternativen besonders groß ist. Häufig wird hierzu ein weiteres Regressionsmodell ausgewählt und der Test so ausgerichtet, dass er gegen dieses Modell eine besonders große Macht aufweist. Ein solches Modell wird konkurrierendes Modell genannt.

Im Folgenden wird nun etwas detaillierter auf den Inhalt eingegangen. Die Arbeit beginnt mit einem Überblick über einige in der Literatur vorhandene Modelle und über verfügbare Modellüberprüfungen. Anschließend werden die Tests entwickelt. Hierbei wird mit der asymptotischen Verteilung der Teststatistik gearbeitet, die sich ergibt, wenn die Anzahl der beobachteten Individuen gegen unendlich strebt. Zunächst wird ein einfaches additives Modell betrachtet, das so genannte Aalen Modell, in dem die Parameter Funktionen sind, die linear eingehen. Hier vereinfacht sich die Teststatistik des Anpassungstests sehr stark und die Modellparameter müssen nicht einmal geschätzt werden. Anschließend wird die Annahme, dass die Funktionen als Parameter linear eingehen, fallen gelassen. Dies ist deutlich aufwendiger, da nun die Modellparameter geschätzt werden müssen. Als drittes wird der Test auf semiparametrische Modelle erweitert, wobei die Teststatistik dahingehend modifiziert wird, dass sie ausnützt, dass einige Parameter zeitkonstant sind.

Wie bereits erwähnt, führt die geschickte Wahl der Gewichte zu einer relativ einfachen asymptotischen Verteilung der Teststatistik. Dadurch wird es möglich, optimale Gewichte gegen lokale Alternativen im Sinne der approximativen Bahadur Effizienz und gegen feste Alternativen im Sinne der Pitman Effizienz zu ermitteln. Die optimalen Gewichte lassen sich mit Hilfe von gewissen gewichteten Orthogonalprojektionen darstellen.

Um den Test gegen konkurrierende Modelle auszurichten, müssen Parameter der konkurrierenden Modelle geschätzt werden und die Konvergenzraten der verwendeten Schätzer bekannt sein, selbst wenn das Modell nicht zutrifft. In der Arbeit wird für das Cox und das Aalen Modell gezeigt, dass die Standardschätzer selbst dann mit der parametrischen Rate konvergieren, wenn die Modelle nicht zutreffen. Diese Ergebnisse gehen über das in der Literatur Vorhandene hinaus.

Setzt man in die optimalen Gewichte geschätzte Parameter von konkurrierenden Modellen ein, so kann unter gewissen Umständen die asymptotische Varianz der Teststatistik Null sein. Um dieses Problem in den Griff zu bekommen, wird für einen speziellen Anpassungstest eines Cox Modells in der Arbeit ein sequentielles Testverfahren entwickelt. In einem ersten Schritt wird die Hypothese getestet, dass die asymptotische Varianz Null ist. In einem zweiten Schritt wird dann der Anpassungstest durchgeführt. Neben diesem sequentiellen Test wird in der Arbeit auch ein Bootstrap-Ansatz vorgeschlagen.

Die Arbeit beinhaltet Simulationsstudien. Diese zeigen, dass die Tests für in der Praxis vorkommende Stichprobenumfänge anwendbar sind. Außerdem sieht man, dass durch die richtige Wahl der Gewichte die Macht der Tests höher ist als bei anderen Verfahren aus der Literatur.

Darüber hinaus konnten die neu entwickelten Tests auch erfolgreich zur Analyse von realen Datensätzen herangezogen werden. Es wurden drei verschiedene Datensätze betrachtet. Diese stammen aus der Softwarezuverlässigkeit und aus medizinischen Studien. Hier können formale Aussagen über die Güte der Anpassung gewisser Modelle gemacht werden, was mit bisher bekannten Verfahren nicht mit gleicher Präzision möglich war.

Abschließend wird kurz auf einige besondere Aspekte und mögliche Erweiterungen der neuen Testmethodik eingegangen. Es wird gezeigt, dass in einem parametrischen Modell die Tatsache, dass die Parameter geschätzt werden müssen, nicht die Macht des Anpassungstests reduziert. Außerdem wird auf Tests gegen mehrere konkurrierende Modelle eingegangen und gezeigt, wie die vorgestellten Anpassungstests auf andere Modelle erweitert werden können. Schließlich wird noch ein Zusammenhang zwischen den Anpassungstests und Schätzverfahren in allgemeinen semiparametrischen Modellen hergestellt.

Zusammenfassend wurden in dieser Arbeit neue Anpassungstests für Modelle der Lebensdaueranalyse entwickelt und ihre asymptotischen Eigenschaften analysiert. Dabei zeigen sowohl Simulationsstudien als auch Anwendungen der Tests auf reale Datensätze, dass mit den neuen Testverfahren Modelle der Lebensdaueranalyse weit besser überprüft werden können als das bisher der Fall war.

Acknowledgement

First and foremost, I want to thank my advisor Prof. Dr. Uwe Jensen. I would like to thank him for his support, encouragement and valuable comments. Furthermore, I would like to thank him for exposing me to new ideas and allowing me to take part in several conferences. Actually, on one of those conferences, a question was raised which lead to this work.

Furthermore, I would like to thank Luitgard Veraart and Ralf Gandy for comments and discussions. Thanks also to Constanze Lütkebohmert for comments and to Dr. Harald Bauer for some discussions.

I would also like to thank Prof. Dr. Volker Schmidt for his support and his interest in my work. A big thank you also goes to the colleagues from the Department of Stochastics at the University of Ulm.